THE CITY

URBAN SOCIOLOGY IN AN URBANIZED SOCIETY

SOCIOLOGY OF THE CITY

URBAN SOCIOLOGY IN AN URBANIZED SOCIETY

J.R. MELLOR

Routledge
Taylor & Francis Group
LONDON AND NEW YORK

First published in 1977

This edition published in 2007
Routledge
2 Park Square, Milton Park, Abingdon, Oxon, OX14 4RN

Simultaneously published in the USA and Canada by Routledge
605 Third Avenue, New York, NY 10017

Routledge is an imprint of Taylor & Francis Group, an informa business

First issued in paperback 2013

British Library Cataloguing in Publication Data
A CIP catalogue record for this book
is available from the British Library

Urban Sociology in an Urbanized Society

ISBN13: 978-0-415-41822-5 (hbk)
ISBN13: 978-0-415-41931-4 (subset)
ISBN13: 978-0-415-41318-3 (set)
ISBN13: 978-0-415-85190-9 (pbk)

Routledge Library Editions: The City

Urban sociology in an urbanized society

J. R. Mellor

Department of Sociology
University of Manchester

Routledge & Kegan Paul

London, Henley and Boston

First published in 1977
by Routledge & Kegan Paul Ltd
39 Store Street,
London WC1E 7DD,
Broadway House,
Newtown Road,
Henley-on-Thames,
Oxon RG9 1EN and
9 Park Street,
Boston, Mass. 02108, USA
Set in 10 on 11 Times New Roman 327
by Kelly and Wright, Bradford-on-Avon, Wiltshire
and printed in Great Britain by
Lowe & Brydone Ltd

British Library Cataloguing in Publication Data

Mellor, J. R.

Urban sociology in an urbanized society.
1. Sociology, Urban
I. Title
301.36'0941 HT151 77–30022

ISBN 0 7100 8683 0

Contents

Preface vii

Introduction 1

part one The British experience 17

1 The British experiment 19

1 Uneven development in cities and regions 19

2 The modern big city 48

1 Regional cities: a suburban diaspora 48

2 The urban problem: the inner cities 65

3 The housing question 88

1 The house as home 88

2 Architects and housing 95

3 Housing finance 107

4 Housing in urban sociology 120

4 The practice of planning 127

1 The town-planning movement 127

2 The town-planning profession 133

3 The sociologists' critique 156

part two Theories of urbanization 167

5 The critique of metropolis 169
1 Urbanism and Marxism 169
2 Tönnies: Community and Society (1887) 178
3 Simmel: 'Metropolis and mental life' (1902–3) 182
4 Weber: The City (1911–13) 189
5 Wirth: 'Urbanism as a way of life' (1938) 194

6 The Chicago School: urban experience 204
1 The ecological city 204
2 Community and freedom 216
3 A continuing tradition? 230

Epilogue The city or the town? 238

Notes 253

Index 301

Preface

> only in the subjective world and in theory and imagination do we deal with identically similar units. . . . We shall realize that all this talk of the organization of sociology as though presently the sociologist would be going about the world with the authority of a sanitary engineer, is and will remain nonsense.[1]

When I was a child places fascinated—war-damaged Tyneside, windswept, maritime and depressed; suburban Cheshire, Edwardian redbrick, canals and ponded countryside eaten away by new estates; overspill Swindon, petty in scale, shopping centre swamped, green downs beyond; the backwater of the Rhondda, a world of memories held in bare hills, terraces strung along the valleys. Then Oxford, then London. Could there have been a more marked contrast in environment, architecture, culture, politics and basic economy? How could this diversity be aggregated in national statistics, or the localism of experience be denied in the rubric of 'society'? Nor could their reality be denied. Industrial conurbations or mining townships might have no common centre or definite boundary, their men might commute 20 miles or more in search of work, their young folk leave for metropolitan areas, their very survival might be a question of national aid policies, but nevertheless these places were there, the context in which people lived out their daily routine. The questions and answers put by Castells[2]—real object or scientific object, the scientific status of urban sociology—are ones which English students reared in a tradition of pragmatic empiricism find hard to stomach.

Very little of this fascination with the varieties of habitat, the nuances of dialect or political life, can be conveyed in teaching urban sociology. The latter shows all the weaknesses of an academic subject practised from a professional career base in the universities.

Although interest in sociology as the new science was stirred up by a diffuse concern for the fate of the masses in the great cities, there was no involvement, no sympathy, with the people; the feel of a place, the intricacies of culture and language were inadmissible evidence. Sociology was something to be administered from *haut en bas*, prescribed in the dosage appropriate to systematic disorders. Sociologists were expected to categorize, to delineate key processes, to give guidance to policy-makers with unambiguous theories and straightforward research reports. Theories were not to be presented as provisional resolutions of class-related uncertainties, nor was research seen as confirmation of commonsense understandings.

Urban sociology originated in the last decades of the nineteenth century as systematic empiricism harnessed social outcry to government reform. The protest literature of Mayhew or Dickens was succeeded by surveys like those of Booth or Rowntree; these poverty surveys together with the later Mass Observation reports were to constitute sociological research until after 1945. The principal concern was that of poverty and the associated issues of housing and community development. Local and regional surveys were mounted by universities and their extra-mural departments to bring home some of the realities of life in an unequal society. Not until the 1950s did sociologists move into other areas of investigation, principally education, social mobility, industry and the trade unions, and the competence in community investigation linked to reform movements in the Welfare State. Interest in local affairs *per se* consequently languished, and only in the mid-1960s was there a resurgence of interest in a truly 'urban' sociology with interests distinct from those of sociology.

Post-war, urban sociology has had three phases: the first a continuation of the native tradition of Fabian empiricism strengthened by an interest in social anthropology; the second, from the mid-1960s, saw a flow-back into urban studies of the sociology hybridized in the universities. The achievements of American sociology had been set against the writings of the European liberals, in particular Weber and Durkheim, and these imported thoughtways embedded in the distinctive British intellectual milieu. The third phase is recent: the rejection of urban sociology for urban and regional studies and a recourse to Marxism to fill the theoretical void. In this last move inspiration comes from abroad—radical community politics in America, and the engagement of revolutionary socialists in the urban question in France.

Of the first phase little need be said. The social surveys and community studies are now invaluable social documents, if sometimes limited in scope. They are revealing both for their ideas and the presuppositions with which they were conducted, and for the life experience they documented. They have already become documents in social history. In a generation the second phase will have been forgotten. There are remarkably few research documents which will be of value to future historians, and despite the frequency and tortured quality of debates on the seminar and conference circuit, little to hand on. A tremendous amount of energy has been expended to very little purpose; teaching time taken up with elaborate modification and refutation of ideas readily swept aside by those responsible for their circulation in the cultural air in the first instance. The intellectual activities of the last ten years tell us little about town life and urbanization in Britain, much more about urban sociologists as a professional group.

Castigation of this phase is now acceptable. The following remarks were prepared for the position paper for the York conference,[3] and there were no dissenting voices. There it was argued that British urban sociology had been characterized by the methodology of positivism in that it had sought to discover the constraints on life chances as environmentally expressed, in definite measurable terms. The urban world was a world of buildings. Consideration of class experience, local culture, and language, political and economic history, was largely absent. As with all branches of urban studies the emphasis had been technocratic. It was assumed that control over urban problems could be achieved if more rational use were made of existing knowledge about environmental technology. The focus of concern had therefore been the professions within the hierarchies of local and national government which have front-line responsibility for the allocation of resources within urban areas. The main weight of attack had been against the planners (thus endorsing their own estimation of professional authority). Change was sought through the established institutions of British political life, redress has been considered possible if only rationality were to prevail. The political bias is evident in the construction of theory and the direction of research towards those who draw up the plans and allocate community resources, rather than those who experience such management. The sociologists have sought to influence the élite, not the mass of urban residents. The medium for the redress of inequality was deemed to be environmental—a redistribution of facilities in space, and the solution to the 'urban problem' was held to be administrative, professional and educational. Accordingly urban sociologists have sought new or modified urban agencies, and socially committed planning.

Further, in that urban sociologists had responded to the pressures of immediate circumstances, reacting expediently and without sense of priority, urban sociology belonged to the liberal pragmatic conception of science. Urban sociology has followed the shifting consensus as to the nature of urban problems. The post-war reconstruction of the British economy threw up new strains in society, and new demands on the state. All of these were necessarily expressed on the ground. The contradictions in the society had territorial definition and so were incorporated into the sociological debates. Discussions about new housing estates were extended into those about urban renewal and community development; the issue of poverty was joined by that of race; to that of housing was added land and property; questions of housing allocation were succeeded by consideration of housing finance; wide-eyed bemusement at mass urbanization was succeeded by shocked appraisal of inner city neglect; satisfaction in the 'new' urbanization was rapidly overtaken by awareness of widespread stagnation in the regions and then the development of their underdevelopment. The endemic nature of these issues has been overlooked as sociologists chased from the banner headline of 'new town blues', to that of 'crack-up of a community', from the seven-day wonder of participation in planning to the novelty of community action, from predictions of new life styles in the year 2000 to gloomy appraisals of the possibilities for action on the urban environment in a zero growth economy. The proud boast of urban sociologists to be in touch with everyday issues, those of pressing concern to the man in the street, is seen to carry its price. The moves to a political economy of urbanization derive in part from awareness of the ephemeral nature of so many themes in urban sociology. Employment structures, financial organization, political hierarchies, flows of capital and of population, have an enduring quality not easily overtaken by events. They can be charted, pinned down (and used for next year's lectures).

Wide as the field appears, the range of issues considered is curiously narrow. Although there is an environmental studies focus which aligns urban sociology with geography or planning rather than history, political studies, linguistics or cultural anthropology, many environmental issues of public and professional concern do not appear. In neat reversal of the attitudes of planners, neither the town centre, nor the issue of roads, transport systems and traffic seem to merit discussion; similarly, only a narrow band of social issues are included. Educational problems, for instance the question of community schools, despite their effect on residential mobility, go unmentioned. Despite the shifts in style there is a continuity of tradition in urban sociology, hence a tacit restriction of the area of discussion.

Urban sociology is firmly identified as having to do with issues of land use, the built environment and the organization of space. It is therefore frankly materialist in conception, the materialism of the engineer rather than that of the businessman. The concern is how men 'act under definite material limitations, presuppositions and conditions independent of their will', and not 'the specific way [they] enter into these definite social and political relations'.[4] This interest in the world around us derives from nineteenth-century currents of thought, an extension of the interests of the geologist, or natural scientist or explorer. As the world was opened out to the imperialist nations, and as their emancipated middle classes used their new affluence to investigate their surroundings, environmental studies took shape. Geography, for instance, is a Victorian product—a groping response to the depredations of the Empire, and the collapse of the old world cultures. The intellectual ancestry runs from Malthus through Livingstone and Darwin, Mackinder or Semple, to Roosevelt and the 'New Deal'. This is a current of thought more influential than that of Marxism or sociology, more widely held among the professional middle classes who shape public debate and academic discussion. Issues of environment take precedence over the question of where the job or market is. 'The orientation on consumption . . . is characteristic of those whose way of life does not make work or output the central problem. Their peculiar social situation (or aspirations) are generalised to explain the nature and problems of society as a whole.'[5] It is their doubts as to the environmental depredations of capitalism, and their moral uncertainties, which have outlet in urban, regional or rural studies.

Three specialist disciplines have a stake in this field—geography, planning and urban sociology. Each of them has undergone trenchant self-criticism in the last decade. In geography, for instance, there has been a search for a scientific basis to a humanist-inspired subject. In planning there has been a return to neo-ecology via cybernetics, a flight to theory in a world moving uncontrollably from the plan. Sociology, whose literature has been combed by the other disciplines, has simply floundered, veering to Marxism in a desperate uncertainty and dissatisfaction with the guidelines of liberalism.

Consider geography first. The movement from the synthetic and regional approach derived in the uncertainties of the inter-war years to an analytical geography reached a high peak in the 1960s. This technologism, spearheaded in Britain in Cambridge and Bristol, relied heavily on utilitarian assumptions and presupposed a naïve behaviourism. The implications of studies carried out in this style, if not neutral in impact—the elaborate demonstration of what was already known, as, for example, the association of single-room furnished tenancies with high rates of occupancy, many small children and immigrant households in London's twilight zones—could be reactionary. The explanation of the presence of poor black households in the American ghettoes in terms of the inelasticity of their demand for transport is the best known example. Now there is a radical geography in which, on paper at least, is set out a heavy Marxism which justifies the study of geography as the distribution of the 'surplus' on the ground. This is loosely and uneasily allied with community action movements as spontaneous and uncertain as the community settlement movements of high Victorian London.

Planning, increasingly recruiting from geography, has moved from the empirical formulae instilled by Geddes and his disciples, to a neo-ecology with strong resemblance to the leading thoughts and loose formulations of the Chicago School. True, the apparent language is different—cybernetics systems analysis has replaced the bio-social language of organism and superorganism, and there is a greater precision in the models derived and urban studies conducted. But the effect on the young professional has been the same. A zest for service has been harnessed by an ideology of professionalism couched in language whose obscurity prevents too close an inspection of the agencies of change. In this way shopping models are substituted for an analysis of shop-owning interests and the decisions of the multinational food corporations, housing market models are formulated independently of either the finances of the house-building industry, or those of the households looking for homes. If nothing else, the adventitious movements of the oil companies and their subsidiaries have exposed the fallacies of the old empiricism (which assumed that what was already there would determine future planning) and rendered suspect this optimistic importation of scientific methodology.

If the scientific reconstruction of geography and the reorientation of planning practice in the 1950s and 1960s is now suspiciously bankrupt, what of sociology? This, too, has not been able to escape from its origins in the ideological currents and counter-currents of nineteenth- and early twentieth-century Europe and America. The British group have merely rewritten a view of the world which Kingsley, or Ruskin, proclaimed as social or humanitarian gospel, Wells or Masterman or Geddes as evolutionary optimism. Each of these disciplines is shackled by its designation. If geography is to do with the interdependencies of man and the land, planning is 'town and country planning', or '*aménagement du territoire*', so urban sociology is similarly bound to the relationship between man and nature. The city is 'a product of nature'.[6] Nor have the recent Marxist formulations escaped the trap of ecologism or urbanism. The Marxist dressing to urban and regional studies is all too often a veneer for traditional concerns and conventional methodology.

The tendency is to exclude certain issues from discussion. Just as the classical sociologists, and after them the Chicago School, excluded in entirety the world of production (the major employment centres of Chicago are never mentioned in the latter's writings, much less the clashes between employers and labour force), so there is still an emphasis on issues of community and consumption. Hence the emphasis on the state whose impact is so much greater on these issues than that of production. Housing, land and now commercial property, community politics (but not industrial politics), population movements, residential segregation and neighbourhood associations, these are still the recognized issues for discussion. The locational decisions made by the controllers of industrial, commercial or financial capital have not been integrated into analysis. The new political economy of urbanization has centred on issues of environment and use of land. Regional studies have perhaps been closer to the concerns of Marxism.

Similarly the formalism of analysis which characterized the development of urban sociology has remained. Discussion still turns on statements of systems as measured by living conditions and material deprivation. The chimera of a universal theory of urbanization is still there, and despite dawning recognition of marked differences among cities of the industrialized societies, has disallowed consideration of the differentials in urban experience within these societies. And the methodology is very much the same. True, we have moved from the street gangs of Chicago or Boston to the street barricades of Paris or Milan, and from the language of process and evolution in ecological systems to that of the spatial articulation of the social structure, but the focus of attention is still the urban environment *sui generis*. Events and activities are discussed as they immediately affect the performance of the urban system. Students of urbanization have 'stopped the time-machine', and limited discussion to externalities.

Marxism is ill-equipped to deal with urbanization. Marx or Engels gave no systematic thought or consideration to the phenomenon of the modern big city and the turmoil of mass urbanization between 1844 and the later fragments of the third volume of *Capital*, always excluding Engels's pamphlets in *The Housing Question*. To them urbanization was demonstration of the irrationality of capitalism and the city was but the stage-set for the drama that was to be played out in the streets. Now the doctrine of Marxism is brought sharply against another vocabulary, another tradition of research and a different sphere of interest. Consumption, planning, the circulation of the surplus on the ground, bring in issues not wholly resolved for Marxism by relating the productive process to the region and that of reproduction of the labour force to the urban complex. Castells takes the urban out of the productive process into that of the management of collective consumption.[7]

The attempts to incorporate the issues of urbanization into the traditions of Marxism are admittedly thought-provocative. For nineteenth-century Marxists as for their liberal contemporaries the agglomerations were the products of industrialism and technology, the evidence of London or Paris to the contrary. They did not allow for distinctions between the metropolitan centres, locus for ruling élites, against the industrial towns which housed the novel productivity of capitalist industry. While it is possible to assimilate Lenin's discussions of imperialism, or Gramsci's suggestions on hegemony to the literature on urbanization and urbanism, the classic Marxist formulations on the production relations of industrial capital are not helpful in understanding the trajectory of urbanization and the role of cities in the economy. There is an implicit break between nineteenth- and twentieth-century strands in Marxian thinking which the urban literature will further emphasize.

No attempt has been made to review the literature of this 'new' political economy of urbanization, Marxist and non-Marxist. It is both too recent and too diffuse to categorize readily. Discussion would be overlong. Urban studies has ramified into regional studies and incorporated rural questions; it is moreover an internationally oriented profession with one world of scholarship, ideas and working material.

The focus of this book is on the urban sociology criticized by Marxists and despised by radicals as an eclectic mixture of case studies, platform-thumping and theory; in short the material expected of urban sociologists as sociologists. This is not material that could be presented in itself, to be taken away, cut up and inserted into research projects, monographs, design exercises and lecture courses. It has to be considered as part of the long run of British or American urbanization and the intellectual responses thereto. Certain responsible groups have reacted in certain ways, so that only some things have been seen, some problems brought out for solution; the response has been selective and partial. Urban sociology is one element in this reaction and move to control.

The emphasis is on British urban sociology. The theories of urbanization reviewed in the second part are in a sense an appendix. In *fin de siècle* Europe and uprooted America theories were derived which, while they articulated sentiments common to middle classes throughout the industrialized or industrializing societies, were constructs specific to those configurations at that time. They are included here for two reasons—first, they did articulate and to some extent resolve the problematic of urbanization met by British sociologists, planners, *et al.*, in pragmatic step-by-step fashion. And second, those theories have been the professional armoury, a thicket of words ensuring respect and autonomy. For all their archaic quality they were in their day respected theories and cannot be summarily dismissed. To understand our own attitudes better, to appreciate the direction taken by urban sociology, it is worthwhile asking just how it was these theories were found necessary, and why they commanded respect for so long. And perhaps still do in some circles.

This is not a book about urban phenomena—there are no maps, photographs, charts or diagrams. It is concerned with ideas, principally those held by a small and specialist group. There is a professional interest here, for ideas are the lecturer's stock in hand, also a personal discomfort when ideas are taken at face value and bandied about unquestioningly. Louis Wirth's edict 'Knowledge of the unstated assumptions and premises of our own and other people's premises comprises the foundation of our intellectual house'[8] can only be endorsed. There is also a belief that intellectuals do have a role in their societies, and that the ideas pushed out into the cultural air have some significance. While they may not determine policy they can legitimate policies exercised for other reasons. But given current economic circumstances perhaps even this argument is specious. It is hard to envisage a sociologist's report that might sway public opinion on government spending on, say, housing, when it is notorious that new housing construction is crippled by interest charges and inflated building costs. Arguments for positive discrimination or community planning belong to the heady days of growth and affluence. It is perhaps no accident that the lean years of the 1970s have seen political economy superseding sociology.

What is represented here is the sociologist's view of the world, analytical, yes, but concerned to establish cross-references, to present patterns and similarities at which the uninitiated may grasp. In the process the subtleties, the variegation, the interest of being an urban resident are lost. If ordering principles are sought then the emphasis has to be on national housing policies and professional creeds of international persuasion which seek to marshal productive forces into efficient units on the ground. So, perhaps paradoxically, an interest in local differences, quirks and oddities, which set one place off from another, took one into the discussion of the 'universals' of urbanization, inner city blight and peripheral expansion, for instance, and the policies of government in oversight of property investment. This despite a conviction that unevenness in development can never be ironed out, and that the contrasts experienced by the itinerant household post-war will be there, in different kind, at any future date. Even if history were erased, as in Morris's Utopian prospectus for the year 2003,[9] England could not be a garden all 'trim and neat and pretty' in which geographical interdependencies did not set in train regional inequalities.

It may well be that we all argue too directly from our own experience. For sociologists enmeshed in the metropolitan regions of urbanizing Europe, street confrontations over inadequacies in housing, or health facilities, or schools, are a facet of capitalist development which has to be fed into the mainstream of Marxist thought. For others directly involved in planning the major urban regions, issues of poverty, race and housing may take attention. Equally, from provincial Britain, as other peripheral regions, different issues are perceived. On the one hand there is the strength and continuity of local traditions with which any newcomer has to come to terms, and there is also the pressure for modernization which might jeopardize the very social nexus on which economic development is to capitalize. There is a detachment from the metropolitan centres fostered in resentment of the one-sided relationship. In that sociology is concerned to establish the norm of development its limitations will be sought out by those not party to the 'decision-making process'. The greater part of this book has been written while living in Hull, a place known to few who have not had cause to work there. It is a place of some character, even on casual acquaintance very different from London, or Tyneside, or Oxford, or the Rhondda. And yet so much of what happens there is a matter of national policy. It is now a matter of personal regret that I did not attempt to map out some dimensions of its class relations, economy and culture, to thread through the interrelationships between national and local in this one town. It is information and analysis of this kind that we lack.

July 1976

University of Manchester

Introduction*

Sociologists of the nineteenth century were in no doubt as to the importance of urbanization in the transformation of Europe. Some of the key concepts of classical sociology—class, *gesellschaft*, anomie, individuation and rationalization—derive from their analysis of the experience of the new urban populations. The emphasis differed; for Marx and Engels, for instance, the towns were an indispensable environment for the generation of revolutionary consciousness, whereas for Tönnies the metropolis was *gesellschaft*, a state of being that would prevent there being a genuine communism; but for them all, the city was a source of economic, political and cultural change. But no 'urban' sociology came out of this interest and concern. To these men the modern world was becoming an urban world, and the new science, sociology, was thereby an urban science. Indeed, it was Weber, in the lecture-papers subsequently published as *The City*,[1] who effectively dismissed a sociology of urban communities.

Weber's argument would seem to be this. The modern city had its origins in the medieval community of burghers, a free association of citizens. This community was marked by relative autonomy and a completeness in its institutional spheres. Here, for a unique moment in time, at the watershed between feudalism and capitalism, public and private interests fused, in that each individual's interests depended on the defence of the walls of the community, So Weber terms this total social system a 'community-association'. But these communities of traders had only a limited independence and autonomy, rapidly became the basis of the nation state, and succumbed to a supra-urban imperialism. They became the residence of kings and emperors, standing armies, national and international bureaucracies, and were themselves divided by class divisions.

With this loss of autonomy came the breakdown in the integration of the different aspects of social life that had characterized the guild city. Residence, occupation, religion, politics, legal and military obligation, ceased to be aspects of a total social experience, and had to be considered as orders of a social structure extending far beyond the urban area. As the populations of the cities no longer shared in this total experience, and as the aggregate of buildings making up the city no longer had the same, or any, significance for them, the city ceased to be a relevant unit of analysis for sociology. In the modern world, the dimensions of class and status, and not locality, were the basis of community formations within the framework of the nation-state. So on these criteria of community autonomy, social integration and individual relevance, the age of the city was at an end.

In situations like that of France, where international labour migration has coincided with an upheaval in the rural economy, or in any of the countries of the underdeveloped world, sociological interest in urbanization remains strong. But in Britain, usually considered a fully urbanized society, in that, in addition to the high proportion of the population actually resident in cities, few areas have escaped incorporation into the urban economies, most sociologists would probably share Weber's attitude to locality as a frame of reference. It is regarded as a primitive category of only occasional relevance in a world of instant communications, near total annihilation of space, international flows of capital and resources, and impressive centralization and control. Students of communities have accordingly been pushed into defensive positions, seeking justifications for community studies and searching for the 'urban' dimension to social existence.

Most urban sociologists now are themselves convinced of the explanatory inadequacy of 'urbanism' and 'urbanization' and the ideological inference in their use as categories of analysis in sociology. There would be little dispute with Castells's statement: 'Toute évolution de la dimension et de la différenciation d'un groupe social est elle-même le produit et l'expression d'une structure sociale et de ses lois de transformation.'[2] The urban, the local, the immediate context of experience, is seen as derivative from structural changes in the society as a whole to which analysis must always be referred. Cities in the contemporary Western world appear as 'a mirror of abstract impersonal forces—of history, class structure and culture'.[3]

And yet there has been growing awareness by social scientists of the local and regional differentials in society. Post-war development in Britain has widened the regional disparities in wealth, prestige, employment opportunities and environment,[4] so that a life-span spent in one of the provincial towns of the North shows only a passing resemblance to one spent in the 'Golden Circle' of metropolitan England.[5] We are not a uniformly urbanized society with the same endowment of resources, but a society comprising many communities, some diffuse, others compact, some very large, others hamlets and villages, with different opportunities for their populations, and with differing sense of community and culture. In acknowledgment of these differentials new specialisms in the social sciences have emerged in recent years: regional and urban economics, for instance;[6] the study of local government and regionalism in political studies.[7] Only in sociology has there been little interest in the comparative study of communities within the national society.

At the same time there has been a growing public concern with the direction of change in the cities. The 'urban crisis' has become a cliché of the political platform as investment in community resources has altered in pattern and quickened in pace. People in British cities have experienced relatively rapid changes in their environment post-war as the economy has been regeared to growth and mass consumption. Immigration, the consequent competition for housing space and educational facilities, slum clearance, decay in the twilight areas, town centre redevelopment, road programmes, land and house prices, have all become 'political' issues, i.e. those considered of direct appeal to the mass of the population. It seems as though the failings of capitalism as a productive system are most readily apparent at this level, the level of the local community, the immediate environment with which we are all confronted. Although few groups of the population are any longer permanently or continuously resident in the one community, local places are substantially the environment of our everyday lives and as such an important basis of our experience and our consciousness of social reality. They are, in short, the world we live in. From this derives so live a sense of an 'urban problematic'.

It is evident that the concerns of the citizenry are bound up with everyday experience of urban living in ways which Weber did not envisage. Weber swept away the ecological argument as it had been phrased by Simmel, and in so doing overlooked the possibility that the resources of the urban environment might themselves become the basis of class divisions, and the city defined for its citizens by the locally restricted allocation of urban resources. The everyday world of the local community has a significance quite other than that of citizenship. A possible justification for community studies might therefore be their focus on 'the concrete problems of life' (Reich) and on 'the welter of daily experience' (Mills)[8] and a specifically 'urban' sociology directed to the exploration of the situations with which individuals are confronted in the urban context. Only if the connection is established with 'the banal, primitive, simple everyday life and wishes of the broadest mass of the population in all the specificity of their situation in society' and 'the objective sociological process'[9] can this urban problematic be restated.

Urban sociologists are therefore in a dilemma. The preoccupation of their colleagues in the dimensions of class, prestige and power, and their general disinterest in urban affairs, have to be set against a widespread statement of the problem of our times as an urban problem. One response has been a movement away from the mainline sociological issues into the concerns of urban policy and community planning, thereby accepting that indeed our problems are 'urban' problems, and as such are questions of administration and reorganization of the cities themselves. More commonly, urban sociologists would answer to their colleagues' arguments that the adoption of an urban perspective only disguises the authentic source of inequalities lodged in the imperatives of the capitalist system by pointing to this widespread awareness of the defects in the urban world. They would claim that the provision and allocation of community resources has so blurred the inequalities deriving from the relations of production that a field of study can be delineated on this basis. The confusion is there in the real world: individuals do not directly experience the objective contradictions of class society, instead they encounter the irrationalities of local and national bureaucracies, the particularism of local politicians and the special circumstances in the local provision of community resources.

And finally there is the view that the issue of the cities is an ideological statement, 'a last protecting illusion in the crisis of our time, that it is not capitalism which is injuring us, but the more isolable, more evident system of urban-industrialism',[10] and as such is worthy of exploration in a critical sociology. Only now is there awareness of the entrenchment of the attitudes of the liberal middle classes of Victorian England to their cities, in the environmental professions and, with them, urban sociologists.[11]

However, at a time when urban research commands increasing investment from the state, and other social sciences are increasingly turning to urban studies, urban sociology is neither respected in sociology, nor with authority in this research. The subject disappoints in that its theoretical statements seem to have only limited relevance to an understanding of the impulses behind urban change. Some, indeed, so contradict common experience of city life as to amount to total mystification.[12] Little attempt has been made to connect the urban realm with definite relations between societies, and within the one society, between its constituent communities. For too long urban sociology has been dogged by the chimera of a universal urbanism to be defined in purely sociological terms.

Urban sociology had its origins in the translation of European sociology into the context of the American colleges and universities in the early years of the century. There was a cross-fertilization between the ideas and assumptions of Social Darwinism, pragmatism, and the sociology of Tönnies, Simmel, Weber, Durkheim, in Chicago under the direction of Small, Thomas and Park, which was to yield a distinctively *urban* sociology. Three features of their approach concern us here:

(i) Their adoption of the city and urbanism as a frame of reference. Urbanization was considered as an autonomous process and source of social change, and urbanism as the expression of modern culture. Change in society was initiated by the cities through their diffusion of the culture engendered by the association of individuals in certain conditions of existence.

(ii) An acceptance of a dualism between country and city, rural and urban. Despite acknowledgment of the controlling activities of the city over its hinterland, the two were considered to be radically different cultures, two ways of life. The urban – rural typology, subsequently modified by the rural – urban continuum, was a theory of development parallel to that derived from Parsons's pattern variables, and in stark opposition to any Marxist-based theories of development.

(iii) The conceptualization of the city as 'an externally organized unit in space produced by laws of its own' (Park, 1915), an organic entity of sufficient unity to command its own future. In effect, changes in the urban environment, a context so governing individual experience as to set the pace and direction of change in society, were explained by the process of urbanization itself.

So was initiated a line of analysis which Marxist sociologists have not hesitated to denote as ideological.[13] The urbanism thesis deriving directly from the working assumptions of classical sociology, empiricism, and the study of the 'social', meant the 'concentration of study rather on definitely bounded units of experience' (Small), and the 'urban', as the immediately apparent context of individual association, became the realm for the investigation of 'the actuality of concrete life' (Simmel).[14] Cities, therefore, could be studied in themselves, and urbanization as a process independent of capitalism as a productive system.

Not only was sociology as an academic discipline narrowly defined so as to exclude the issues of power and scarcity, but the 'social' became both 'communal' and 'personal'. Weber's concept of social action—'oriented to the past, present and expected action of others'; Tönnies's 'relations of affirmation', and Simmel's concept of 'sociation', linked with the concerns of the American pragmatists for consensus and communication, meant that sociology became the study of relationships 'which involve both mutual recognition and the sense of something shared or held in common'. In consequence, the realm of urban sociology was restricted to the study of 'community', and 'community' became the world of women and children. In assuming communities to be traditional complexes of relations of natural will there was a refusal to acknowledge that 'urbanization ought more logically to represent a peculiar form of community formation rather than community destruction'.[15]

The exclusion of the issues of scarcity and power from inquiry has had devastating consequences for the study of cities deriving from the effective exercise of power in the centralization of resources, and themselves creating a new scarcity, land.

British urban sociology

At the time Glass published her critique of the parochialism of domestic urban sociology (1962), and advocated a convergence between the study of urban diffusion and the sociology of development, there was full acceptance of the ecological tradition as the basis for urban studies. Glass's critique was itself premised on the assumption that the urban was denoted by a contrast with the rural, as environment, which in a long-developed society had become a trivial distinction. As she acknowledged, 'it is the regional distinctions which are clearly noticeable at first acquaintance, it is the class distinctions which are paramount'. Urban sociology in an urbanized society had become an irrelevant pursuit, and instead attention had to be turned to 'cities both as instances of and contributors to social change'.[16]

Subsequently, however, there has been a reorientation of British urban sociology towards the study of the urban community as a politically defined arena of conflict within which competing interests dispute the allocation of a limited range of resources. These resources include urban space and environment, and the full range of community services. The statement 'there is a class struggle over the use of houses, and . . . this class struggle is the central process of the city as a social unit',[17] deriving from a study of race relations in the twilight zone of Birmingham, became the reorienting principle for urban studies in the Welfare State.

The starting point for Rex's designation of housing classes had been the observation of the differences in access to resources of housing and education facilities, between immigrant workers and their local counterparts. Seemingly equivalent status in the labour force did not open up the same opportunities for the two groups in the urban context because of housing policies, which, directly or indirectly, gave preference to the established and local households. There seemed, therefore, grounds for a distinction between the class struggle deriving from workplace relations and that emerging in the conflict over housing space within the city. In particular the local working class had gained access to housing of a standard incompatible with their control over the means of production, while the immigrant worker was excluded from new housing and restricted to the twilight areas. There is here a recognition that access to community resources, what Castells terms 'collective consumption', does not derive straight from the relationships of production.

The themes indicated by Rex for urban sociology—that of the allocation and use of scarce resources in the urban environment, and the constraints operative in the urban system—have been amplified by Pahl. Urban sociology becomes the sociology of constraint: 'The basic framework for urban sociology is then the pattern of constraints which operates differentially in given localities.'[18]

Marked inequalities in claim over investment into community resources have meant unevenness in their distribution; this, taken together with the tax of distance, has meant differential access to facilities by different groups in society. A low level of command over space, that is, an income too low to open up access to a wide range of community facilities, imposes yet another handicap on the poor in society. Urban sociology has to focus on the politics of the allocation of collective resources in space. In particular, attention is to be directed towards the system managers, the local technocrats and social gatekeepers, who mediate in the allocative processes of the 'socio-spatial or socio-ecological system.'[19]

Clearly the sociological tradition associated with Weber, whose themes are those of inequality, class, power and conflict, has replaced the alternative tradition on which ecology and community studies had been based. The small-scale focus on neighbourhood and domestic life castigated by Glass in earlier urban studies is dropped in favour of political analysis of the urban community as a social system defined through the conflict over resources that are intrinsic to that locality—its land and property. Although the parameters for class conflict are prescribed nationally, the city remains an arena for struggle in so far as these resources are allocated locally. A local social unit, or community, is therefore defined by conflict, not consensus, a political struggle in which local resources are the subject of negotiation and threat and temporary truces, and in which the professions' intervention may be crucial. The city is no longer the autonomous community-association as defined by Weber, but, as at Birmingham, there is demonstrably a local social unit which has meaning for its inhabitants through its being a material resource.

The British sociologists, therefore, would claim a middle ground between Weber's rejection of the urban community as a unit of sociological interest, and the ecologists' dismissal of society beyond the urban limits as irrelevant to urban analysis. They acknowledge the impact of nationally determined policies for intervention into local markets, and yet refuse to overlook the particular circumstances of allocation of resources in the community. In a measure a balance is held between the local and the national, and between the premise of class as the base category of existence, and that of community, or place.

And yet, in reacting against the aimlessness, and possible triviality of the community studies initiated by the Chicago School, an unnecessarily restrictive approach may have been adopted. This sociology of constraint, with its emphasis on material conditions and indices of inequality in life chances, loses sight of 'the interactions of men's experience',[20] the world of the men in the street. The main emphasis has been on the distribution of resources and facilities at the interface between the ecological and political systems; that is, it has been directed towards those who control or allocate rather than those who experience such control. Should we not be looking closer at the ways in which people, not planners, adapt, respond to, and make use of their community? There is a greater emphasis on the system, and its management, than on the men and women within it. We need to bring the people back in, to deal with 'the specific way [they] enter into these definite social and political relations'. Although the situations in which individuals must act are predefined for them in situations of constraint, they must realize the situation in their actions, apprehend on their own terms the action possible in given circumstances.

But British sociologists cannot yet escape the charge of parochialism. The use of systems analysis is indicative: there is an inference that the urban system may set the parameters for explanation as it did for the Chicago School. Halsey's comment, 'There is no city which is either autonomous or internally homogeneous. It cannot be the analytical unit for all life chances',[21] indicates the unease prevalent among sociologists. Moreover, the concern is still with domestic issues—poverty in the inner cities, race in the twilight areas, community development in the inner cities, the failure of the local executives in their professionalism to redress inequality on the ground. But all these questions are issues which must raise further questions about national policies for investment and growth, and the metropolitan society's command over international resources. This is particularly so in the case of housing. The personal world of home and family life is at the mercy of international swings of fortune in that policies for housing expenditure are governed by national status in world markets. The most important of 'urban infrastructural investment' issues is not in this respect 'urban'.

What is lacking is a working theory based on a historical analysis of the British experience of urbanization which, necessarily, must relate diverse urban communities to their sources of capital and supply. Urbanization means dependence—on other workers, on other centres, and the hinterland of nation and world society. It also means dominance, as resources, capital and labour are drawn into the centre—first from country to town, then from satellite territory to metropolitan society, and concurrently from the towns, to the metropolitan regions themselves. It is these relations of urban community to nation and world society that have been left unexplored and yet it is these that must underpin any more localized sociology. This investigation must proceed before ever internal differentials in the allocation of resources are considered. And these differences between places are not random, nor can their derivation be ascertained by empirical inspection of their condition now: they are firmly rooted in the specific relationship each local area has had, and must have, to the national and international disposition of capital, labour, control and resources, through time.

The theory of metropolitan dominance

There is a widespread view in sociology that local studies are not a legitimate method of sociological analysis, and in general Weber's argument is accepted. Loss of autonomy and completeness of social experience, with mass urbanization and the growth of the nation-state, means that the latter forms a more relevant aggregate of analysis. They are, however, to be used as 'the source for overall images of society . . . as reference points for doing other research, and for . . . commentaries on society at large'.[22] A survey of family life in Hull therefore has wider applicability, and a community study in Bethnal Green has findings generalizable to 'working-class families' irrespective of their local circumstances. The local variations in resources, environment, opportunities and culture within the society are disregarded. And yet in any examination of the British experience of urbanization, it is easy to demonstrate how each place has had a special occupational history, and has stood in a specific relationship to 'society' throughout its life cycle. The uneven distribution of the resources of labour and capital has meant wide variations in the conditions of existence, in experience, and in culture. Some of these differences in culture are subtle—as between Manchester and Leeds; others are brutal, as from Glasgow to Surrey. The specific and regional cultures of Britain have maintained a relative autonomy despite the small size of the country and the centralization of authority.

It is all too readily assumed that we live in mass societies in which local differentials in culture are survivals, soon to be obliterated, and not societies which might better be analysed in terms of metropolitan dominance and provincial dependency, in which pre-existing variations deriving from the relational position of one community *vis-à-vis* the others are continually being reactivated. Even in a country as compact and highly centralized as Britain, the continual recreation of locally-based cultures cannot be overlooked—the 'Mersey Beat' from Liverpool, the gangs of the Glasgow housing estates, Welsh and Scottish nationalism, Geordie pride and truculence—the peripheral regions and provincial communities outside the Golden Circle of metropolitan England are generating cultures as pervasive as those from the heart of metropolis—whether West Indian Reggae, Cockney skinheads or Hampstead socialism. If culture is taken to be 'a system of concepts which stand in symbolic relationship to our experience, as well as fashion it for us',[23] then it must be considered in the context of the differential opportunities for social experience derived from the shifting relationship between peripheral community and metropolitan centre—between localism, or provincialism, and (national) hegemony.

The idea of metropolitan dominance is not new, but the use made of it by Frank[24] is. Stein had revived the term (attributable to Park) to denote the impact of urbanization, industrialization and bureaucratization on the communities of America, but he uses it within the context of the mass society theme.[25] But for Frank, strongly influenced by the experience of Latin America post-war, dominance from the centre becomes the process by which the peripheral communities are restructured; concurrently, the centre, metropolis is itself transformed. He sums up his own thesis as: 'That in chain-like fashion the contradictions of expropriation/appropriation and metropolis/satellite polarization totally penetrate the underdeveloped world *creating* an internal structure of underdevelopment.'[26] (Emphasis mine.)

Not a quiet disintegration of the world into an amorphous undifferentiated penumbra to the metropolitan centre, but a polarization which is replicated down the chain of expropriation, world, national, provincial, local—at each level, there is a community which replicates the contradictions of capitalism. The global relations of imperialism are recreated with increasing intimacy as beads of a chain from world metropolis to remote hamlet. And at the same time, metropolis and satellites are in a state of tension and antagonism as metropolitan centre draws further away from provincial community. Frank poses the same political issues so tortuously considered by

Tönnies. But whereas Tönnies could only foresee the dissolution of the older social groupings, including local communities, and their substitution by the institutions of the modern state centred on the metropolis,[27] Frank's model embraces the re-creation, not the dissolution of community formations, and reverses the impulse of change from centre (metropolis), to periphery (rural-satellite-province).

However, only the looseness with which Frank employs his master terms 'surplus expropriation/appropriation' and 'metropolis/satellite polarization' allows him to speak both of spatial relations and the relations between social classes in the same terminology. This weakness was early seized on by Mexican critics, and it can rightly be argued that 'the collapsing of social relations and spatial relations into the same vocabulary denotes a lack of concreteness, or in other words, that a substantial residue remains to be properly explained'.[28] Equally, the flaccidity of the analogy of colonization and underdevelopment, as used by, for instance, Clark and Carmichael for the black ghetto, has subsequently been exposed by writers as diverse as Oppenheimer and Tabb.[29] Unlike a typical colony, the ghetto has no existence prior to capitalist urbanization, and no possibility of territorial control. The appeal is emotive; the designation of the ghetto as colony impedes progressive analysis of the context by which blacks are bound to limited neighbourhoods in the cities.

And if the language of dominance/dependency, metropolitan/provincial polarization is to be carried from the experience of the underdeveloped world back into the metropolitan societies, which superficially at least has attractions for insurgent planners, further correctives may have to be enjoined. There is another argument, which can be validated by the many community studies undertaken by sociologists in both America and Europe. The centralization of control over resources in the exercise of dominance from the centre, prevents awareness of this wider social reality, in that capital and control both become invisible hands, not evident in everyday experience. Metropolitan dominance in Britain, at least, has acted to truncate the locally-based provincial class system, and so for the antagonisms of class are substituted the small-scale gradations of provincialism. The head of the local status ladder is seen as the head, and his position is seen to be realizable, in that many of the present occupants of a high position in the local status hierarchy are known to have had humble origins. And so an image of boundless opportunity is built up. Only in rare instances has the expropriation been live, as in the nineteenth-century Oldham described by Foster or mid-twentieth-century Featherstone,[30] and a sense of common identity overwhelmed these claims of status. In fact the global antagonisms cannot be replicated. The local class structures are only pallid versions of the global antagonisms, and so status becomes more real than class.

In metropolis the argument must be turned. As Williams, echoing Simmel, writes:[31]

> In the twentieth century, metropolitan experience has been dominant. Yet no city, no metropolis is a thing in itself. What the modern city concentrates, both physically and in its dimensions of experience, is still a society, a whole society which in fact extends beyond the city even while it is dominated by it.

At the very least this indicates that in studying the cosmopolitan existences of metropolis, the immediate reality with which one is confronted is insufficient as explanation. To paraphrase, the appearance of things does not signify their real being. The slum populations of the world cities are in no better position to understand the circumstances of their expropriation than their provincial counterparts. For example, Stedman Jones's analysis of the casual poor shows how socialism failed to take root in nineteenth-century London. London did not lead the social revolution, but lost the initiative to the provinces—places on the periphery such as Glasgow, South Wales and the North East. He concludes, 'The law of uneven development had worked cruelly against its creators.'[32] In the language of Althusser, the economic dialectic is never active in the pure state. 'The contradiction between capital and labour is never simple, but always specified by the historically concrete forms and circumstances in which it is exercised.'[33]

Analysis of the national society in terms of the differential allocation of resources brought about by the exercise of control from the central metropolis shows that this contradiction will not be uniformly experienced within the one society. That is, class consciousness will not have the same dimensions from one community to another, and classes can never be national phenomena. So the comparative study of communities within the national society is vital to the understanding of class formations, and cannot be dismissed as secondary to the study of the relations of exploitation between classes.

The comparative study of settlements

Since the pioneer work of Christaller, and its elaboration by Lösch,[34] the concept of 'hierarchy' has been integral to locational studies in geography and planning. Order, both in the distribution and ranking of settlements, within a region, is seen as inherent, indeed rational, to the conduct of society, in that activities must necessarily be centralized; accessibility determines the level of control, or 'service', a given settlement provides. To a geographer, therefore, the uneven distribution of resources between communities and their discrepant growth rates comes as no surprise. Marxism, however, has not yet, except incidentally, been concerned with this dimension of inequality within societies. The refreshing quality of Frank's statements is that there is an acknowledgment of a spatial component to inequality, and an attempt, however opportunist this may be judged, to assimilate technically correct observations of spatial inequalities within societies to a Marxist theory of uneven development.

From this orientation it is possible to reopen the debate as to the continuance of comparative studies of settlements with the following suggestions:

(i) In principle no distinction need be drawn between metropolis, towns and villages: they are all local clusters which present their members with certain conditions of existence and bases of experience, and whose history is made up of their interdependence one with another. *Urban* sociology then becomes a specialism in the wider context of settlement studies.

(ii) A comparative, specific, and concrete methodology in which the position of the particular community formation in the total productive process is taken as the starting point should be adopted. Local communities are to be studied as definite articulations of the social structure, contexts specific in their life chances, their class systems, their forms of political association, and their cultures, but always in terms of their relational situation in a 'field' or matrix of other settlements.

(iii) This matrix of settlements is not random, nor are the variations between places beyond scientific analysis. Factors of geography, the distribution of the raw materials of economic activity, for example water, coal, oil and relationships of proximity, as for instance between the Thames Estuary and Europe, cannot be discounted. But neither can the control and centralization of resources from specific urban centres throughout history. For this reason it is suggested that a unifying theme for the study of communities could be that of uneven development or, more specifically, metropolitan dominance.

(iv) As the basis for a typology, it is suggested that the dimensions of 'dominance' and 'dependence', 'metropolitan' and 'provincial', be explored. While cities comprise our society, in that in the developed world urbanization has become the way of life, any one city is not society; some local communities are wholly dependent, others exercise dominance while themselves dominated.

As local studies proceed, the need for a typology becomes increasingly evident. To take two examples: the housing class model advanced by Rex and Moore in Birmingham, and the criticisms made by Dennis of planning and housing policy in Sunderland.[35] The Birmingham model is not only questionable within its own context but cannot be extended to other places where immigrants were confronted with rather different housing situations, as in either London or Bradford, and have had sharply contrasting opportunities for occupational mobility subsequently. And what happened in Sunderland was in a measure special to the depressed industrial towns of the peripheral regions. National housing policies were being implemented by professional groups whose criteria for judgment came from metropolitan standards of living, and were being applied to a population in a region whose income levels were only 80 per cent of those in the South East. Dennis's failure to place Sunderland in the national context weakens his own analysis,[36] in that the professions arbitrating over the future of the inner districts of the town are made to seem whimsically arbitrary and personally culpable, and not as furthering the hegemony of the directive centre. Equally, the local politicians' attempts to redress the region's long-standing inequalities of opportunity and living conditions in face of persistent and bland neglect from national élites, got no credit.

It is not hard to make a preliminary identification of the metropolitan centres of the developed world. They are all characterized by high rates of economic growth, from which derive labour shortages, inward migration, pressure on housing, transitional zones, or temporary housing, and large-scale commercial development backed by massive state investment in transportation. In all respects they are different from the provincial communities of, say, northern England, south-west France, or the southern states of the USA whose low rate of economic growth gives rise to a labour surplus evidenced in their unemployment rates, and the outward migration of their young adults. Population levels may actually fall, housing markets are depressed; the problems are those of environmental neglect, and under-use of the human potential of the local society.

These, however, are the symptoms, and some kind of measure ought to be found for *dependency*, as distinct from *interdependency*. As this crucial analytical problem remains unresolved even for the national economies of Latin America and the Caribbean,[37] it is virtually impossible to make a definite categorization for entities (cities, city-regions), which neither control their own revenue nor set up the priorities for expenditure. Obviously questions should be directed towards measuring the strength of local linkages in the area's economy: are, for instance, new industries being financed locally; is there a local commodity exchange, as in the height of Manchester's command over the cotton trade; is existing industry reliant on policy directives from head office elsewhere; in what respects is the local economy self-generating . . . ? Similarly, it might be possible to make estimations of the net outflow of 'the surplus' from given regions, or even areas within an urban region. But it is plausible that in a metropolitan society the most depressed and seemingly underdeveloped regions might well demonstrate the greatest economic autarchy, and net inflows of capital in the form of subsidy and welfare benefits. In fact these become almost academic questions in an economy where land itself is no more than a counter in the transactions of the major financial institutions, and the multinational and national corporations move almost at will across the territorial board.

Glass, in her review article of 1955, argued for, first, 'a systematic outline of recent and present urban Britain, of the variations in urban conditions—in environment, institutions, and society', and second, a 'concern for the general contours of urbanism'.[38] On the first account, there has been an accumulation of raw data in planning studies and historical monographs, but without any attempt to pull them together into the comparative study she sought. For the latter, the contributions have come from writers outside academic sociology.[39] Both the culture of cities, and the prevailing attitudes to the modern big city, have been ignored in the search for more positive models of urban development. And yet the debate as between 'dependency' and 'interdependency' needs to be discussed with reference to the extent of penetration of the local and regional cultures of Britain by the culture of the dominant bourgeoisies of the national metropole, i.e. that which Wirth denoted 'urbanism'.

Experience of, and knowledge about, the peripheral regions of Britain shows that there is only a limited acquiescence in the metropolitan way of life, and its criteria for private consumption, public investment *and* commercial development policies. At a very obvious level, fashions originating in the metropolis are adapted to local conventions, and some are never adopted at all. In a measure the same applies to local government and local firms, whose investment policies may be governed by local and personal considerations. 'Provincialism' has come to mean this localism and detachment from metropolitan life. The local community comes to have its own scales of deference and esteem, built up on a wide range of local associations, and buttressed by the local media, and there is a widespread but tacit rejection of the metropolitan way of life. The active resentment of metropolitan control, expressed in the nationalist movements of Scotland, Wales or Brittany, is symptomatic of the failure of the persuasive hegemony of the metropolis.

As in 1955 it is evident that a major cleavage in British society is that between the provinces and the metropolitan South East, one which, moreover, is articulated politically, and may well be met by devolution of administrative powers by the state. For the mass of the people, the routine and quiet satisfaction of provincial life is far removed from the restless cosmopolitanism and search for novelty, as well as the assurance of the metropolitan existence:

> in the Golden Circle they are aware, and they know that everyone else is aware, however much some of the provincials may scoff at the idea, that here the national pace is set, and the national trends formed . . . the people living here are part of *the* exclusive set-up.[40]

The chief value of Frank's approach to Latin American development for local studies in the metropolitan societies, is that it refutes, finally, the urban–rural distinction on which the ecological theories of the city were premised, and indicates an alternative basis for community categorization. There may be a distinction between *rural* province and dominant *metropolis* but it is one that derives from the economic and political connections between the two, rather than their ecological characteristics. The artificial dualism perpetuated in the idea of urban and rural sociologies cannot be maintained: far more important is the metropolitan–provincial distinction. Cities depend for their being on a network of activities that far transcend the boundaries of the urban settlement. They are 'the initiating and controlling centres of economic, political, and cultural life that have drawn the most remote communities of the world into their orbit, and woven diverse areas, peoples, and activities into a cosmos'.[41]

On this argument, urban sociology exists in an 'urbanized' society as 'urbanization' is never complete. The whole society is not uniformly urbanized, and communities have differential experience of the contradictions of expropriation and centralization. The traditional concepts of urban sociology take on a new complexion: 'urbanism' can be considered as the culture of the metropolis, and 'ruralism' is associated with 'provincialism', i.e. detachment and exclusion from metropolitan life. Urbanization can be considered either as the transformation of places as their status in the matrix of communities is altered, or, alternatively, as the incorporation of groups in society into the way of life of the controlling metropolis, and their acquiescence in the hegemony of those interests dominant in the metropolitan centre.

part one

The British experience

By 1900 the British élite had already lived with mass urbanization for a full century. This was then a rare experience, for no other society had undergone so marked a shift in way of life. But only as satisfaction in the new dispensation clouded in the last decades of the century were there purposeful moves towards the organization of the swollen towns. For guidance the urban bourgeoisie and the new middle classes turned to the Americans, French and Germans, all of whom had less familiarity with urban problems, but who already had taken initiatives in directing capital resources into unprofitable urban infrastructure. The characteristic features of British urban policy in the twentieth century—restrictions on the growth of existing towns, the encouragement of suburban development, controlled rebuilding of old urban districts, all without direct intervention by central government—took shape in the period before the First World War. In the decades 1890–1913, the ruling élite of the most urbanized society in the world were impelled to take stock of their towns, and the failures of private capital in their construction and maintenance.

The balance struck between the state as central government, its agencies the local authorities, and private investors, was, however, peculiar to Britain. The very earliness of mass urbanization, the relative isolation of urban aggregates, and their independence of existing concentration of power and wealth, had resulted in a diversity of communities resistant to direction by central government based on London. Arrangements for local government had to take account of these interests and attitudes. In the same manner, state intervention over investment into land or buildings had to respect existing arrangements for the circulation of capital. State control over urbanization in Britain has, therefore, been indirect, standardization of policy slow, and a considerable diversity retained. Although this is in statistical terms an urbanized society and in law subject to remarkable centralization, past experience and present practice contribute to a notable unevenness of development.

Urban sociology in Britain has its origins in the same period as the consolidation of policies for urban development by government. Traditionally it has been concerned with the environmental aspects of Welfare State policy, i.e. housing and land-use planning, and the issues for debate have reflected the shifting consensus as to the scope and directions of state intervention into development decisions. Of necessity, therefore, a discussion of British urbanization must embrace its urban sociology. It would be misleading to extract the academic discussion from this political context for the latter has given the vocabulary as well as the subject matter of urban sociology. Its frequent blandness of language and lack of specificity in analysis, which the account in this section reflects, are indicative of its role in this governing consensus. Its weaknesses are of more than academic significance.

As this section presents British urban sociology as integral to the response by the ruling élite to the problems experienced in the towns, the conventional headings—suburbs, slum, twilight area, housing, urban renewal, planning—have been used. These have been the debating issues of urban policy in this society. From this it should not be inferred that they are the basis for any future development of urban sociological study. As the language of liberalism and the Welfare State fall into disuse a different range of issues—industrial investment, labour migration, community politics, as well as questions of class and class culture, language and ideology—will come to the fore. A programme for the development of either urbanization or urban sociology in Britain cannot be derived from this survey of experience over the last century.

1 The British experiment*

1 Uneven development in cities and regions

There are in Britain, as in other developed societies, marked regional distinctions. One in particular is conspicuous, that between the stagnant economies of the provinces—Scotland, Wales, South West and Northern England—and the prosperity of the Midlands and South East. It is as noticeable in British society as the cleavages between slums and suburbs; of greater salience perhaps than the familiar division between town and country. The major differentials in way of life apparent between these two zones extend to the towns. Just as the countryside has been remodelled over the centuries to meet the requirements of international trade and urban markets, so now existing urban settlements in the provinces find their economic basis removed, reorganized or replaced, their fabric neglected or restructured, and their population induced into alternative forms of commodity production. 'Urban' as a category of experience pales in significance as relations of dependency assume new patterns.

In any assessment of British society in the past ten years certain geographical changes are evident for which there is no adequate sociological language. There has been a decentralization of productive activity from the major metropolitan regions, a devolution of administrative control from the capital, and a general deconcentration of population. New areas are subject to development pressures in the search for alternative sources of energy and raw materials, ever distant country districts are being penetrated for homes, for holidays and for work, detached from the major centres, and formerly remote regions are being reindustrialized. Underdevelopment has become a resource, as in Latin America. Cheap labour, cheap land, minimal congestion costs, and living conditions which attract the managerial class precisely because they do not raise issues of the 'urban problematic' are the inducements to capital investment away from established urban centres. Peripheral communities outside the metropolitan regions, some subject to persistent neglect for a half century, some longer, now find themselves, their labour force and their resources, reappropriated.

Urbanization, in the sense of concentration of productive activity into densely populated settlements, was an important phase in the transformation of agrarian Europe into capitalism. The antagonism between industrial or commercial town and the countryside was in the nineteenth century stark; day-to-day experience of work, housing, social life, political opportunity, confirmed the impact of the towns on old ways of life. The European sociologists could therefore argue that there had been a major break with the past (so shifting attention from class formation to the retrieval of community), and postulate a dualism between town and country. Urbanization was to become the shorthand symbol for the impact of capitalism as a fully developed mode of production on social and political life. 'The ideas [of town and country] mediate human interests and purposes for which there is no other immediately available vocabulary.'[1]

For example, in the writings of Tönnies, it is clear that he used forms of community organization—village, town, metropolis—as illustration of the underlying reorientation of social relations. The rural/urban distinction located his argument, but on his prediction 'the rural organization is doomed to dissolution',[2] its usefulness would soon be outlived. A generation previously Marx and Engels had also singled out the town/country separation as 'The greatest division of material and mental labour . . . here first became manifest the division of the population into two great classes . . . the antagonism of town and country . . . is the most crass expression of the subjection of the individual under the division of labour',[3] but had related it to specific class relations at a moment of history. The town signified the beginning of property having its basis in labour and exchange.

Urbanization once seemed an inexorable process: once traditional sanctions on rural migration had broken down governments were powerless to stem the movement from the land to the towns. The most decisive edicts against new building in the major cities were disregarded by all classes. The concentration of population and centralization of activity associated with the development of capitalist society were seen as inherent to the new mode of production, and not to be checked by direction, or reform in administration. So thought Marx and Engels, hence the exhortation to achieve the proletarian revolution before 'the abolition of the antithesis between town and country'.[4]

To those familiar with the capital cities of Western Europe, or its new manufacturing centres, in the nineteenth century, this was a natural conclusion. Urban growth was traumatic, for both rulers and ruled. The stagnation, even decline, of smaller centres could be overlooked, and interruptions in the previous expansion of the large cities disregarded. Now, however, it is the interruptions, the changing relationships between towns and cities that intrigue. Growth no longer impresses, the inevitability of urbanization is in doubt. Is it strictly accurate to speak of 'the accelerating concentration of the means of production . . . monopolistic state capitalism . . . progressively concentrating large masses of the population . . . establishing vast communal units to organize daily life'?[5] Empirical evidence from most developed societies can be assembled to demonstrate the converse: the deconcentration of productive units and the associated aggregates of population, and the reforming of the 'vast communal units' on a more personal scale.

The voices of dissent to this 'urban view of history' come from the regions. There it is apparent that concentration in control of productive forces in society does not necessarily entail centralization in location. A new industrial plant, a government office removed from the primate city, may generate renewed development in a peripheral region without any diminution in the web of controls radiating from corporation headquarters of central office. The spatial analogies in sociological vocabulary—hierarchy, lower/upper, centralization, centre/periphery, convergence, centrifugal/centripetal—have confused the debate so that issues of social structure have been subsumed in questions of geographical location. From a regional vantage there is no question as to the maintained concentration of power in the metropolitan centres, hence the efflorescence of nationalist movements, but equally the impact of development in the peripheral regions is inescapable. The environmental issues, the planning pressures, the social problems, and political movements, the turmoil in the social map of the region and its intellectual world all demand systematic investigation. Regional rather than urban issues come to the fore; in ideas, as in politics, the dominance of the metropolis is challenged.

Given the sensitivity of regional populations to the proclaimed hegemony of the metropolitan centres, it is not surprising that there has been a resistance to received wisdom in the social sciences, and a search for alternative vocabularies in which to articulate the changing situation. The dissatisfaction general among students of academic economics, geography, planning, politics or sociology, in this context, has been focused on development studies, in particular the neo-Marxist theories derived from experience of societies with apparently similar development issues, i.e. Latin America. Dependency theory, as expressed by Frank in his statements on underdevelopment, was derived in reaction to the subordination of the Latin America economies to those of the imperialist nations, and was advanced to explain both the distortions in their pattern of economic development and the nature and balance of political power in the subcontinent. The aspect of Frank's statements which has been seized on is that of the relationship of metropolis to province, one which is replicated in a chain of exploitation within the regions and subregions of a country, prevents autonomous development, and aggravates economic stagnation, the 'development of underdevelopment'.[6]

Superficially at least, there are tempting comparisons between developed and underdeveloped societies. Most of the former show marked internal disparities in development between one dominating region and its satellite regions: France, Italy, Japan, USA, are in this respect similar to Brazil or Nigeria. But this observation is no more than recognition that in all these societies the national élites are concentrated in regions appreciably more prosperous than the rest of the country. Frank's own statements were grounded in a careful examination of historical material especially for Brazil and Chile. They are to be understood as 'a specific aspect, or a mentally dissected part of a historically concrete of existing bourgeois society'.[7] His arguments as to the further economic and political development of regions within these satellite states can be extended only with difficulty to a society, Britain, with a much longer, and more intricate, history. The British Isles were once on the very edge of the developed world, and yet they achieved dominance over international markets and independent industrialization. Remote provinces became the industrial power houses of the world despite the concentration of capital in London and the opposition of agrarian capitalist and industrial bourgeoisie. Many of the conspicuous strands in political, scientific and intellectual life took shape and colour in the provinces. It is arguable that the hegemony of the metropolis was broken in the nineteenth century and has never been reassumed. And in this century centralization of power has not prevented economic diversification elsewhere and with it renewed political challenges to the state. At the very least the thesis of metropolitan dominance needs to be examined closely against the historical evidence.

Development studies, like all area studies, have an eclectic quality. Not only have the conventional distinctions between the social sciences in subject matter been disregarded, so that elements of anthropology, sociology, economics, geography and politics are all pressed into service, but the theories around which these disciplines have been built have been pulled apart to be reassembled in unfamiliar ways. Frank's own critique of sociological theory is well known; there are also restatements of classical economics from the West Indies,[8] of academic geography from North East England and elsewhere.[9] Marxism has met the same fate. Perhaps because 'straight' Marxism was only indirectly concerned with most of the issues of regional (and urban) studies, perhaps in reaction to its metropolitan characteristics, Marxist categories of analysis have not been arbitrarily adopted. The quest for an $n + 1$ science once met by sociology, now frequently by Marxism, is answered in development studies by a patchwork quilt of concepts, theories, empirical studies and polemic. It is doubtful whether a 'political economy of cities and regions' is a sufficiently precise specification of what is required to confer scientific status on area studies.

Since the era of Ricardo, Malthus or von Thunen, regional questions have been missing in 'political economy'. (The tremendous attraction of Henry George's book, *Poverty and Progress*, may well have been its novelty in relating popular concern about land to economic distress and political discontent.) And few academics now would have the audacity to tackle such wide-embracing questions as the consequences of regional disparities in use of resources and rate of growth for the economic development and political institutions of a society. Conversely, how could one relate regional disparities in development to a society's social structure and its status in the international economy? And yet it is questions of this level of generality and range to which answers are sought. Modern scholarship is more cautious, its field of study more circumscribed than that of writers such as Semple or Mackinder whose interests were in continents not islands, the comparison of societies not regions and cities.

Despite the interest of the Chicago School in the urban region as the natural area for study, sociologists subsequently have paid little attention to regional questions. Formality in conception allowed urban groupings to be lifted out of their regional matrix and preoccupation with issues of the environment emphasized the phenomenal distinction of urban and rural. The synthesis sought in regional studies was in methodological terms alien to the analysis directed by the sociologist at the city. The learned quasi-metaphysical appeal of the former has had short shrift from zealous students of urban systems. If the city has, in sociological concept, become detached from the region, equally urbanization has been treated as a process quite apart from issues of regionalism.

It is the contention here that the sequence and form of urbanization in Britain can only be discussed in the framework of a political economy of regional development. That there is a definite sequence to urbanization in this society cannot be in dispute: the contrasts between the medieval landscape of small towns, the three centuries, *c.*1500–1800, in which London was the only large city, the reversal of the rural urban balance in the nineteenth century, and the ebbs and flows of investment between regions and their centres this century, are accepted features of Britain's historical geography. The relationship of this sequence to the development of a mode of production and its class relationships is the field for debate. 'Political economy' has been construed too narrowly: in the hands of Adam Smith, or Ricardo, it was the master science: in the hands of Marx it was a sociology in the best sense of that word. To restrict discussion to the role of the state in the allocation of the surplus is to accept the liberal democratic definition of the realm of politics in a mixed economy/Welfare State.

Metropolitan dominance

In medieval times England was neither a developed nor an urbanized society: it stood on the periphery of the main trade routes between Europe and the East: its principal export was raw wool to the manufacturing centres of Netherlands and Italy, and its imports were principally luxury commodities. The towns were small, primitive places, with few substantial buildings, limited trade connections, and a provincial outlook. The organization and prosperity of the cities of the Hanseatic League were absent, as was the cosmopolitan vitality of the Italian city states. By 1377 it is estimated that London had a population of 35,000, while that of York, the second city, was 11,000. And although London had been the country's first city for 300 years, it accounted for less than 2 per cent of the then population of 2 millions.[10] The overwhelming predominance of London in the national economy was to come later, in the sixteenth century.

After 1500 London's growth dominated the internal balance of trade and capital accumulation. By 1600 the city's population was estimated to have been over 200,000, in a national population of some 3.5 millions, i.e. some 5.5 per cent of the total. By 1700 the estimates indicate a population of 550,000, something like 9 per cent of the national total. By 1801, then resident in the capital city, described by contemporaries as 'The metropolis of England, at once the Seat of Government, and the Greatest Emporium in the known world',[11] were about 12 per cent of the population of England and Wales. The only comparable city in Europe was Paris, with a population of some half a million, but in a nation whose population was then three times that of England and Wales. And despite the industrialization of the provinces throughout the nineteenth century, London's growth continued unabated, accounting for 15 per cent of the population in 1851, and 20 per cent in 1901. Nothing, it seemed, could shake the supremacy of the national capital in which wealth, power and prestige were concentrated.

What then happened in the sixteenth and seventeenth centuries? London 'took off' with the consolidation of the nation-state under the Tudors, the growth of foreign trade, particularly with the New World, the tightening grip of London's merchants and financiers on that trade, and the increasing capitalization of agriculture. London grew by plunder of the resources of the New World, absorption of the primitive accumulation of the national territory through monopolies of trade, expropriation by the state, and the establishment of London as the centre for conspicuous consumption. The opening up of new markets for agriculture meant prosperity for some in the countryside (evidenced in the substantial seventeenth-century farmhouses across the Midlands) and urban fortunes were lavished on the great country houses, but existing urban centres were stifled by London's monopoly of trade, commerce and industry. Throughout the sixteenth century accusations that London was causing 'the loss and decay of many [or most] of the ancient cities, corporate towns and markets within this realm, by drawing them to herself alone, say they, both of all trade of traffic by sea, and the retailing of wares and exercise of manual art also',[12] had been prevalent. Even by 1500 the cloth trade was controlled by a monopolistic corporation upheld by the state, the Merchant Adventurers, based on London. A 'distributive grid' was set up across the country bringing cloth to London for export. The traditional suppliers, the towns, found their trade undercut by rural cloth manufactories set up by the London cloth dealers; centralization on the national capital was supplanting the older complementary ties of provincial towns and local regions. Bindoff writes:[13]

> But in this aggregate of local market patterns there was being imposed a new national one, in which the provincial town was replaced by a metropolis which took the whole country for its province. Much of the economic and social—and not a little of the political and religious—history of the Tudor century can be written in terms of the growth of London; of a London which fed its vast population . . . from an ever-widening area of the country, which called into existence new rural centres of industry and doomed old urban ones, and whose wealth brought the nation increasingly under its domination.

In this way a new balance was struck between town and country. Instead of a reciprocity in exchange of goods and services on a local basis, the economies of remote regions were reoriented to commodity production for the London market—the woollen industry of the

West Country, and coal-mining in the North East being the two conspicuous examples.

Such a leap forward in one sector of the economy clearly meant strains in the national society. London was absorbing a disproportionate share of the nation's wealth, and for much of the seventeenth century an estimated half of the national increase of population.[14] The disruption in traditional society was greatest in the South East: this region became the most heavily commercialized, the most closely incorporated into the market economy focused on London, and the least traditional in its attitudes, and was to become the stronghold of modernism in the Civil War. It appears that the areas of Parliamentary strength in that conflict were London itself, the East, the Midlands, and the South East, while the North and West were the strongholds for the Royalists. That is, the areas closest to the metropolis were strongest against the Crown, and those remotest, most drawn to support. Barrington Moore comments[15]

> the result is almost what one would expect in a society where capitalist and generally more modern ways of thinking and acting were forcing their way up through an older social structure. This new world had its centre in London, from which its influence radiated out most strongly to the South and the East.

There was then evident a major antagonism, which was not that between classes, nor between town and country, but between the remote provinces and those regions subordinate to the major urban centre. The first stages in the bourgeois revolution were forced through by those most aware of the restrictions in their economic activities by traditional authority, an authority which none the less had secured the basis of economic modernization, and its corollary, urbanization without industrialization.

The pattern set in the sixteenth century was not seriously altered until the last quarter of the eighteenth century. New trading outlets, particularly with the Americas—sugar and slaves—encouraged the growth of the ports of Bristol and Liverpool;[16] the development of industry in the provinces enabled new urban growth—by 1750 towns such as Leeds, Bradford and Manchester were already sizeable trading centres, and the general prosperity of the country had lifted many existing towns out of their earlier stagnation. But the disjuncture between London and the provincial towns was still great—in size if nothing else—and there was a great gulf between the wealth, prosperity, wit and culture of the metropolis and the quiet self-satisfaction of provincial life. And yet it was the provinces that were to generate the new industries by which Britain's supremacy in world markets was consolidated. The new industrial centres were in areas remote from metropolitan influence, and developed on local initiative and with local finance. So, for example, the mining, engineering and cotton manufacturing industries of Oldham in the first decades of the nineteenth century were dominated by a few families with investment in local land.[17] The importance of cotton in the industrial revolution was precisely its low capital requirements —its machinery 'could be installed, if need be piecemeal, by small men who started off with a few borrowed pounds, for the men who controlled the great accumulation of eighteenth-century wealth were not greatly inclined to invest large amounts in industry'.[18] The great industrial towns developed apart from the national centre and its ruling élites, without their financial support, or state backing.

Nineteenth-century urbanization

In 1801, 85 per cent of the population still lived in country districts: it was industrialization and the development of the Victorian cities of the North and Midlands which transformed the balance between town and country, and brought the British population into the cities. The reality of these industrial towns is with us yet: the factory chimneys, industrial devastation in spoil-heap and subsidence, square mile on mile of cramped terraced housing, and the proud monuments to civic confidence and local pride. The reaction of established opinion was puzzled, amazed, and in a measure hostile to the barbarism of the new industry:[19]

> the chimneys smoked, the ceaseless roar and mighty beat and dizzying whirl of machinery, struggled and strove perpetually. Senseless and purposeless were wood and iron and steam in their endless labours; but the persistence of their monotonous work was rivalled in tireless endurance by the strong crowds.

But 'Manchester man' believed himself to be the equal if not the better of his metropolitan equivalent. The new cities became focal points of affection and loyalty, demonstrated in rivalry among themselves, and in antagonism to the claims of the national capital. In their view the metropolis was a parasite on the national growth, its cultural pretensions disputed, and interference in economic affairs bitterly resented. In a measure initiative did slip from the metropolis: wealth accrued to the new industrialists, finance and trade were not so firmly centred in London—Manchester too had its stock exchange, and controlled the country's principal export industry, and the political initiative of the provincial élites had to be reckoned with.

Kiernan raises the question of the long-term significance of the geographical demarcation between the older and the newer ruling classes, and suggests that it 'reflected a drawn battle or compromise between them, an arrangement by which the millowners instead of ousting the landowners allowed them to go on running much of the State apparatus',[20] a demarcation which later in the century was to be redrawn between provincial industrialists and London's financial plutocracy. If, earlier, Manchester and Birmingham liberals had fought London's Whig aristocracy, for political representation and free trade, late century the battle lines were drawn between conservatives seeking imperial protection for provincial industries, against metropolitan liberals intent on maintaining unlimited frontiers for trade and financial transactions: Chamberlain from Birmingham was to oppose Rosebery, a Rothschild son-in-law. The implication is that the predominance of the interests of finance capital in the conduct of the economy in the closing years of the Victorian era, and subsequently, has an underpinning in the long-standing hegemony of London.

As the profitability of manufacturing industry declined in the last quarter of the nineteenth century, and London gathered control over the money market, trade and the Press,[21] there was a gravitation to the capital of informed and critical opinion. Whereas in the early years of the century it was to the new cities of the provinces they had responded and recoiled, 'Everywhere barbarous indifference, hard egotism on the one hand, and nameless misery on the other',[22] later it was the metropolis which governed the attitudes of the professional classes from which the intelligentsia was drawn. London, with its tight-packed alleyways and courts, its crowds and its unknown poor, was the source of moral conscience and political turmoil, and not the factories and working people of the industrial towns. Simon, working in London, had succeeded Chadwick as the instigator of public health legislation,[23] and the great social survey of the end of the century was that initiated by Booth (a Liverpool businessman drawn like so many others into London) in 1886. Whereas the models for new forms of environment had been due to provincial initiative, now they came from the heart of London: Owen's vision at New Lanark, based on his experience of Manchester's turbulence and squalor, is superseded by that of Howard, oriented to the colonization of a poverty-stricken countryside by London's poor. And Engels, fifty years on, had shifted his attention and hopes for the future from the factory proletariat of Manchester to the masses of East End London. 'The revival of the East End of London remains one of the greatest and most fruitful facts of this "fin de siècle".'[24]

The situation and condition of London in the nineteenth century, economically and politically, merits consideration. There is the seeming anomaly of the world's major city (6.5 million in 1901), capital of the first industrial nation, which was itself largely non-industrialized, and had at its heart the greatest pool of unemployed and underemployed in that nation. Economic activity in London revolved round its role as the nation's first port and major distribution centre, its position as the major single domestic market, and its status as centre of court, government, and the administration and control of the affairs of international trade and empire. The first of these (taken together with the seasonality of building work) helps to explain the significance of casual employment in the labour market, quite unlike conditions in the factory towns; the latter meant a demand for specialized rather than mass-produced goods. Consequently at mid-century, in industrial terms, London was primarily a finishing centre for consumption goods, clothing, furniture, engineering, printing and precision equipment.[25] Stedman-Jones estimates that in 1851 86 per cent of firms employed less than ten men, only twelve concerns 300 or more.[26] And even at the end of the century less than 20 per cent of the industrial labour force could be said to be in factories—the small workshop predominated. The impact of industrialization in the North and the dissemination of consumer goods by rail had meant the collapse of traditional London industries, and the creation of conditions in which industry could only survive by minimization of overheads—i.e. cutting of expenditure on fixed equipment—buildings and machinery, and on wages. The infamies of outwork and the sweatshops documented by Mayhew were the result.

Unlike the factory towns, therefore, where the social divisions were only too apparent, and the motive power underlying these divisions readily discernible, there was no such clear pattern to London. The upper class viewed with contempt the pretensions of the new middle classes, professional men, clerks, shopkeepers; the latter knew little of the world of the artisans, still closely tied to their workshops in the inner city, even less of the lives of the mass of the casual poor. They were shocked and stirred by the revelations of the missionary explorers to this underworld to their city. Nor was the productive basis to this phantasmagoria evident—no factory chimneys and roar of industry; in their stead the paper transactions in city offices, quiet conversations in the clubs, and the gentle movement of shipping on the Thames. It is not, therefore, surprising that the problem of London was defined in traditional terms as that of the 'poor', not the 'working class', and their plight was seen as soluble through charity and enlightened guidance by the middle classes. In Victorian London 'for many of the middle class reformers, the poor were—and are—a problem of restricted consumption, not the failure of the productive machine to provide jobs':[27] with good reason—there was no evident productive machine to the metropolis.

Given these conditions it is surprising that Engels, while acknowledging the difference between the capital cities, the ports, the manufacturing cities and the industrial towns, could write in 1844, 'What is true of London, is true of Manchester, Birmingham, Leeds, is true of all great towns. . . .'[28] Both in description of living conditions, in selection of the factory proletariat as front runners in the 'social war', he tacitly recognized the gulf between 'the modern manufacturing town', epitomized in Manchester, and the established centre of nation and empire. But by 1872, London, together with Paris, are taken as the models of the 'modern big city'.[29] There is a failure to distinguish between the urbanization that followed the concentration of productive activity at railhead and canalside, and that which accrued to the major centres of finance and commerce; the centres of commodity production were only exceptionally seats of bourgeois power. The significance of the gap in experience and political awareness between the new factory proletariat of the provinces and the submerged poor in a metropolitan economy dominated by finance, trade, and the production of luxury commodities is not considered. The great mills, ironworks and coalpits of the northern towns concentrated the working population of the whole town in a tiny compass; there was no equivalent experience for London.

With industrialization of the national economy, working conditions for the mass of London's population had deteriorated through the century, as had their housing conditions, and yet their numbers continued to increase—this 'urban residuum' formed a reserve army of labour as marginal to the society of the city as to its economy. Their world was the world of the rookeries of central London—St Giles, Holborn, Clerkenwell, Shoreditch, and the vast extent of the East End, described by Park as the 'most remarkable of these cities within cities . . . of persons of the same social class'.[30] How was it this huge labour pool was not utilized, why did London continue to grow, why did it not 'get on the industrial track'?

London was remote from the major sources of energy, the coalfields, but it had imported coal from the North East for three centuries; land was at a premium at the centre, but cheap on the outskirts, as was evidenced by housing speculation throughout the century; there were the advantages of a large market, large pool of cheap labour, and the world's major financial exchanges. Yet these last two conditions were precisely those which impeded industrial development. The financiers of London were notably disinterested in industrial investment, and the very presence of a seemingly inexhaustible labour supply was a deterrent, in these circumstances, to mechanization and factory production. There is therefore the paradox of small-scale workshop production, operating on minimal fixed capital, in the financial centre of an empire, and, also, the continued movement of a pauperized population into a city unable to give them a livelihood. London, as the only major urban centre in the heart of the agricultural counties, whose labour force was pushed from the villages and small towns with ever greater severity of working and housing conditions, took the brunt of the final stages in the dispossession of the rural population from the land in the closing decades of the nineteenth century. The slow death of agriculture marked the supersession of agrarian interests in the English ruling class, the victory of commercial and industrial capital over landed capital. Investment in the countryside virtually ceased from the 1870s, and the rural areas, if not the playground of the rich, were left to stagnate for a century. The plight of inner London was directly related to the immiseration of the countryside.

Contemporary observers from both Europe and the New World considered the concentrated misery of the population of inner London to have no equivalent elsewhere in the world. The world's greatest city had its most extensive slum, a constant threat to public order and, with enfranchisement of the municipalities, to democracy itself. It was, moreover, a slum that seemed to be without solution. The strategy of civic improvements wielded with such draconic effect in Paris by Haussmann, in London had left the East End virtually untouched. The worst of the rookeries of central and western London had been eliminated, as for instance with the cutting of Victoria Street through those of Westminster, and although the East End had been affected by the encroachment of the docks and associated warehouses along the Thames, these had only exacerbated the problems in the tracts behind these frontages. While the very newness as well as the dynamism of Chicago or New York's growth was to wipe out the inner slums, and *étatist* strategies were politically acceptable in France, in England the question of the slums was to vex a generation of discussants and legislators. The answer finally was to depend on the provision of housing outside the existing city limits by the municipalities, and cheap transport. The working-class suburbs of London, as elsewhere, were to depend on cheap energy—coal from the pits of the provinces, and cheap land, land which seemed to have no long-term value for the production of food.

Frequent comparisons were drawn between the experiences of Paris and London, the two great cities of Western Europe; Paris in a state of continual political turmoil, a constant reminder to the ruling classes of Europe of the explosive potential of urbanization, London always growing, inexorably expanding, but without co-ordinated administration. The executive power held by Haussmann under the direct aegis of Napoleon III, which enabled him to provide a new sewerage system, water supply, network of public parks, and transform inner Paris from a sordid medieval city into a city of orderly boulevards, was the admiration of the civic reformers.[31] Chamberlain's reconstruction of central Birmingham, and the institution of the London County Council were inspired by the spectacle of the reconstruction of Paris between 1852 and 1870.

In that period some 40 per cent of the houses in inner Paris were demolished, and 85 miles of major highway constructed; in the 1850s, 20 per cent of employment in Paris was in the building industry.[32] Population rose from 1.053 million in 1851 to 1.970 million in 1870, largely in the new suburbs, for, like London, large tracts of the inner city were emptied of population. The economic impact of the project is not easy to assess: the expenditure was huge—in 1868 Haussmann was forced to ask for retrospective approval for loans covering the period 1863–8 which were equivalent to a quarter of the annual French budget.[33] The entire programme had required unorthodox financial practices—new credit institutions, and the persuasion that municipal expenditure was an investment and not an outlay against revenue. In this respect capital circulation had probably been speeded up in that property investment tapped small savings which had no other outlet. The policy of the reconstruction of Paris as the prestigious centre for a modern nation, a fit setting for finance, trade and large-scale industry, incidentally providing employment for the restless unemployed of the city, was an integral part of Bonapartist policies for the modernization of France. But perhaps as important was its impact on the social structure of the provinces. As yet[34]

> The capitalist part of the French economy was a superstructure erected on the immovable base of the peasantry and petty bourgeoisie. The landless free labourers merely trickled into the cities, the standardised cheap goods which made the fortune of the progressive industrialist elsewhere lacked a sufficiently large and expanding market. Plenty of capital was saved, but why should it be invested in home industry?

Unlike that of England, the rural population had rights in the land, and were not so easily displaced; building in Paris accelerated urban migration, in that it tapped traditional skills, for example masonry,[35] and eroded peasant, and provincial, self-sufficiency.

The development of Paris at that period, as subsequently, is not comparable with that of London. Though ostensibly their position in their respective nations was similar, dominant centres of finance and exchange and specialized commodity production whose interests were often opposed to those of the provincial towns and rural hinterland, one was a centre of a fully developed capitalist economy, and the other that of a partially modernized society, with a very much more limited empire. So London was a city whose pattern was shaped by business interests, with only partial intervention from state agencies, and Paris a city whose economic, social and physical development was directed by the state. London was a city growing beyond its capabilities with the movement of landless labourers to its streets—the culmination to the long centuries in the elimination of the English peasantry, and the subordination of country to city; Paris was still the centre of an agrarian society based on a peasantry who were to be impelled into the cities. The differences in the political economy of the two societies is manifest even in the physical appearance of the two cities.

The obvious comparison is between the orderly appearance of Paris—in layout, height of buildings, and regularity in their design, and the haphazard development of London. More specifically, Paris, like other major continental cities, was densely built up, with high plot coverage and a density in housing unmatched in London. The low density profile of the latter, the provision of houses rather than apartments, was to lead Rasmussen to term it the 'Unique City'.[36] Conventionally the explanation for these differences is phrased in terms of the vulnerability of foreign cities, the need for fortifications, and the restrictions on building outside these defensive lines. More important, however, were the fiscal divisions: a sharp distinction was enforced between town and country to safeguard the collection of urban taxes, an important source of revenue to European governments. So the eighteenth-century wall round Paris, the 'Wall of the Fermiers-Généraux', was built to ensure the collection of the levy on goods entering the city, and development outside this boundary was strongly discouraged. Important also, eventually, was the control by the municipalities over land in their boundaries and their desire to maximize revenues through intensive development. Also significant were the differences in land tenure between England and the continent. In the former landholding had early become commercialized and treated as an income-yielding investment on which leases rather than yearly tenancies were granted. For urban land these leases might be granted for long periods—often ninety-nine years—so that the realization of ground values for the landowner took place infrequently. This attitude to land as an investment indicates both the consolidation of land in a few large estates, and the close connections of the landed classes with commercial interests. As Rasmussen argues for the nineteenth century:[37]

> the owner could afford to wait. In London which was a commercial city and where it was now possible to build as

> far as one wished, speculators thought in terms of building speculation rather than land speculation. . . . The money was used to produce something and the investor was not interested in building as many houses as possible on a plot of land, but only in building houses as attractive as possible.

By contrast, in situations of fragmented ownership, and shortage of alternative investment outlets, building would be delayed until the greatest profit could be obtained by the most intensive development permissible.

There were in Britain strong incentives to develop land outside the cities after 1870. Agriculture was increasingly hit by the penetration of domestic markets by foreign suppliers, and the price indices for agricultural produce fell steadily. For wheat the index dropped by 40 per cent between 1870–4 and 1910–13, and for wool by 50 per cent; livestock production was less affected. In addition to economies in labour as farmers converted land from arable to pasture (the agricultural labour force fell from 986,000 in 1871 to 665,000 in 1911), there was a severe fall in rental income. Estimates vary, if only because of regional variations. In general there seems to have been a reduction of 25 per cent in rents between 1870 and 1900 (and a stabilization at this latter level subsequently), with falls of up to 50 per cent in the south-eastern counties. In many of the latter areas 'the bottom dropped out of the market'.[38] The heavy clays of Essex in particular became virtually worthless.[39]

Given phenomenally cheap money, 'Britain was enjoying such a surfeit of liquidity that the market rate of discount was below 1 per cent',[40] the speculative developers of the London suburbs in the building boom of the 1890s were able to take full advantage in the collapse of the rural land market. This new suburban sprawl was seen as a marked departure from previous development of London; not only was the green heritage of the Home Counties being swallowed up at an unprecedented rate, but the new developments broke with the conventions of orderly estate development and the long-term maintenance of property values.

> Estate management in terms of good design, in terms of well balanced land-uses, and especially in terms of long-term profitability . . . was fundamental to early concepts of urban and rural development. Formal statutory town-planning attempted to regulate development where good estate management was absent, and to create and regulate development through the municipalities where the ancient patronage failed.[41]

The lobby for town planning at the turn of the century came not only from the countryside preservationists, but also from the surveyors, a profession as concerned as the architects with the new institute of town planning.

By the time that Unwin's influential tract, *Nothing Gained by Overcrowding*, was published in 1912, conditions in the landmarket were such that he could argue that there were considerable advantages to landowners from low density development as more land in total would be converted from rural use (at values of *c.* £40 per acre) to housing (at values of *c.* £300–500). He gives the example of development at thirty-four houses per acre (standard by-law layout), which would require 100 acres, and that at fifteen houses per acre which would require 227 acres for the same number of dwellings. In the first case a development increment of £45,000 would accrue to the landowners, in the second, that of £102,000. He concluded 'in spite of the fears of the land-owners and the speculative builder there does not seem to be any reason why town planning should not prove to be of great benefit to both parties'.[42]

The incentive to extensive suburban development for the builder was straightforward—cheap land, and building costs kept to a minimum in two-storey development. For the new suburbanites, as Unwin emphasized, the advantages were those readily appreciated by every prudent family man—space for house and garden with relatively little increase in rent or purchase price. It is hard to interpret this movement as an anti-urban reaction: the middle-class families moving to the suburban estates were precisely those who played the greatest part in the activities and associations of their city, and took greatest pride in its achievements, as, for instance, in Manchester. That status did accrue to suburban living is indisputable—the centuries of convergence of agricultural ruling class and urban bourgeoisie, the traditional conversion of urban wealth into country house splendour, particularly round London, ensured the suburban villa a social *cachet*. It is perhaps therefore ironic that it was the collapse of landed interests which permitted mass development to overtake the selective suburbanization of the more affluent. The invasion of the countryside by the towns, most marked for London, and least evident in the industrial townships, marked a transition in the social structure. It symbolized the eclipse of landed interests in the ruling class, and the pauperization of the rural districts, the growing importance of the new middle classes and a domestic market for consumer durables, and was one factor, among others, in precipitating intervention by the state in urban investment.

In the opening years of the twentieth century the lines of demarcation in British society seemed well set—a small upper class increasingly integrated as provincial industrialists sought to assimilate with London élites, the growing but still numerically insignificant middle classes, and the great mass of urban workers; agricultural workers were an unconsidered group. On the ground these distinctions were to be read in a segregation in the towns as pronounced as that encountered by Engels in Manchester in 1844. The conspicuous lines of cleavage territorially were not then the regional distinctions which have since become the subject of so much debate, but those of the towns, between slum and suburb, and that between town and country. Howard in his opposition of town and country magnets, the one of social opportunities and places of amusement, highways and unemployment, and the other long hours, low wages, lack of society, no public spirit,[43] was expressing the known conditions of two ways of life.

Regional underdevelopment

Unevenness in development is inevitable given the natural distribution of resources and the differential attributes of location. The question at issue is that of underdevelopment, development whose form is ordained by interests outside the area, which once set in train prevents its ever achieving autonomy. At worst the region is denuded of resources, its ecology destroyed, its people rendered paupers, at best economic advancement is dependent on outside financial interests, the social structure modelled on that of the dominant metropole. The stagnation and actual decline of some regions in Britain this century can either be interpreted in terms of technological innovation or with relation to the specific features of the political economy of twentieth-century capitalism.

For instance, the national grid and the combustion engine both had important effects on industrial location and the use of urban land: it would be foolish to pretend otherwise. The first has freed industry from locations in proximity to coalfield and railhead, and the second has released warehouse distribution from the imperative of a central location in the urban market, allowed industry to move to sites in the suburbs, and widened the universe of choice for a private household in search of a home. The introduction of new materials, for instance, aluminium, cement, plastics and artificial fibres, similarly entailed new production processes, and different criteria for plant location. But equally important have been the conditions of production. The key words here are rationalization and mechanization, i.e. higher capitalization, higher productivity per man hour and larger units of production.

A crucial period in the reorganization of the economy was that between 1919 and 1939. In conditions of international recession, industrial production was increasingly geared to domestic (mass) markets. The production and scientific management techniques pioneered in America in the early years of the century became wide spread in Britain in that period. Previous skills were devalued and the use of untrained labour in industries other than textiles became widespread. At the same time new forms of industrial organization became more conspicuous—joint stock companies, trade associations and combines such as ICI.[44] This process of rationalization and the extension of monopolistic market conditions proceeded with the backing of the banks and the direct encouragement of the government, in reversal of previous disinterest by finance capital in industrial production and a switch of previous attitudes towards monopoly and free competition. And no restrictions were placed on the location of new industries.[45] The Royal Commission of 1940 reported that the movement of industry had 'proceeded with little or no regard to the fact that it necessarily involves heavy expenditure by the community for . . . new roads, housing accommodation, water supply, sewers, gas and electric mains, schools, churches, increased transport . . .'.[46]

The consequences for industrial location in the inter-war years, and subsequently, are beyond dispute: rationalization of production in large units, accelerated by merger, take-over and 'asset-stripping', meant the closure of plants in small centres, as well as the inner districts of the towns; and the concentration of production in new plants in proximity to major markets, and with access to new sources of labour. There was a national redistribution of commodity production. London and its suburbs underwent an 'industrial renaissance' and the region of acute labour surplus became the centre for industrial growth. Industrial towns declined, the great manufacturing cities stagnated, and modern industry transferred to small towns (Banbury), historic cities (Cambridge or Oxford), or the countryside peripheral to London, all areas without an industrial tradition, no background of organization in the labour force, and within reach of the centre of power, London. As this industrial reorganization and relocation was accompanied by a considerable growth in distribution, and white-collar employment, also concentrated in London, the dominance of the metropolis was consolidated as never before. Commodity production, in the particular conditions of Britain inter-war, required a docile labour force, willing to accept assembly line and shift working, which the metropolis and its poverty-stricken rural hinterland provided, closer connections with the institutions of finance, and the state, as well as access to the established market for consumer goods, the middle classes of suburban London.

In absence of control over industrial or housing investment, the dream of Howard's contemporaries—the solution of the problems of the metropolis by the resettlement of its unemployed in the countryside—was realized in suburban sprawl rather than the controlled integration of industrial production and community development. New industries located in the suburbs for mass-production techniques, continuous assembly lines, large-scale production, all required extensive sites with good accessibility, all conditions fulfilled by suburban rather than inner city locations. Initially, at least, this entailed outward commuting from the inner districts to the new plants—at Longbridge, Birmingham, in 1938 half the work force had to travel more than five miles.[47] At the same time there was a rapid increase in city centre employment with the proliferation in the selling and distributive mechanisms of offices and shops. Meanwhile, new housing in the suburbs was largely for the white-collar market, unrelated to proximate industrial development, and commuting increased.

The impact on the regions can be demonstrated variously: unemployment rates in the 'Special Areas' were consistently double those of the South and Midlands throughout the 1930s, that in South Wales averaging 30 per cent from 1929 to 1936; the differences in terms of protracted unemployment were even more marked. Between 1931 and 1938 London with the South East absorbed an estimated million out of a total increase in population for England and Wales of 1.2 million. The provincial townships were left as the nation's twilight areas, industrial graveyards, with a population that stubbornly defied conditions of state-administered poverty to remain in their communities. Despite their youth as communities—the greatest incursion of population into the Rhondda, for instance, came only after 1870, the boom in shipbuilding and engineering on Tyne and Wear only after 1880—industrialization had forged strong ties that were not so easily broken as those of their rural forebears. And yet there was a persistent refusal by government to encourage alternative employment, until after 1935, and the cartelization of industry blocked the establishment of new plant, as at Jarrow.[48]

The cleavage between metropolitan and provincial regions which is still so conspicuous a feature of Britain's social geography hardened in the inter-war years. The relative decline of urban industrial communities as well as rural districts can be related to certain features of twentieth-century capitalism—the engagement of finance interests with commodity production, the establishment of monopoly in capitalist organization, the involvement of the state in regulation of terms of trade and conditions of production, and the reorientation from an export-based economy to the production of consumer goods primarily for the home market—at their most blatant, unmediated, in the 1930s.

It is not often appreciated that regional inequalities have widened post-war, for a more rapid rate of economic growth, as indicated in the doubling of real wages in the twenty years after 1945, has blurred the more obvious disparities. But in terms of employment opportunities, wage levels, housing conditions and availability of social services, the provincial regions have dropped behind the 'Golden Circle' of London.[49] For example, unemployment rates in the North are consistently double those of South and East, wage levels in the North East are estimated to be only 80 per cent of those in the latter region, despite the pockets of poverty in inner London; again with the exception of inner London, housing need is most acute in the regions, and in terms of professional or social services, the South and East score on all indices of availability. The disparities in conditions of existence and life possibilities remain, and are reflected in rates of population growth. Between 1951 and 1969 it is estimated that the Midlands and South East absorbed 78 per cent of the population increase of England and Wales.[50]

The consequences of this imbalance for the regulation of the economy were not openly debated until the late 1960s. Then it was argued that the differentials in demand for labour, which in the South and Midlands periodically led to labour famine, high wages and demand for consumer goods, had had inflationary consequences which until then had only been checked by national 'stop-go' policies which had pushed up unemployment in the Development Areas to politically unacceptable levels. Selective Employment Tax was introduced as a regulator which could take account of these regional variations.[51] Subsequently, the endemic shortage of labour for service industries in London has threatened nationally-based income policies. In effect a premium has to be paid for the continued functioning of London as a world city.

In this period, however, two counter trends to industrial location have been evident. One is that, as in nineteenth-century Britain, the raw materials required for continued growth are to be extracted or processed away from the major urban centres. So the Highland Zone of Britain which once provided its timber, coal and iron ore, now is to provide its oil, and other raw materials—phosphates, tin, copper, limestone and water. Areas remote from the main urban agglomerations—Teesside or the Trent Valley, or Wales, or the Western Highlands of Scotland—provide its energy and process key raw materials. Here are conditions of industrialization without urbanization—in this epoch the demands on labour are small and the labour catchment area scattered.[52]

And second, the concentration of industrial investment in the key regions of the South East and Midlands, taken together with the ever spiralling growth of the tertiary sector, carries its own external costs. Shortages, and consequent high cost of labour, land and housing, and congestion costs both in communications and amenity, lead to the decentralization of industrial production beyond the metropolitan region. Whereas London once exported its poor to the factories of Lancashire and the mines of Wales, it now exports its assembly plants and process divisions to the unemployed and under-employed of the nineteenth-century towns. Control by London head office over short-term management is of less importance, the banks and financial institutions are themselves decentralizing, as are government agencies, and for the multinational corporations with international markets, access to home markets may be inconsequential. The costs of the over-urbanization of the metropolitan regions, and the relative benefits of underdevelopment in the regions—cheap labour, land, housing and services—now override the disadvantages of decentralization. While the promotion of this movement by national government and local interests cannot be discounted, its gathering momentum has to be related to Britain's failing position in the international relations of exchange—the once dominant metropolis is now a dependent economy forced to make full use of its own resources in labour as in minerals and land.

Uneven development—the new suburbs

Extreme unevenness in regional development in the 1930s was matched by growing disparities within the urban centres themselves. Not only was there a mass influx into the 'coffin' of England from the Depressed Areas, but an accelerating migration from old residential districts into the new suburbs. The very condition which led to the stagnation of the peripheral regions stimulated suburban expansion to the towns. A more rapid rate in growth of GNP than hitherto, based on protection and Imperial Preference, cheap money (the bank rate pegged at 2 per cent between 1931 and 1951), industrial rationalization, and diversification, spurred on a massive investment in housing outside the existing towns. This was the suburbia which horrified contemporary observers, and whose encroachment on traditional landscapes lent emotional fervour to the case for statutory planning. It represented the life ambitions and savings of millions, undoubtedly gave further impulse to the recovery of the economy,[53] and confirmed the new pattern of urbanization. While some regions became economic backwaters, their towns starved of investment, cities elsewhere became regions.

It is with everlasting amazement that one looks at the inter-war suburbs. Crippling continuous unemployment, as on Tyneside or South Wales, was the experience of some regions; mass suburbia, as in the Home Counties, that of others. Widespread poverty, related to unemployment, irregular work and low wages could be matched in the same town by the cosy 'Welwyn', 'Cranleigh' or 'Thorsgarth' of the new suburbs. In a town like Hull where unemployment rates in 1938 were still at 11 per cent (the national average), the built-up area of the town virtually doubled between 1919 and 1939 as 27,000 houses were constructed (16,000 private, 11,000 public), without an increase in population. The suburban explosion, houses, shopping centres, schools and factories, meant a new way of life for the housewife, looking to her roses and the Townswomen's Guild for diversion outside the semi-detached home. In total 4.3 million houses were constructed inter-war, 3 million after 1930, and of the latter 70 per cent were privately financed. At 1939 one-third of the housing stock was less than twenty years old and investment in housing as a proportion of GNP had risen from rates of 1.2 per cent—1.7 per cent in the period 1900–14—to an average of 3.3 per cent in 1930–8.[54]

The 1930s was a period of unprecedented prosperity for the majority of the population: excluding building, industrial production is estimated to have risen by about 50 per cent between 1924 and 1937,[55] and industrial output per head of population by a similar amount. On any measure, income per head, including wages, increased most rapidly in the 1930s, and real incomes rose, despite widespread unemployment. Living standards overall were transformed in improved nutrition, better housing conditions and a lower proportion of income expended on these necessities.[56] Two important elements in this were an improvement in the terms of trade at the height of the recession 1929–34, which brought cheaper food, and a marked reduction in the cost of living index, and cheaper housing.

Housing production had been primed in the 1920s by government subsidy to both local authorities and private enterprise, as in post-war conditions of inflation the latter would have been unwilling to meet demand, but building was hesitant. In only one year, 1928, did construction reach the levels of the 1930s: 366,000 in that year, the average for 1934–8 being 409,000 per annum. In this latter period cheap money was influential, but the building boom anticipated the fall in interest rates: more important was the fall in costs of construction and the increase in demand as domestic savings were directed to the building societies. Between 1928 and 1934 Bowley[57] estimates that capital costs for a small house fell from £422 to £361, i.e. 14 per cent; the costs of building materials had fallen by 13 per cent, and that of labour in a period of high unemployment in the building industry by 10 per cent. Even for craftsmen in the industry, unemployment reached 25 per cent in 1932.[58] Meanwhile, the number of share investors in the building societies was increasing rapidly (1 million in 1924, 2 million in 1939), their capital rose in the same period from £137 million to £771 million, and new advances or mortgage trebled. And the greater part of this finance came from the small investor—the diversion of large investments to the societies in 1932 after devaluation was soon checked as it would have imperilled the liquidity of the societies.[59] Increased salaries, and wages of those in the growth industries, accompanied by the fall in the cost of living, allowed savings and their diversion to housing. Consumers' expenditure on housing increased by 40 per cent between 1920–9 and 1930–8, and by 1938 repayments on mortgage accounted for 64 per cent of building societies' funds.[60]

The housing boom, and the growth of the building societies, must be related to the shifts in the national employment structure. Between 1920–1 and 1937–8 the numbers employed in agriculture, fishing and forestry, mining and quarrying, and manufacturing, fell by 892,000 (10 per cent); those employed in distribution, banking and insurance, professional and miscellaneous services increased by 1.7 million. By 1938 salaried workers accounted for 25.5 per cent of the labour force, an increase of over 1 million since 1920.[61] It was largely their savings that were channelled by the building societies into the construction of suburban housing estates which the latter, in liaison with the builders, were promoting. The financial uncertainties of speculative building were minimized and, as in the 1890s, the builders could operate in confidence of an assured market: money was cheap, land freely available, labour and materials similarly at base prices, in a period of rising incomes and increased expenditure on housing. It was a period of building speculation for a mass market backed by government in subsidies to private construction and encouragement to building societies.

So was established the characteristic pattern of urban development in England this century. This low intensity in the use of land has, it is claimed, enabled a comfortable, relaxed, easy-going style of life, 'the natural environment for a naturally chosen way of life'.[62] And yet this environment, which placed larger sections of the working population in situations of relative housing affluence, was only achieved at the expense of the living standards of the remainder, including those who built the houses. It depended on cheap labour, cheap materials, land and money. The comfort and cosiness of the suburbs must be set against the neglect of the slums, and also that of the towns in the peripheral regions. Housing investment went to new property, not maintenance and replacement of that already existing. And, further, the huge step forward in living conditions for those moving to the suburbs depended on the international relations of exchange established by Britain in previous dominance over world markets. It can be argued that 'to some extent . . . the imperial heritage of British capitalism gave the government during the 'thirties the flexibility and ability to compromise under popular pressures . . .'.[63] Suburbanization owed much to imperialism. The British working population as labour aristocracy of the world benefited from favourable terms of trade and cheap food, but also from cheap land on which they could achieve middle-class standards of living.

Judged in terms of movement of the labour force to work, inter-war developments had produced highly inefficient towns. White-collar employment had grown rapidly in the town centres, while its work-force had moved in increasing numbers to the suburbs; new investment in manufacturing industry was generally in plant in suburban locations, while the workforce was still drawn from the inner urban districts. The mobility of the labour force was presumed. In this respect new town and expanded town policies post-war have endorsed but humanized the locational decisions of those who control capital investment in production. Otherwise planning policies have been unable to rationalize urban investment, public or private. Inducements to manufacturing industry to leave crowded inner urban districts, and clearance policies, have encouraged suburban expansion. At the same time development control in the central city business districts has been unsuccessful in mitigating the concentration of 'non-productive' activities at the core of the urban complexes. In fact zoning of land use, rationing of space available for office space, hotels, shops, conference centres, *et al.*, has strengthened speculative pressure for property investment.[64] The slow realization of these trends in relocation of work is witness to the inertia built into property-holding. Industrial plants however unsuitable are not readily evacuated—buildings, plant, labour supply, supply lines, and the historic valuation of the site as industrial land, are all assets not easily discarded. In the same manner, commercial and residential users of central city space are loathe to relinquish a prime location. So not until the 1960s did inner London for instance lose manufacturing employment at a rapid rate—400,000 jobs between 1961 and 1971, a further 72,000 in 1972. And only in that decade were fears voiced as to a repetition of American experience in 'social polarization'.[65]

The dispersal of population in the city-regions has been furthered by government intervention in housing policies and land-use planning. Construction of housing by local authorities extended suburban living to considerable sections of the working class. Full government backing in subsidy and provision of capital, as well as cheap money, allowed a massive building programme, largely in the suburbs, between 1945 and 1956 (1.43 million houses), and although subsequently the local authority contribution to housing construction was directed into slum clearance rather than 'general need', the effect was to accelerate the falls in population in the inner districts. Population had to be decanted into suburban estates, or new towns; other households, faced with dwindling choice and higher prices in the inner districts, elected for the cheaper house out of town. Private building has been almost entirely in suburban or country districts throughout the post-war period, a practice enforced in many towns by the acquisition of all building land in the urban area by the local authority. Only in the major metropolitan centres and their satellites has there been privately financed redevelopment.

Although checking the excesses of 1930s development was a main impulse to post-war planning legislation, restrictions on housing development adjacent to the towns have further diffused the city into the countryside. The residential growth areas in the city-regional complexes are the estates attached to small town and village. The intention was to protect the countryside from the ravages of speculative sprawl, regulate land use to ensure long-term profitability, integrate housing requirements with those of industry, and relate development to local services. In that the loss of land from agriculture has been reduced from an average of 60,000 acres per annum in the 1930s, to 38,000 per annum between 1945 and 1967,[66] planning policies have been effective, but many of the traditional settlements of rural England have lost their identity, urban populations have been forced back into the countryside, and the city has spread further afield. The planning profession is strongly identified with this planned dispersal as 'an antidote to the problem of our cities, providing an environment which blends space with humanity, and offering in all probability, the best choice of rehousing the nation more quickly'.[67]

A guiding theme in planning has been that of 'amenity'—in housing, in green space in the cities, in parkways, in replanned city centres and in the greenbelt. Industry too large or too noxious to be camouflaged is located away from the eyes of the citizen, and suburban landscaping brought into the city. The emphasis is on consumption and display, not production. Maintenance of the economy, at a level far removed from subsistence or necessity, means lavish expenditure on 'overheads'—glossy buildings, landscaped parks for industrial plant, and lavish use of space in the cities. 'Amenity' is a luxury—amenity land is not something to be used, but something to be viewed, space by which activities and people are to be kept apart. In urban development terms, amenity indicates more extensive use of land and a further impetus to decentralization. It is an index to the extravagant use of resources which has characterized the economics of advanced capitalism, i.e. the exploitation of the resources of the Third World, in particular, oil, and the lavish use of the basic resource of any country, land.

The combined use of powers over housing investment and land use to promote humane living conditions for the mass of the working population might be termed the 'British experiment'. Tolerable living conditions compatible with a high consumption economy have only been maintained in the major cities by suburban development. This now would seem to have been checked by inflation generalized in the international economy, allied with the long-term weaknesses in Britain's competitive situation in world markets. Cutbacks in public spending, increased costs of raw materials, the premium on agricultural land, the charge on capital, have all had their effect on housing production and its location and design. Older housing is to be renovated, not replaced, and residential densities in suburban locations are to be increased. That comfortable, natural style of life was premised on certain relations of production, internationally, and also nationally.

The costs of uneven development are only appreciated now that the hidden subsidies are removed. Economic activity in Britain was focused on the major urban aggregates, principally London. Now the costs of such aggregation, land itself, expenditure on communication, the price of housing, the limited outlets for leisure, all reflected in higher labour costs, have pushed on the regeneration of regional economies. Suburban development was combined with, and in a sense dependent on, the regional imbalances already discussed. The differentiation of the metropolitan community (which to the Chicago School was the key feature of urbanization), in itself an index to class interests as manifest in rights and interests in land, is an integral component of the relationship between the metropolis and its satellites. 'Over-urbanization' has accelerated the 'development of underdevelopment', so precipitating new tensions in the political economy of the nation. The hegemony of the metropolitan élites is challenged in the debates over devolution and the political consensus stretched paper thin. Suburban expansion—innocent as it may look—has been a precipitant to the diversion of capital investment to the regions. The continued expansion of capitalism has been ensured, but so have its inequalities and costs.

Afterword

The outline of British urban and regional development presented here is no more than that—an introductory sketch of some of the issues which would have to be considered in a political economy of urbanization. Some issues stand out more clearly than others. For instance it is clear that there are episodes in British history in which uneven development has had political consequences, which would repay further study. Similarly the issues of built form and its determinants might be clarified in a framework of analysis wider than that commonly adopted. What is emphasized here is the interrelationship between regional development and the differentials in opportunity within the urban complexes themselves.

In summary, economic growth for many of the provincial communities in Britain was greatest in the period of acutest poverty in inner London, i.e. in the last quarter of the nineteenth century. Then, when capital sought alternative sources of labour for new industries, the regions were left in stagnation and decline, as the South East prospered. Existing investment in 'community infrastructure' was discounted in favour of new locations and new building. Now, as the economies of the regions are restructured and their underemployed offered jobs in the processing divisions of the industrial corporations, the prospects for the poor of the major cities are once more in jeopardy. Does not London have a higher total of unemployed than Wales, with some of the highest unemployment rates in the country in the inner boroughs? And, unlike the regions, these inner urban districts have no identification as communities, no political weight; their interests are subordinated to the functional requirements of the metropolitan region as an aggregation of economic activity. They are the internal colonies, for they are wholly subordinate to initiatives taken by ruling interests, whereas the provincial communities have never been mute subordinates to metropolitan initiative.

It is important to recognize the peculiar conditions of urbanization in Britain. In the formative years, 1780–1840, industrialization was effected in conditions of simple technology in a barbaric productive system. Industry had to move to sources of energy and raw materials, and labour had to follow, hence the urbanization of previously backward and isolated regions, areas almost beyond civilization. That large, densely populated and heterogeneous settlements resulted was due to the relations of production by which the working population in poverty and insecurity were bound to the principal labour markets, closely packed at the heart of the urban complex. But now, the insecurity of the labour force is cushioned by the state, industry is 'footloose', distribution and services are increasingly so, information is internationally disseminated, contacts are global. Is then agglomeration less and less necessary?

The conditions of nineteenth-century urbanization have therefore given Britain a very distinctive balance between town and country, metropolis and province. Compare Italy, which superficially at least in the imbalance between north and south, and the attempts by the state post-war to redress the uneven development of the country, presents a similar regional profile to Britain: but there the disjuncture is between an agrarian peasant economy, and an industrialization grafted on to ancient cities. The historic centres of northern Italy have been industrialized without coal or easy access to raw materials, and small-scale industry has been superimposed on an agrarian society in the countryside. A considerable increase in the industrial labour force is possible without urbanization. Industrialization in Britain created new communities, ranging in size from the small townships of the coalfields to the major conurbations, communities of such substance—as built environments, as social groupings, and as political forces—that their existence cannot be discounted in further industrial location decisions.

It is difficult to incorporate either the nineteenth-century experience of British development, or the subsequent reactions against neglect of once dynamic communities, into a schema of a chain of expropriation from provincial satellite to metropolitan centre, and an associated subservience of local bourgeoisies to that of the metropolis. The extreme sensitivity evident in twentieth-century Britain to shifts in population and activity which are inconsequential by continental standards,[68] is a measure both of the integration of disparate regions into a national political community, and their development as distinctive communities with diverse cultures. The pattern of nineteenth-century growth permitted independent economic growth under the auspices of provincial bourgeoisies and generated alternative cultures in which there is neither acquiescence in the rights of industrial capital over total life experience, nor a consensus in the hegemony of metropolis. The model of metropolitan dominance is too simplistic: it cannot be related to the problems of a once imperialist nation pushed into making full use of indigenous resources in land and raw materials, labour, and existing capital investment. New relationships between regions result, and a modified political balance must be struck if the state is to maintain authority over the national society. The known conditions of geographical inequality within all societies, at all stages in their development, cannot be subsumed under a simple hierarchical model which does not take account of the new tensions and antagonisms as places are reassigned positions in the national matrix.

The accusation levelled at dependency theory, that designations of capitalism in terms of the circulation of commodities and conditions of exchange rather than by its mode of production[69] preclude consideration of the exact character of the relations of production, would seem to be justified. The definitions used are so wide that 'we could conclude that from the neolithic revolution onwards there has never been anything other than capitalism'. The same could be said of metropolitan dominance. Examination of the British experience of urbanization confirms the suggestive qualities of statements as to underdevelopment in the metropolitan society, but demonstrates their inadequacy as theories in the face of the historically concrete. The statements are so general that they can neither be corroborated nor denied.

2 The modern big city

1 Regional cities: a suburban diaspora

The study of cities is governed by the images we as sociologists, planners, architects or politicians hold of them. In any society there is a diversity of perspectives, inchoate and unspecific as to the nature of existing cities and the form of alternatives. If British attitudes have been charged by the long-standing reluctance to invest urban wealth in urban living, and heavily coloured by the squalor of the towns of the first industrial revolution, those of Americans bear the reactions to the disappearance of the small town democracies of the agrarian states, and then the urban migration of Southern blacks. In comparison, intellectuals of the European cities of France, or Italy, or Germany, are caught between traces of an older urban culture in which a comfortable bourgeoisie monopolized the privileged quarters of the towns, and the new experience of mass urbanization, the *grands ensembles* and the suburban villa. The evident diversity of urban experience in one society, at one time, confuses discussion, but perhaps more important is the transmission of differing intellectual traditions, in themselves ideological responses to past changes in society, into current debate.

Prevalent in all these societies there is an image of the country town, a close-packed huddle of roofs round a market place and a church, distinct from and yet integral to the countryside whose activities it co-ordinates. Town and country are seen as in 'reciprocal interdependence'; the town is a haven of peaceful activity, embedded in fields and trees. This was an image of urban life that haunted the generation of the 1880s and 1890s, acutely aware that the old way of life was slipping away. Tönnies, in Germany, hailed the country town as 'the highest, the most complex, form of social life';[1] Morris, in *News From Nowhere*, evoked the picture of a London, then the world's greatest city, as a series of small towns in the year 2003.[2] And the traditional market centres of England now attract all sections of society as homes, to shop and stroll around at weekends, and to preserve as integral elements in the nation's heritage. And this was the governing image behind Howard's prospectus for the garden city in which the fields of rural England were to be brought within reach of the factory worker. The aestheticism and romance of small town living, not the rationalist social engineering of Owen, was to become a main impulse to twentieth-century planning. Both the industrial city and the metropolis were to be restructured into a form which flatly denied the changes in conditions of existence that had already taken place, and were to gather pace in the twentieth century.

Now, we must take account of another view of urbanization, that of the regional city. This is a city of many communities, many centres, some old, some new, a city which spreads far out into the countryside embracing in its tentacles country town and village alike, incorporating even distant hamlets into its rhythm of work and play. The countryside has been incorporated into the living space of the city while the city itself diffuses activity and population away from its core to peripheral and partially autonomous centres. The metropolis encountered by Dickens, Engels, Booth or Simmel has become the 'megalopolis' described by Gottmann as 'the main street and cross-roads of the nation', a 'stupendous monument erected by titanic efforts'.[3] The same is decried by Mumford as 'sprawling giantism'.

The great metropolitan centres have become of so vast a scale and so diffuse an influence as to defy measurement, and far transcend daily experience, so that we no longer have any criteria for what is urban and what is not. This urban explosion, encountered to greater or lesser degree throughout the cities of Europe and North America, has upset previously held images of cities and urbanization leaving the urban student groping after something he knows to exist but which eludes analysis or conceptualization. Urban sociology, premised on the earlier reactions to the modern city, has still not come to terms with this latest mode of urbanization; its working assumptions and concepts, its taken-for-granted world of inquiry, still relate to the seemingly bounded settlements of the nineteenth century. In this it has parted company with planning thought, in which the city-region, and the possibilities of a mobile society, without community identification, shape discussion as to strategy.

A persuasive spokesman for a new image of urbanization has been Webber, in a series of articles throughout the 1960s.[4] In these the city is seen as the 'non-place urban realm', in which the traditional costs of distance which held the older cities together have been eliminated. Cities of the world of advanced technology of communications can dispense with agglomeration and scatter activities and residents over a wide area so that everyone experiences an expanded living space which enables them to realize 'expanding opportunities for learning new ways, participating in more diverse types of activities, cultivating a greater variety of interests and tastes, developing greater capacities for understanding and savouring richer experiences'.[5] As accessibility is removed from propinquity the local neighbourhood becomes of diminishing significance, and other forms of community, the specialized occupational or interest group, take its place.

This image of a free-wheeling society in which diversity and freedom of choice is achieved through a totally mobile way of life has been treated with great scepticism by sociologists familiar with either the new suburbs, or conditions in the inner cities. But as a projection of the emergent pattern of urbanization in post-industrial society, it has had widespread currency in the British planning profession. It has entered into the policies for the latest new towns, for example, Milton Keynes,[6] as well as statements of planning goals for many structure planning documents, and was endorsed by the conference convened in 1969 by the Centre of Environmental Studies.[7] As the compact city form is superseded by the interlaced activities of the regional city,[8] large sections of the profession have been convinced that there is to be a new way of life, perhaps more akin to that established in America than hitherto experienced in Europe.

City dispersal—a suburban diaspora?

> To be your own unique self; to build your unique house, mid a unique landscape; to live . . . a self-centred life, in which private fantasy and caprice would have licence to express themselves openly, in short, to withdraw like a monk, and live like a prince—this was the purpose of the original creators of the suburb. They proposed in effect to create an asylum, in which they could, as individuals, overcome the chronic defects of civilization while still commanding at will the privileges and benefits of urban society.[9]

In the decade up to 1970 only New York City among the world cities is known to have gained in population. Moscow is estimated to have lost up to 1 million residents during 1962–72; London, already by international standards a low density city, a further 600,000 (7.9 per cent). And these large-scale transfers of population are at the root of so many of the contemporary issues of urban living:

social polarization, the inner city crisis of twilight area, ghetto, or slum, suburban anomie, privatization and neurosis. The depopulation, and all-encompassing deprivation of the relict communities of the inner neighbourhoods, the loosening of family networks with geographical mobility, the attenuation of neighbourhood affiliations, even the weakening of provincial cultures, are all to be related to the selective migration of populations away from their communities of origin.

Post-war British cities have experienced a major redistribution of population. There are different ways of demonstrating the shifts in location. One, simply, is to use as a measure the proportion of population resident in the major conurbations. Between 1951 and 1971 this fell from 38.5 per cent to 32.6 per cent, the proportion resident in nominally rural areas rising from 18.4 per cent to 21.6 per cent. Individual urban areas suffered different rates of emigration. Whereas Manchester and Liverpool both lost near one in five of their population in the decade 1961–71, the loss to Birmingham (9.1 per cent) was more in line with that of London. To some extent these differentials are artefacts of administrative divisions between municipality and county, but there can be no doubt as to the upturn in the rate of movement out of the inner city districts for both population and industry.

The new feature to this spread of the cities was its impact on the inner districts. Although the growth areas in the previous decade had been the outer areas of the metropolitan regions,[10] this had relatively little effect on the level of population in the inner districts.[11] Whereas in the 1950s, housing shortages had slowed up movement from the inner cities, and immigration from abroad had offset outward movement of the local population, the 1960s saw the extension of large-scale clearance schemes in the inner districts, the encouragement of new town and overspill movements, and the promotion of owner-occupation in the suburbs by governments and building societies alike.

The acceleration of the drift away from the traditional city complexes is the direct consequence of housing and planning policies in the so-called age of affluence. As Ash commented: 'They are sowing down the grass in the middle of Liverpool . . . vast further areas of Liverpool are derelict, and in their turn, are destined to become green.'[12] There is here a ghastly distortion of the visionary prediction of Morris, in 1890, of a land without cities.

The new pattern of urbanization in Britain goes beyond suburbanization as previously experienced. The *Oxford English Dictionary* gives the definition of a suburb as 'immediately outside and adjacent to the walls and boundaries of a city', and yet the present pattern of development is that of extensions to villages, and towns functionally distinct from the metropolitan core to the region, and separated from the built-up area of the dominating central metropolis by open land. These new developments satisfy none of the definitions of the suburb: they are not peripheral to the urban area, they are frequently not economically dependent, but fully diversified urban economies; they are not part-communities, mere bedroom communities, but offer their residents a full range of services and opportunities. The concept of the 'suburb' conveys the idea of something incomplete, reliant on the more complex city to which it is attached. Presently there is instead a considerable degree of autonomy and interdependence among the constituent communities of the regional city.

Post-war urbanization in Britain has of course been moulded by the exercise of planning controls. Planning policy has steadfastly maintained the aims of containment of urban sprawl, i.e. suburbanization, and the protection of the countryside through the designation of greenbelts, urban fences, village envelopes and the like. So suburbs as such have been limited in extent by the exercise of control over development, and the regional city promoted, both directly, as in the construction of the new towns, and their subsequent designation as growth points, and indirectly, through the spreading of development further from the metropolitan core by restrictions on the use of land for building adjacent to the existing built-up area. It is therefore not strictly accurate to speak of the new pattern of urbanization as that of 'suburbanization': whole regions have been recast as activities and people have dispersed from the urban agglomerations.

In another sense, too, the epithet 'suburbanization' is misplaced. The suburb historically was an upper- or middle-class arcadian retreat from the hazards and discomforts of urban life. Mumford characterizes it as 'a segregated community, set apart from the city, not merely by space, but by class stratification: a sort of green ghetto dedicated to the élite'.[13] The archetypal suburb is a middle-class preserve maintained by cost and access barriers as well. But these green ghettos are fated to be engulfed in the subsequent spread of the city; mass urbanization overtakes the selective suburbanization of the more affluent.[14] The promoters of these select enclaves were unable to guarantee the long-term survival of the very features which had enabled them to market the new communities—seclusion and nature.

Mass urbanization inevitably entails the creation of suburbs in the sense of extensions to the built-up area, and these new neighbourhoods engulf the distinctive community form of the middle-class suburb. The latter must fight to maintain its distinctiveness and social *cachet*, but normally succumbs to obsolescence, and becomes yet another transitional zone within the urban area. The present-day extension of something of the living standards of the middle classes to the mass of the population through greater affluence, the availability of private transport, easy credit facilities, and publicly subsidized housing, has meant an extension of suburbs in their physical form, without an accreditation of their residents with the social status of the once exclusive middle-class suburb. Lowering of residential densities must not be confused with the use of suburban living to maintain or consolidate a high status position in the local community. It is this mass urbanization which has transformed the compact metropolis into the diffused city-region. Local centrifugal movements out of the inner cities are superimposed upon an international gravitation towards the dominant cities, the 'poles of command', so that complexes of living and production equivalent to nation-states in scale and wealth become the urban way of life for ever-increasing numbers.

There has been much discussion in urban sociology as to the reasons for this movement out of the inner cities. Commonplace is the observation that the majority of the most affluent households in the city complexes of North America or Britain occupy suburban, peripheral locations. As these, of all households, can exercise preferences as to location which are not open to lower income households, the indication is that where choice is possible it is exercised in favour of residence out of the cities. Speculative interpretation of the available literature suggests a number of motives.

Frequently the benefits of such a location are expressed in terms of its environment: the air is fresh, free from the pollutants of smoke and chemicals, the atmosphere is tranquil, away from the noise and bustle of the urban area. 'Peace and quiet' is a prime asset of the suburban location; it represents a refuge from the hurly-burly of the city's commercial life, its crowds, its children. In addition it offers seclusion and privacy; neighbours can be kept at arm's length in a way that is impossible at higher densities. Then, too, there is the greater proximity of the neighbourhood to the countryside, the ideal location being one that actually overlooks the rural landscape, so completing the illusion of detachment from urban life. Environment provides a convenient rationale for suburban living.

Less often mentioned are the economic advantages of living out of town. Low rates are an incentive, but paramount is the attraction of a lower expenditure on space than in any location closer to the urban core. The suburban resident exchanges the costs of travel to central urban facilities, activities and job opportunities, for a more lavish use of land: space replaces accessibility in the choice of housing. But there are other less measurable benefits to suburban living. In most urban areas there are inner neighbourhoods of houses lavishly provided with garden space, at a cost considerably less than comparable suburban houses. If utilitarian calculations prevailed, these, not the peripheral estates, would show high concentrations of the middle-class populations for which they were originally built. Something other than economy is at stake. Sociologists argue that those who move away from the inner city neighbourhoods do so with positive expectations of what extra-urban living has to offer. They hold certain images of life outside the big city, its rurality, its friendliness, its community spirit, its suitability for children, even its prestige, which motivate their movement there.[15]

Strongly emphasized has been the 'familism' of the suburbs. In the nineteenth century, when infantile mortality rates in the inner urban districts of the English towns frequently exceeded 175/1000, the flight to the suburbs by the middle classes had as a motive the protection of their children's lives. In similar manner the reluctant suburbanites in Essex housing estates expressed satisfaction in suburban living through the opportunities it offered their children. Instead of damp and overcrowded cottages and tenements, their children had sunlight and space, hot water and sound houses. As one housewife said: 'You've got to put up with things if you want a place for your children. Your children come first.'[16] Suburbs are therefore seen as the environment for child-rearing, providing healthy living conditions, and in their relative homogeneity (in America), a secure social environment. As Dobriner points out, the 'better for children theme' is a socially approved rationale for suburban movement, and its use in response to questioning may mask the parents' own concerns—a better house, a secluded environment, and a higher status neighbourhood.[17]

The theme of familism must be related to attempts to translate economic gains into higher living standards—a more modern house, greater space, a more salubrious environment than that in the older districts. It is hard to sustain the argument that the mass movement from the inner districts is motivated by a desire to achieve status in the community at large. The suggestion made by Rex, that 'all participate in a socio-cultural system in which the middle-class way of life enjoys high prestige and in which the move to the suburbs is a built in aspiration',[18] is a masterly overstatement. It is all too easy to demonstrate that there is no unitary value system: working-class groups, immigrants, the professional meritocracy, all deviate from the bourgeois vision of the suburban idyll. The location of so many council housing estates in the suburbs has not been sought by their residents; circumstances of housing provision allowed no choice. This explanation for suburbanization is enticing in its very simplicity: but it simply rationalizes that which has transpired.

Yet another query must be raised: is it that suburban residence has had high prestige, or is it that their landscaping and proximity to the countryside lent the middle-class suburbs a rustic air? Does in fact rural, and not suburban, living offer the highest status; is the country house and not the suburban villa the acme of achievement? Not only have successive élites built, or taken over, country retreats, but increasingly the broad mass of the population sees the suburb as a half-way house, a limited realization of their dream of a country house or cottage: 'a large proportion of those settling in representative new suburbs, though recognizing that they are now suburbanites, aspire and expect to live in the countryside eventually. . . . Life in the countryside [is] a peculiarly prestigious way of life'.[19] The magnet remains the countryside, remote from urban pressures.

In Britain there is an active antipathy towards the cities. Although the main shift from the land pre-dated the mass extension of the suburbs, unlike therefore many European and American cities where first generation urban migrants have moved direct to the suburbs,[20] every town has many families that have never lived in the close-packed conditions of the inner cities, and every region its communities in which a close link has been maintained with the surrounding countryside.[21] There is also a deep-seated nostalgia for the pre-industrial past. Popular archaeology and urban history stops with Georgian England, there is a search for rural living and rural traditions as representing a way of life older, and somehow truer, than the culture offered by industrial-urbanism. There is a 'myth of a happier past',[22] a world that has been lost, and which can never be retrieved in the conditions of the cities. Retreat to the country cottage in the village or, at worst, the garden suburb, is an affirmation of a conviction that an environmental solution to the problems presented by capitalism can be found. The middle classes are drawn out of the cities towards the countryside in their refusal to accept the circumscriptions on living presented by urban capitalism. In similar vein Riesman argued that the movement of Americans out of the cities represented a 'tremendous but tacit revolt against industrialism'.[23]

For those who have choice, the odds are stacked against permanent urban residence except in the special circumstances of the metropolitan city. The central areas of cities such as London, Paris, New York, are sufficiently lively, and their ex-urban sprawl so far-reaching that residence in an inner district becomes an attractive proposition once more. So ensues the process of 'gentrification' in which the normal urban distribution is reversed in that the richer households oust the less affluent from the central districts; rich and very poor co-exist in the central city, while the less affluent households are pushed out to the suburbs or beyond into districts remote from central area attractions and services. Is this then a sufficient explanation for the deconcentration of cities and their transformation into the diffused regional city—a search for seclusion, a deliberate withdrawal out of the urban environment except where central city glitter conceals the drabness of urban existence?

There is in these explanations a sense of inevitability: this new environment derives from the very nature of man, brought to an intolerable state of being in a strictly urban environment and now, with affluence, released into a way of life in harmony with his, and his family's, needs. The discussion has been conducted on voluntarist, nominalist principles, a position in which it is assumed that the individual can make decisions about future actions which are not taken under duress, under situations of constraint, and can implement those decisions freely. In this focus the sociologists and the urban economists have written up the urban process as the logical outcome of human needs, irrespective of the circumstances in which those needs are defined, and oblivious to the barriers confronting many households in their search for a desired way of life. They ignore the constraining matrix of the urban land market, and the limited availability of housing, but it is from these conditions that the demand for suburban housing derives.

The urban land market

The urban land market presents special problems for economists. Land is a peculiar commodity in that the possibilities of increasing supply in relation to demand are limited: either the intensity of use can be increased, or with improvements in accessibility the area available to the community can be extended. So by the end of the eighteenth century, with road improvements, many English towns were losing their élite of merchants and professional men to the country districts.[24] The railways were to accelerate the promotion of these upper-class retreats (often by the large landowners themselves), but large-scale speculative developments depended on a mass market, i.e. a section of the population able to take advantage of new forms of transport. For Britain this happened in the second half of the nineteenth century. Only when sufficient numbers of the population are lifted above subsistence level can the compact city be exchanged for a diffuse sprawl into the countryside as once rural land is incorporated into the urban land market in a random scatter of development. Accumulation of wealth from property can now be achieved in a more open fashion by the subdivision and marketing of undeveloped land by the speculator/dealer, rather than by the intensification of use of existing urban property, buildings or land.

All theoretical discussions of the urban land market start from the differentials in value between developed and undeveloped land. Hurd seems to have been the first to translate Ricardian rent theory into the urban situation, so initiating the main line of urban location theory.[25] In this land values are related both to accessibility, and to the sequential expansion of the urban area in which the value of already developed land is automatically pushed up by yet further expansion of the urban area. In this way a further impetus is given to speculation on the urban fringes, as the differential between urban and non-urban land values is widened. Since the nineteenth century, in Britain as in America, the suburbs have been built up on this speculative momentum.

For Britain, most is known about London's expansion. From the seventeenth century onwards the big estates had been parcelled out into sedate town environments.[26] Nash in Regent's Park, and Cubitt in Belgravia, were heirs to this tradition of development. In these circumstances, careful planning and promotion were necessary to attract an upper-class market as 'the profit from just building is extremely small. . . . [The] main profit comes from raising the ground rent, from careful selection and skilled utilisation of the building terrain'.[27] But for a mass market, building was less carefully regulated and estate planning less important. Dyos calculated that in Camberwell (1878–80), some 416 firms and individual builders were involved in the construction of 5,670 houses. Of these, half built less than six houses each.[28] This fragmentation of interests, a diffuse nexus of private investors, solicitors, estate agents, building societies and builders, and their subcontractors, was to characterize the development of London, as other urban centres.

In the twentieth century suburban development has been regularized in that finance for housing has been channelled through building societies and local authorities, more house building is in the hands of large, established, construction companies, and land use is itself regulated under building and planning acts. But the incentives for suburban development remain the same, and are common to private and public development. They can be summarized under four headings:

(i) the cost differential between rural land (with planning permission) and urban land remains in a ratio of something like 1:10, at minimum.

(ii) Land can be acquired in large plots, suitable for economies in mass production. (One estimate is that the cost of building a single housing unit is halved once the scale of production exceeds fifty dwellings.) 'Infill' housing is costly.

(iii) There are no difficulties in rehousing and relocating the existing users of the land. In the urban area the cost of 'recapturing the site' deters all but the provider of luxury housing (or government subsidized agency) from undertaking redevelopment.

(iv) For the private developer there is the expectation of profit in realizing the development value of the land.[29]

Financial policies and fiscal incentives add yet further impulse to suburban movement. Not only were the British building societies a main influence in suburban development during the 1930s when they acted in collaboration with the builders,[30] but their lending policies now encourage purchase of housing in remote districts. Not only is the borrowing ceiling raised on new development (to 95 per cent of valuation), but the calculations as to ability to repay the mortgage do not take into account the household's travel costs. Therefore 'one way in which potential borrowers overcome deficiency of current income which unacceptably restricts their borrowing, is to substitute higher travel costs for loan repayment charges to find cheaper properties in more distant locations'.[31] For young households, with anticipation of increasing income, there is considerable incentive to maximize their present standard of housing by moving to the fringes of the urban region. A situation in reverse of that observed by the Chicago School then pertains, in which relatively low income households occupy the 'commuter's zone'—Zone V in Burgess's model.

Fiscal policies in Britain, as in the USA, since the 1930s, have been oriented towards the encouragement of borrowing for house purchase. Favourable treatment of funding institutions, whether credit banks or building societies, and tax relief on interest payments by the occupier, have given every incentive towards maximum expenditure on housing, and hence a seemingly insatiable demand for new housing. Clawson,[32] in a review of post-war housing policy in the USA, concluded that the direction, pace and extent of suburbanization in that period had been heavily determined by federal housing and income tax policy. The flight from the cities owed as much to political directive over housing finance as to the anarchy of the market. In Britain, similarly, in economists' terms, those who use valuable resources in land and construction materials have not been obliged to 'pay a price at least equal to the opportunity cost of these resources'. In conditions of inflation, it is argued, 'the so-called demand for space in part results from the cost of space being little or nothing'.[33]

Local authorities are now in a different situation, caught in a pincer between governmental policies favouring private enterprise construction in the new residential districts out of town, and the reluctance of their tenants to join the urban game of leap-frog, as successive generations of new households overstep each other out beyond the urban fringe. Willmott quoted one respondent from Dagenham as saying:[34]

> It's a ridiculous situation. The LCC put its overflow into Dagenham. Dagenham puts its overflow of necessity into

> Canvey Island. Where they go from Canvey Island in the next generation, God knows. We can only assume that they'll put them on to rafts and set them adrift.

Housing policies, the land market, and planning strategies, may converge to push young households even 60 miles from their families and friends (as in Milton Keynes). The alternative is costly reclamation of land within the urban area: goods yards, docks, marshes (Thamesmead), mud flats, industrial sites, or areas of low density housing. This is not simply a question of vested interests in property pushing up land costs, and hence densities, and so building costs. By very reason of their nature as sites, development costs must be high.

In this analysis it is the availability of housing alone that draws the urban resident out from the existing built-up area into new districts. What these households seek is a house, and they move where they do simply because it is there they can find a house they can afford. The move away from the city has very little to do with a quest for a rural idyll, or with a search for status, but simply relates to the nature of the mass markets in housing. And yet, of all those who undertake the move to the new areas, few do so reluctantly; most have some belief in the images of ex-urban residence already discussed. Familism, prestige, rurality and the return to the small community, all play their part as rationalizations of what must be undertaken.

It is possible to argue that these images of extra-urban living are 'myths';[35] the simple pastoral image of life out of the city, the rural setting, the small community, the establishment of identity round the family unit, all are elements in an image that disguises the real inconvenience and cost to the individual householder of the new form of settlement. Even the proximity of the countryside is illusory in that there is no greater access to it than from the town; a farmer's wife from Suffolk complained that if she wanted to walk on grass she had to come up to a London park. The escape from the city means a sacrifice of the urban assets of diversity and choice in exchange for rural environs, and a new house. Is not this myth of 'rural' life an ideology—using that word in the specialized sense of an idea which serves vested interests in society? The belief in the intrinsic superiority of life out of town provides a motivation for movement there, a decision in which there is in effect very little choice, and a legitimation for those who continue to create new suburbs.

The implications of suburban movement

The post-war extension of the mass suburbs provoked a near

hysterical reaction among social commentators in Britain and America alike. In America popular journalism and sociological investigation built up an image of suburbia, differing from both traditional working-class and middle-class cultures in its 'other-directedness' and conformity, its frenetic socializing and lack of self-confidence, and its quest for status. 'Status becomes an autonomous motive and mode of life', concluded Stein in his review of the literature of the 1950s.[36] On the English side most attention was directed to the new suburban council estates in which working-class families were being resettled in the post-war housing drive. The reactions here were equally extreme: the new districts were described as the 'breeding grounds of the privatized worker',[37] or as 'the classically anomic society'.[38] It was argued that the matrix of traditional working-class life was being destroyed, and its localism broken down. Working-class families now separated from their kin, and reserved with their neighbours, were retreating into an isolation of family life.

> In a life now house-centred instead of kinship centred, competition for status now takes the form of struggle for material acquisition. In the absence of small groups . . . people think they are judged, and judge others by the material standards which are the outward and visible mark of respectability.[39]

Opinion was by no means unanimous on the consequences of the enforced relocation from the inner districts: Mogey, for instance, compared the strong if circumscribed loyalties of the old-established district with the more open society on the new estates, in which 'The inhabitants of Barton [had] lost their ties to the neighbourhood and gained in return citizenship in the wider and freer atmosphere of the varied associational life of the city',[40] but there was concurrence on the fact of change. Even in Gans's later study, *The Levittowners*, designed to disprove the suburbanism hypothesis, there is substantial evidence that the new community had had an unanticipated impact on the adults moving there.[41] In that case the implications of a new pattern of consumption were found to be more wide-ranging than the movement to the suburbs, as such. It was the house that had drawn them to the district, and once there it was the house that gave the greatest satisfaction. It is 'the isolation of the house as object'[42] that forms the central theme to life in the new suburbs.[43] Geographical mobility had enabled a break with tradition, and the new house encouraged an orientation of interests towards the family unit, and consumption.

Evidence from the suburban studies was fed into the 'embourgeoisement' debate concerning the convergence between blue-collar and white-collar workers in advanced industrial economies, and the collapse of traditional working-class solidarities. The privatization of existence at home was related to instrumental attitudes to work and trade unions alike.[44] On the first count, various studies have demonstrated the maintained barriers between classes, and the tensions in a neighbourhood when households of disparate life-styles and expectations for themselves and their children are forced to co-exist.[45] Despite affluence, despite suburban living, there was little indication that working-class households were adopting middle-class ways of life. Suburbs were 'new homes for old values', 'old communities on new land'. And there is now no evidence to confirm the suggestion that greater affluence and a suburban environment have meant the collapse of working-class militancy. In at least one interpretation, movement out of a traditional community into the suburban council estates had acted to revitalize political activity, and while hastening new patterns of family life and sociability had in no way broken down working-class loyalties.[46]

The suburban solution

The suburban solution to the problem of housing the working classes was the one advanced by the slum reformers of the late nineteenth century. If only the new lands of the suburbs could be opened up to the working classes, particularly the skilled men, the craftsmen, a filtering process could be initiated in which the overcrowding of the very poor, the urban 'residuum', in the central slums, would be relieved. Cheap, efficient transport from the central city to the suburbs was seen as the key to the solution of the housing question;[47] the lengthening of the working day which resulted, the increasing separation of city from country as the suburbs extended, the perennial cost of transportation, were not seen as in their turn creating new problems. Movement to the suburbs, not the redistribution of incomes, was the answer to the crisis of the modern city.

No counter-proposal came from the socialists, far more concerned with organizing a mass party than with providing (Utopian) solutions to the housing question. The only programme to which Engels was prepared to commit socialism was the 'abolition of the antithesis between town and country'. Not only was it hopelessly unrealistic to hope to solve the housing question within the pattern of land values set up by large-scale urbanization, but it was Utopian to begin to 'prescribe the form in which this or any other of the antitheses of present-day society is to be solved'.[48] The only challenge to the suburban solution has come from the garden city movement which would seek to reintegrate town and country, and solve the land problem of the large cities in the one programme of controlled resettlement of the working population in communities where work and home were once more brought together.

But, despite the humanism and social commitment of the members of the movement, the garden city could offer no challenge to the capitalist city. It failed in two respects: one, in its diagnosis of the rationale of urban, more specifically metropolitan, growth; two, in its presentation of a restricted, environmental solution. In Howard's analysis the basis of urban growth was industrial production and the attractions of industrial wealth, so further growth could be checked by industrial decentralization. Like his contemporary, Geddes, Howard saw mass urbanization as resulting from the application of the new technology, a technology which could be controlled, planned, and so liberated; but the controlling activities of the metropolis, from which derives its magnetism, were ignored. No answer was presented to the centralization of power in society; the network of controls, financial, administrative, political, radiating over national society and international empire, on which was based the dominance of metropolis, went unrecognized. The garden city was seen as a productive rather than a controlling unit. Nor could an appropriation of development values for the community offset the effects of the accepted allocation of wealth between capital and labour. The garden city programme differed only from suburban extension in its provision of a community infrastructure out of its own resources. In all respects the garden city, and its realization in the new towns, is part of the suburban solution to the consequences of living under capitalist conditions; both garden city and suburb present environmental solutions to the issues of capitalist urbanization. Their creators 'proposed in effect to create an asylum, in which they could . . . overcome the chronic defects of civilization while still commanding at will the privileges and benefits of urban society'.[49]

The transformation of living conditions achieved by thinning out, deconcentrating, urban populations and providing new housing away from the urban core is indisputable. The answer by Bauer, advocate of planned suburban housing, to Engels, was succinct: 'It would be difficult to dispose of several million houses as mere visionary utopianism.' Use of cheap land, and the treatment of housing as a public utility, along with the controlled use of modern methods in the planning, design and construction of housing, has 'abolished the distinction between my better and your worse',[50] at least for certain sectors of the population. As, too, first industry, then offices, have moved out of the cities to suburban or new town locations, the position of the worker, industrial or white-collar, has yet further improved. Life is comfortable, relaxed in house and garden, and for the major metropolitan regions largely independent of the central city.

But the cost-benefit equation for the suburban solution cannot be calculated in terms of the comfort and satisfaction of the suburban residents alone, for the true costs of suburban extension are felt by other groups of the population, who are not eligible for suburban living. The long-term concentration of community investment into the new areas of the suburbs or new towns, and the continual drain of population from the inner areas to the periphery, have as a consequence, the twilight areas of the inner cities, depressed grey areas or transitional zones, housing the elderly, the poor, the unskilled or semi-skilled service worker, on whose labour inner city activities depend. While certain categories of activity remain so heavily concentrated in the core to the central city there will be competition for accessible housing in which the poor and the poorest are forced into the areas furthest from the suburban ideal. High standards of living for one section of the working class were only achieved at the expense of the living standards of the remainder. The comfort and cosiness of the suburbs must be assessed against the neglect and squalor of much of the inner city districts.

In the largest metropolitan centres there is now evident a more effective segregation of an 'urban residuum' from other sections of the working class than anything experienced in the nineteenth century. In fact it was the forced coexistence of the artisan, the 'respectable working class', with the poor and the very poor, in the same streets even, that spurred on housing reform then. Now the service workers on whom the metropolitan economy depends must be retained in the inner city, in the barracks of local authority housing or, at less expense, in the rehabilitated structures of the nineteenth-century speculative builder. 'Social polarization' means not only a heavy burden on the rates for the municipal authorities, but a division within the ranks of the working class. As it is these same urban centres that attract the main influx of overseas immigrants, this cleavage between suburban and inner city working population is underscored by racial divisions.

For the smaller provincial centres, the distinctions are less marked, for a number of reasons. One, certainly, is the slower rate of economic growth which has slowed up industrial decentralization, and held back demand for suburban housing. Important also is the lesser impact of congestion costs: movement into, across and around the inner areas is still possible, so that the town centre remains the social and economic hub of the urban community. And planning controls have been exercised to prevent any diminution in importance of the town centre as the focus for business, shopping and leisure. Equally important have been housing policies: local authority estates are juxtaposed with private housing, rehousing from clearance districts places inner city dwellers of all ages and occupations in suburban areas in inner city estates, and pressure for cheap housing forces many young households into inner slum areas, alongside the elderly. Underlying all these aspects of provincial life is scale: social contacts can be maintained despite suburban movement, and any one workplace draws on a labour force from all parts of the town. Proponents of the garden city, which was to be a 'balanced community', were intuitively correct in their imperative of a fixed limit to the size of the town, but to maintain this implies economic stagnation, lack of diversification, and limited opportunities.

In conclusion

The implications of this new pattern of urbanization for urban sociology have not yet been fully appreciated. The classical theories of urban sociology are derived from experience of cities before their diffusion into the countryside, and before their transformation into regional cities made up of many interlinked, but partially autonomous, subcommunities. In this situation 'urban' sociology has no meaning. The images of urbanization held by Tönnies, Simmel, Weber, Park and Wirth[51] must be related to their experience of cities as distinct from and in their terms even antagonistic to the countryside; of cities as having a unity of existence epitomized in the concept of the 'urban community'. Park in his recognition of the separate worlds of the metropolis, its many subcommunities, comes closest to present-day conceptions of the city as presenting a fragmented, diverse, diffuse and segmental community existence, but even he presents the urban community as one organic entity focused on the central city.

Perhaps surprisingly something of the same image is present in more recent attempts to provide a theoretical framework for urban and regional studies. In reading Rex, or Pahl, or Castells,[52] there is an impression of an 'urban system' having a distinctive community power structure, not of the fragmentation and interleaving of community life, the diversification of interests epitomized by Long as 'the ecology of games',[53] found in the metropolitan regions. In these there is not one, but many, labour markets, not a single 'allocative structure' but a plurality, a system of subcommunities contending between themselves for investment and space. One has only to think of London where access to community resources is mediated so variably even between adjacent local authorities to realize the futility of a simple model of urban social structure.

There are, however, circumstances where the older models of urban social structure may apply, where the city concept is not outmoded. Although the provincial communities are also diffuse in outline they have not undergone the functional differentiation and multiplication of constituent communities that characterize the metropolitan regions. The simplest urban situation for study is that which shows the greatest dependency and where there is closest identification with a local culture. The provincial community has a common history and an identity of interest in its dependence; the very lack of autonomy makes for an identification with the local community. Sluggishness in economic growth indicates a slow rate of change in the environment, a low rate of mobility, and lack of affluence, all of which put a brake to the fragmentation and complication of community structure encountered in the metropolitan regions. Continuity, not change, simplicity in community structure, characterize the provincial community, whereas the regional city has broken with the past—the city is now a region. In these circumstances 'the city concept is an anachronism, and even far-reaching reforms cannot prevent its decay and final disappearance'.[54]

There is currently a lag between policy reactions in planning and social theories as to urbanization, in that planners have responded more openly, and less critically, to the exigencies of the modern big city. So among professional planners, the 'Los Angeles in Hampshire' image for the new urbanization, the open-ended formlessness of structure planning, oppose the older ideas as to compact tightly integrated towns, and the fixity of the plan. For both planning and urban sociology the guiding assumptions were derived in the context of the *fin de siècle* metropolis (to which in their formalism of approach the Chicago School were able to assimilate their observations of a very different urban situation), and yet the conditions of living that have succeeded those cities in crisis generate ideas that flatly contradict the earlier premises as to the nature of 'the urban'. Entrenched beliefs and responses to current situations war uneasily, and ideas are extrapolated from one context to another without critical assessment as to their fitness for action.

2 The urban problem: the inner cities

For the past century celebrations of progress in the societies of industrial capitalism have been accompanied by a grudging awareness of an 'urban problem'. Living conditions for certain groups in society have repeatedly forced attention on the injustice and partiality inherent in the structure of access to resources, in this way maintaining a constant living indictment of the urban-industrial order. It is not the working conditions of capitalism that have generated uneasiness in the ruling consensus about failings in the mode of production, but its living conditions, the squalor, misery, pollution and sheer inefficiency of its urban centres. This awareness of the living conditions of the urban poor dates from the nineteenth century, although there had been no marked deterioration in the administration or sanitary conditions of the largest towns. Although cities were notorious as centres of plague and pestilence, foetid environments that swallowed up the lives of men, women and children, and their poor were cramped into back courts or houses subdivided by floor, room and partition, there had been no evident concern for their planning as more humane places in which to live. It is therefore significant that awareness of urban conditions did alter as the pace of economic and demographic growth quickened in the nineteenth century. Was it that 'the industrial town was something new . . . which seemed the very reverse of being pre-ordained and inevitable'; did it 'seem natural that the inventive powers of man and the strength of machines would be able to change the course of the situation which they themselves had created'?[1]

It is this urban problem that has preoccupied urban sociologists from the days of Geddes and Park. For the former, cities in the 'paleolithic' civilization stood condemned by their 'slums, semi-slums, and super-slums', and the task of the urban scientist was to outline the conditions, causes and solutions to bleak urban deprivation. Parkways, gardens and flower-boxes were to signify the release of urban citizens from the old order in a new consciousness of urban destiny. And it was the inner city that preoccupied Park and his colleagues at Chicago. The main task of sociology there, as at other American universities,[2] was to elucidate the process of urbanization for a bewildered public, and discover the mechanisms by which the traumas of existence in the inner cities could be resolved.

In Britain, as in America, over the past fifteen years, this urban problem has once more come to the forefront in policy discussions. Inner city decay and deprivation are the underpinnings to a growth industry in urban research. But the phrasing of the problem has undergone significant shifts. For example, where once the issue was defined as that of the slums, that is, of problems of physical decay and dilapidation which once rectified by redevelopment would be resolved once and for all, the new definition of the issues of the inner city is less certain: in America the ghetto commands attention, whereas in Britain it is the twilight area debate that covers concern for the future of entire inner cities. Economic growth and technological prowess have freed most of the population from their confines, but have hastened their deterioration as environment. Planning has shown itself incapable of creating conditions of living which overcome these defects; the new is excoriated as inhumane, creating wastelands in the cities as in the suburbs. Nothing, it seems, can remedy conditions in these inner city districts—housing, education, welfare and community development projects can only stave off their incipient collapse as communities. There seems little alternative to their slow decay and their use as reservoirs of cheap labour, as the encampments of an under-class debarred from integration into society by their residence in these areas.

In the debates, many aspects of developed capitalist society are called into question, and the content of discussion is political. The analytical tools of the social scientists coexist uneasily with the doctrinaire statements from political platforms as to national identity, race, poverty and class differentials in education and intelligence. The language, the concepts, the very intentions of research in the inner cities are only comprehensible as part of this political debate. And as the debate shifts ground, in response to changing definitions of the situation as much as changes in the urban situation itself, so do the concerns of social research. In urban sociology there has, for example, been a shift from issues of cultural assimilation to those of the political economy of the ghetto, in America; and in Britain, the discussion has shifted from the demoralization of the urban poor, to the conditions of the slum, and then to the problematic of the twilight area. In these latter two the interests of sociologists, planners and other 'urban managers' converge, and it is on these issues that the sociologists have had a ready hearing from their audience in the professions.

The slum: myth and counter-myth

The slums once had power to shock: the accounts of squalor, misery and inhuman living conditions in the nineteenth-century cities shook the middle-class public of Victorian England, scarcely willing to believe the fearful degradation of life in the slums, so close and yet so remote from their own secure homes. The very alliterative qualities of the word 'slum' are calculated to rouse feelings of pity and disgust —pity for the misery of others, disgust at their fecklessness and irresponsibility. As Abrams comments, 'Slum gives its meaning the moment it is uttered. From the day the word entered the language of social reform, its mere mention was enough to revolt the good citizen, win the support of the crusading press, and dedicate official action to its extinction.'[3]

There was then a clear image of the slum: it was an area where the misfits, the dregs and the outlaws from society accumulated, 'a bleak area of segregation of the sediment of society',[4] a social residue daily joined by newcomers to the city, and left behind by the more energetic and ambitious. The slum was the locale of vice, crime, delinquency and disease, a disorderly gathering of people beyond society and without community. Such a classic slum, for instance, was the 'Jago', a notorious huddle of courts and alleys in Shoreditch, demolished in the earliest of the clearance schemes undertaken by the London County Council.[5] Sunken below the level of the surrounding streets, it was entered by narrow alleyways down which no respectable citizen dared enter. It was the home of fugitives from justice, the pariahs of the urban world; a submerged and detached area, with a shifting slum population. It was this kind of area that the housing reformers were dedicated to extirpate.

Presently, however, the image of a slum is more hazy: neither physical nor social criteria isolate the slum as an entity with definite characteristics. In Great Britain, physical unfitness is marked out by categories of repair, damp and sanitary facilities. To warrant clearance the dwelling has to be in such a state as to be 'not reasonably suitable for occupation in that condition'.[6] The stringency with which such criteria are interpreted varies with the resources and willingness to act of different localities, and changing economic circumstances. Neither can the sociologists offer a firmer definition. Typically they have emphasized the slum's conditions of social degradation, disorganization and exclusion, while emphasizing the subjective and relative quality of such labelling.[7] Not only are the physical and social criteria for a slum acknowledged to be of limited objectivity, but the two indices to a problem area may rarely coincide. The modern slum may be the substantial rooming house of the twilight zones, or the council house estates used as the dumping ground for 'problem' families and difficult tenants.

The image of the slum has been diffused in changing circumstances. The crippling total poverty that characterized the slums described by Mayhew, Mearns or Booth[8] has been alleviated by the Welfare State. No longer do the families of entire streets live as animals without beds, bedding, or food, their children half-starving and near naked. The slums are still there, but their incidence, their harshness, their concentration and their visibility have been reduced. At the same time the areas designated as slums have changed over the last century. In the nineteenth century, housing associations and local authorities tackled the areas of total dilapidation, overcrowding and squalor, the 'exploitation slums'. With few exceptions these areas have been eliminated, and the clearance programmes have turned to the working-class cottages, areas of dilapidation and neglect, 'blighted slums',[9] starved of investment since their construction.

In this there are marked regional distinctions. In areas of growth, the metropolitan regions, conditions of housing stress are found in very varied physical structures, so that solid, modern housing may be brought into disrepute and dilapidation through intensive occupation. But in the provinces, the problem is quite different. The unevenness of national growth has meant the long decline of their towns, the ageing of their population, and the decay of their housing stock. Their slums are the slums of underdevelopment, whereas those of the South East or Midlands are those of over-urbanization. If in the nineteenth century a distinction could be drawn between urban overcrowding and rural neglect, an analogous comparison can be made now between the slums of the metropolitan region and those of the provincial communities.

But the clearance programmes roll on, despite widespread anger at the destruction of familiar neighbourhoods, and the personal upheaval entailed. Unlike the planners who see each project as the realization of a long-term strategy for the reconstruction of the city's fabric, the urban resident sees only displacement of householders, shopkeepers, small businesses and workshops, the removal of low cost housing, and the creation of wastelands. From 1955 to 1973 alone, 1.7 million dwellings were demolished in Great Britain. Urban renewal is seen as yet another pressure towards monopoly and concentration: housing concentrated in the hands of the municipalities, small firms deprived of premises, independent entrepreneurs ousted. Possibilities of autonomy at home and at work are thereby further reduced under the slogan of slum clearance, or urban renewal.

When Dickens or Engels were writing, the slum problem was one of urbanization. The great cities were growing more rapidly in population than their ability to provide the social capital necessary for the reproduction or maintenance of the labour force. Slums resulted from the overcrowding and exploitation of the poor and very poor, scratching a living in the hurly-burly of the Victorian city. The problem was akin to that faced by the cities of the Third World now, where peripheral shanty towns complement the inner city slums. In the last quarter of the twentieth century the more typical 'slum problem' is that of the blight and obsolescence of the houses of the inner cities, and the endemic poverty of their residents. The social disorganization and marginality which so concerned the earlier writers was a problem primarily of urbanization, and not of bad housing. As the problem has changed, so have the attitudes to it. The main concern has ceased to be the welfare of the population, and become the state of the housing stock; urban renewal has replaced slum clearance as the professional statement of aims.

And yet the designation of an area as 'slum' may still legitimate the furtherance of urban renewal policies which are unwanted by the populations affected, and viewed with suspicion by many others. The slur and stigmatization attached to the designation of an area as 'slum' not only convinces its population that they have no rights—as regards their housing they are beyond civil society—but also convinces informed opinion of the fitness of this strategy, regardless of the exact conditions of the area so categorized, and the interests involved. All kinds of housing situation are bracketed together, and the diagnosis and remedy for the problem predetermined, for to designate an area as slum is to prescribe a certain policy—clearance. The causes of the slum have been forgotten, and a myth built up that conditions of social pathology are most acute in the slum, that a slum could be designated in terms of housing conditions, and that to remove this housing was to remove the slum.

In Victorian England the debate on the slums was initiated in the context of more wide-ranging discussions among the middle classes as to urbanization, demoralization and poverty. 'Slums only become a social problem when a large enough group of important enough people decide that poor people ought not to live in such places.'[10] London attracted the greatest attention, both as the largest city with the greatest concentration of the poor, and that with a chronic situation of unemployment and casual labour. Conditions in its labour market were quite unlike those of the northern towns, where a factory proletariat was housed in an unregulated and planless sprawl of mean houses. London was extreme not only in its worsening employment situation, but in the deterioration in living conditions as housing for the poor diminished in supply as London's commercial expansion, offices, warehouses, docks, railways, commanded housing land.[11] Overcrowding increased throughout the century.[12]

The slums had originally aroused concern as health risks as cholera and smallpox epidemics overran most of the industrial towns in the 1830s and 1840s. The lack of 'infrastructure' to the towns, water supply, drainage and sewerage, as well as services such as street lighting and rubbish collection, was seen as one part of the problem, and housing conditions in the poorer districts, the other. Accordingly the measures taken were two-fold: first, authorization to, and surveillance over, the municipalities, in the provision of basic services for the towns;[13] and second, control over housing conditions in the slum districts. Always the rack-renting landlord was seen as central to the problem. Regulation of the overcrowding in common lodging houses, powers over landlords to maintain adequate standards of repair and basic facilities, were elements to the attack against the slums from the 1840s, and only gradually was it realized that these were unavailing. Powers for the compulsory clearance of slum districts were available in 1868, and for the replacement of working-class housing by the municipalities in 1875, and again in 1890.[14] Through the century the task of remedying the chronic evils of urbanization had devolved to the municipalities (under government direction), and their main task in the housing question had been defined as that of removing the unfit or unhealthy districts.

This then was the first stage in regulation of living conditions for the working class: a reluctant adoption of powers against landlords and the provision of urban services, benefiting all sections of the population, at the expense of the property-owners themselves. In this statement of the urban problem, the slum was seen as the key, it was an intrinsic element to 'the mental landscape within which the middle class could recognize and articulate their own anxieties about urban existence'.[15] If the area that was the seat of moral pestilence and degeneration threatening urban civilization could be removed in entirety, either by charitable associations or by the municipalities, then the future was assured.

Very soon, however, it had to be recognized that slum clearance was not the solution. It was expensive in that it took place in the most crowded and intensively used areas, and its removal only exacerbated the problem of cheap housing, enabling larger profits to be made by slum landlords. In effect there was a double subsidy to the owners of the slums—first through the compulsory purchase at market value of the site to be cleared, and, second, through the forcing up of rents, and subsequent acquisition costs, of the surrounding areas. This was early acknowledged:[16]

> It is notorious that the Artisan's Dwelling Act has in some respects made matters worse for them. Large spaces have been cleared of fever-breeding rookeries, to make way for the building of decent habitations, but the rents of these are beyond the means of the abject poor. They are driven to crowd more closely together in the few stifling places still left to them, and so *Dives* makes a richer harvest out of their misery, buying up property condemned as unfit for habitation, and turning it into a gold-mine because the poor must have shelter somewhere.

Even if municipalities provided the new housing, the needs of the poorest households could not be met. In the clearance of the Jago and its replacement by the Boundary Street scheme, only eleven people out of 5,719 displaced took rooms in the new housing.[17]

In fact clearance exacerbated conditions of shortage in the inner cities. Particularly in London, there was a vast pool of labour forced to cluster close to the markets, docks and streets from which they drew their livelihood. But equally in the factory towns, the worker was forced to live in earshot and walking distance of the factory gates. The poor were compelled to live close to their place of work, and therefore had to accept what housing was available to them there. In an account of one of the worst of the mid-century London blackspots, Jacob's Island, Bermondsey, Mayhew quotes the women explaining their situation in simple terms: 'They knows it's handy for a Man's work—and that's the reason why they imposes on a body.'[18] Those who worked long and irregular hours, such as market porters, hotel and restaurant staff, dock workers, were particularly dependent on central area housing and therefore had to compete for space with the shops, offices, warehouses, public buildings and luxury housing. Inner London in Booth's poverty maps was a crazy patchwork of good and bad housing, the poor relegated to the back streets and courts, the rich in the squares and wide thoroughfares. If these sordid slums were removed then the slum problem would simply be displaced elsewhere.

But by then, the principle of state intervention was accepted, and there was no dispute that measures had to be taken to safeguard living conditions in the cities if for no other reason than ensuring the reproduction of the labour force. The alternative of forcible removal of slum properties was to open out the cheaper land of the suburbs, first through the provision of cheap transport and then, eventually, by subsidies to the builders of new houses. The first steps were hesitant, parliamentary power to impel railway companies to provide cheap trains in 1893, municipalization of bus and tram services from the 1890s (and quite inadequate for the poorest households, who only very slowly benefited from the 'filtering up' of more secure members of the working class from the inner cities), but they formed the basis of what can be termed the 'suburban solution' to the housing question. Certainly, by 1900, there was a well-established argument that population had to be decentralized as had industry already, if cities were to function: 'The Housing Problem is not to be solved in the slums of Camberwell or Whitechapel, but in the green fields of Harrow and Hendon . . . in the suburbs of the south.'[19] This essentially was the policy promoted by the radical liberals, and enshrined in the Housing Acts from 1919 onwards. Slum landlords were to be put out of business by the provision of alternative cheap housing.

In face of this new appraisal of the housing problem, the proponents of slum clearance had to make their case with even greater stridency. They had to demonstrate that bad housing conditions had an effect on health and welfare over and above conditions of poverty,[20] which was sufficiently deleterious to justify expenditure to the ratepayers, punitive action against the property-owners, and the compulsory upheaval of the population. The issues were necessarily simplified in the preparation of a 'charter for action', and the conditions and circumstances surrounding particular areas of poor housing had to be overlooked to put the message across in unmistakable terms—clear the slums, eliminate the housing problem. There was a vulgarization of the issues at stake which was necessary in political terms if the slums were not to decay as the suburbs expanded, but which had its basis in the cleavage in life situation and culture of middle and working class. Only the separation between classes in the city permitted the stereotype of social pathology, disease, deviance and moral turpitude, common to all areas of dilapidated housing, to achieve credibility.

The slum, then, became a reformer's myth. As Dennis has argued, the reformers had to resort to easily understood slogans and propaganda statements which simplify the issues and render clear cut the solutions, in order to bring the needs of the worst housed to public attention, and into the sphere of legislation. '. . . research, detailed knowledge, and the fastidious weighing of alternatives must give way to the slogans of propaganda, and to the enthusiasm, which stems from clear and uncomplicated statements of right and wrong.'[21] Inherent to the statement of the issues is an assumption that slums exist as identifiable groupings of people and dwellings, which have common features in a variety of urban contexts. The principal features to the slum are those of social pathology, which are soluble if the physical environment is reconstructed. This image would have been a distortion and oversimplification of conditions in the inner cities even in the nineteenth century, and subsequently it has acted to stigmatize households forced by conditions of shortage in housing, and their own subsistence wages, to live in houses which have long outlived their lifespan.

Nor has the slogan of slum clearance been directed as its spokesman intended. Throughout the century, the earlier solution, removal of the older housing, has been seen as an alternative to the policy of construction of more cheap housing in the suburbs, and not as an integral aspect of a combined strategy. Government directives, in 1933, and then again in 1956, specifically restricted local authorities to the costly task of urban renewal, leaving the green fields of the suburbs to private enterprise, except where rehousing commitments forced expansion. Insufficient resources have been allowed to the public sector to provide new housing over and above replacement requirements, and the humanitarian pleas of the housing reformers have served to justify restriction of the local authorities to the tasks which private enterprise could not undertake.

Sociological approaches to the slum

All studies of areas defined as slums have been coloured by the reaction of the writers to the reformers' policies. There has been acute resentment expressed on behalf of the neighbourhoods affected, at the stigmatization and high-handed treatment of their populations. It would seem that sociologists have been concerned to correct simplistic arguments concerning the relation between poor housing and 'social pathology'. They have certainly tried to demonstrate the validity of the culture, or subculture, in these areas, by dwelling on the life experience of the residents. 'Slum' areas have therefore been defended as socially optimal environments, and the planners castigated as 'clumsy giants crushing underfoot things of whose value they have no conception'.[22] They have argued that either the existing social network should be maintained on redevelopment, in 'the movement of street and kinship groupings as a whole, members being transferred together to a new setting [which] would enable the city to be rebuilt without squandering the fruits of social cohesion',[23] or the houses should be rehabilitated rather than demolished.

The antagonism between sociologists and planners derives from their different interests, the one in social organization, the other in the environment, and also in the slum myth itself. Sociologists have distinguished between the 'urban jungle', an area which 'attracts the criminal, the mentally ill, the socially rejected, and those who for one reason or another have given up the attempt to cope with life',[24] and the 'urban village', i.e. areas with a relatively self-contained, slowly changing, highly personalized social environment, in absolute contrast to the mobile shifting world of the 'urban jungle'. As there is real difficulty in studying the latter, in that its population may put a high premium on anonymity, and independence of surveillance,[25] the sociologists have concentrated their attention on the urban village. But the planners, operating in the assumption of homogeneity of the slums, and in acceptance of physical dilapidation as an index to 'social pathology', have been oblivious to the differences between areas, both in their position in the housing market and the populations they house.

A definite image has been built up, largely on the basis of local community studies, of life in the urban village. The relative lack of mobility is emphasized, as is the focus on the family circle, that is, a unit rather larger than the nuclear family of parents and children. Once working-class families had moved into the towns from the land, mobility often ceased for two or three generations, and a strong identification developed with one particular neighbourhood. On the basis of this localization, it is argued, there is a rejection of urban culture, in favour of a more personalized way of life in which status is conferred on criteria other than those of education, wealth and occupation. Suttles terms this 'provincialism'.[26] Particularistic assessments of merit are rooted in close group identifications, the networks of housewives, neighbours and relatives in Bethnal Green or Sunderland for instance,[27] the pub- and club-based worlds of the miners in Ashton,[28] or the street gangs of Liverpool.[29] All of these groupings are based on ascribed categories of sex and age, family ties, or on residence. Individuals do not have to find out who they associate with, as the relationships are given by their position in the community. The social value of the urban village therefore lies in its provision of security for every resident. By the very simplicity with which its members are allocated roles, the poor are protected from anomie.

In this image there is again an inference of homogeneity, and the internal stratification and segmentation of social life is overlooked. The localization and segmentalization of life permits considerable diversity of social groupings within the same neighbourhood, so that, for instance, in two American studies, both in Boston, the research workers, relying on personal contact and participant observation, found themselves restricted to certain groups in the neighbourhood, Whyte to the street corner gangs, and Gans to the Italian working class.[30] Access to one group in the neighbourhood did not lead to association with others. The vulnerability of working-class households in the housing market is revealed in their inability to resist the invasion of undesirable groups. The sociological portrait of the urban village has encouraged the planner to discount the possibility of divergence of interests and aspirations between different sectors of the population.[31]

In most inner city districts in Britain there is considerable diversity of population as judged by age, stage in household cycle, income and ownership of property. Areas of blight and decay are frequently associated with elderly populations, but in many towns the clearance areas also have a high proportion of families with children, living alongside older households.[32] Modernized and smartly painted houses are intermingled with the brown paint and shabby fronts of the older population, and the population divided on aspirations for their housing and their attitude towards the future of the area. And as renewal programmes extend into areas not previously considered as slums, so may the range of households included widen.

There are also studies which present a different interpretation of the slum culture. Mogey's study of St Ebbe's, Oxford,[33] must be set alongside the better known contemporaneous study of Bethnal Green. In this it is clear that Mogey welcomed the opportunity given to the inner city population to break out of the traditional community, based on deference and assent, into the new housing estates. The lack of privacy, the conservatism, intolerance and restriction of outlook, were exchanged for a more open life in the new areas. Even in Bethnal Green, it was noted that scholarship places to grammar school were not taken up, as 'This was a working-class community, and those who tried to become something else, were not behaving as they should'.[34] Similarly, Coates and Silburn, in their study of St Ann's, Nottingham, found that 40 per cent of their sample were living in poverty, that there was considerable turnover of population, and a marked heterogeneity. They write of the breakdown of community feeling, and the continual friction between neighbours, especially over children 'having to be careful who you're friends with, and who your children can play with',[35] in an area where social tensions were at least as strong a source of discontent as the lack of physical amenities. Two-thirds of their sample were 'very glad' that the area was scheduled for clearance, and half wanted to leave the area completely.

These differences in assessment may relate to different circumstances. Bethnal Green was an area of considerable stability, and relative homogeneity, a relict community at the apex of the working-class sector of London extending into suburban Essex, deserted by successive waves of the more ambitious. Post-war rent controls and housing shortage had reinforced this immobility, and local occupations were in general poorly paid. Oxford, however, was a town experiencing a new affluence, and expansion, with the growth of high wage industries, in particular, cars. Similarly, St Ann's, Nottingham, and Millfield, Sunderland, are evidently very different communities. The relatively rapid rate of growth in Nottingham had put strains on the inner city housing stock in that immigrants to the city employed in its low wage industries came to live alongside local households in dilapidated low cost housing. While in Sunderland, an area of endemic unemployment, there was a strong identification with the one neighbourhood by a local population strongly resistant to a clearance programme that had continued in advance of their demand for better housing. Slum clearance was being used as a legitimation for urban renewal policies intended to modernize the town, that outran the community's preparedness for change.

It is clear that the slums of one town are not the slums of another, even in the same society. Each urban area presents a different housing profile in that it will have inherited a housing stock representative of the historical situation of the town in the development of the national economy, and the demands placed on that housing stock will depend on changes in the local labour market. Generalizations as to 'the slums' and policies for slum clearance throughout the country are therefore virtually impossible to make. While the residents of Cardiff, or Manchester, for example, may be campaigning against the continuation of clearance policies, young housewives in Hull are blockading their streets in attempts to draw attention to the fate of their children in rat-infested and damp houses.

The contemporary debate—the future of older housing

The term slum is of limited usefulness, referring as it does to a definite ecological unit. The 'slum' problem now is more fragmented, the young household forced to live in its car, or doubling-up with relatives; the basement flat in an otherwise modernized house, the caravan site in a rural setting, or the tenement block used for homeless families. Nor is the slum the current problematic; the issue is that of the future of older housing, and not the fate of their populations. The problems set in train by slum clearance schemes, or 'urban renewal programmes', as they should be termed, are treated as of secondary consequences to the modernization of the housing stock. Urban renewal, 'the substitution of new social capital for old . . . in a programme for providing twentieth-century amenities in nineteenth-century towns',[36] is the concern of the planners. Housing programmes are not set by reference to the needs of the present occupants of the housing but in terms of some universal standard for the condition of the housing stock in years to come. Homes are for tomorrow. The decision whether to clear, rehabilitate or leave alone an area of housing becomes a technical and not a humanitarian question.

Concurrent with this important shift in attitude to areas of older housing have been changes in their social structure. In many cities there is evidence of community demoralization as the former enclaves are broken up by outward mobility, and subsequent invasions by low-income households.[37] In Blackburn, for instance, like many other northern towns in decline, but because of the operational requirements of the textile industry attracting a new labour force, the older residents felt themselves marooned in the inner districts, abandoned by their own youngsters who had moved away from the town, if not to the suburbs. An increased rate of economic growth, post-war, has enabled relative affluence for the younger generation, increased personal expenditure on housing, and the break-up of neighbourhood groupings. So the old community is 'at the mercy of the rising expectations of the young, which in felicitous conjunction with a construction industry with a need to build, ensures the flight from the town centre of all but the old and the poorest'.[38] Affluence has been as effective a destroyer of old communities as the bulldozer.

Sociological concern has therefore shifted from fears for the destruction of community groupings, to a concern for the promotion of community associations, and local identification. Community action and community development programmes must be seen in the context of breakdown in neighbourhood cohesion as strangers with instrumental attitudes to housing and locality move into the inner districts to take the place of the younger and more ambitious local families. Ostensibly these areas are the least likely to generate urban social movements because of 'the inadequacies of the poor community's political socio-economic infrastructure and knowledge base, plus difficulties of organizing a representative and accountable citizen's group in the face of futility, alienation and mistrust',[39] and yet there are increasing efforts to mobilize these groups in face of the planning process which may culminate in the destruction of their homes.

The mass popular backlash against the erasure of whole neighbourhoods and their replacement by new environments has been taken as the opportunity for creating awareness of the world beyond the close-up scene of job, family and neighbourhood, so arousing the people out of their parochialism and apathy into a consciousness of their own rights and a sense of their position in society. This, it is claimed, is a starting point for 'political activity anchored in life'.[40] But, despite community action, protest and participation, the future of older housing is not a matter which will be locally determined. Decisions as to the level of investment in housing are primarily national policies taken with an eye to the desired level of public expenditure, the rate of housing investment considered desirable in the national economy, and the conditions and requirements of the building industry. Local initiative can have little effect so long as decisions as to the allocation of housing finance and the distribution of wealth are nationally determined, and in that respect public opinion is of little consequence.

From the public, the sociologists, and increasingly the planners, there is widespread criticism of the housing policies of the last decade. It is argued that these policies represent a philosophy of housing affluence, 'the belief that housing standards ought to rise in step with gross national product because we are members of a rich and technologically advanced society'.[41] It is argued that these policies can only create situations of hardship for low income households, in the extra expenditure involved, and that public health criteria are running ahead of demand for modern housing. It is considered that older housing is needed as a reservoir of low cost accommodation, insufficient attention is paid to the consumers' preferences, and to the benefits of neighbourhood, and location within the urban area.[42]

In this debate there is a consensus that the market, and the consumer, should determine future housing policy, rather than general conceptions of social need. Modern housing is no longer seen as a public utility available as of right, and instead the social scientists have turned to micro-analysis of the consumer's preferences and attitudes to their housing. So Dennis, echoing Gans, argues for the participation of the public in the decisions that affect their future and the organization of social surveys to ascertain their preferences. In this he shares common ground with orthodox economists who maintain: 'In general only households know their own needs, means and preferences, and so only they can choose the price, location, quantity and quality of housing that suits them best.'[43]

The current mood in discussion of housing policy is for a 'new empiricism committed to an induced understanding through research and public information gathering: this replaces logically deduced and insensitive dogmatism'.[44] The supposition is that the population is able to exercise preferences which are a better indication of their needs than any predetermined social policy. If, however, it is accepted that choices of housing are decisions taken under circumstances of constraint, then the assumptions of this new approach to housing policy are questionable. Empiricism is a measure of present circumstances, present conditions of resource allocation, and cannot give the answers if conditions of existence were to change. What, for example, would the demand for clearance be if higher subsidies were available for new housing? Should housing be treated as a market commodity, and not excluded from the calculations as to consumer preference and cost-benefit analysis? The reactions against compulsory upheaval are as much a reaction against its cost, as the manner in which it is conducted, and the end-product of reconstruction.

Nevertheless, legitimation for the shift in emphasis since 1968 from comprehensive redevelopment to rehabilitation, improvement areas, and cellular renewal, has been sought in this popular reaction against the planners' policies. The effect of changing conditions in the inner city housing markets, distorted by the clearance programmes, on household attitudes have been ignored. The mounting costs of new housing have to be set against the increased investment required in the diminishing supply of old housing. Similarly disregarded are the economic and political conditions governing public expenditure on housing. The switch in policy from new development to improvement of existing houses was necessitated by restrictions on public expenditure in conditions of national economic crisis, and the restrictions on subsidy, which further penalized public housing, were intended to shift the costs of urban renewal from national exchequer to private purse.

Into this debate, too, has been fed the sociologists' portrayal of the slum as the urban village, the cosy, secure community ravaged in the name of comprehensive neighbourhood planning. The sociological analyses, designed to expose the tenuous social bases of the reformer's myth of the slum, have themselves acquired the status of myths in validating changes in public policy for which there was little alternative on economic grounds. Where the earlier image of the slum justified the bulldozer, and legitimated the sweeping decisions taken by the urban managers without reference or consultation with the populations concerned, the new image justifies piecemeal improvement and protracted exercises in participation.

Real as are the hardships of redevelopment, the new policies should not become substitute slogans for slum clearance. In some situations housing conditions may not warrant drastic change, but elsewhere the situation may be different. Despite the onslaught on the inner cities since 1956, there were in 1971 2.245 million households lacking or sharing one of the basic amenities of a bath or shower, W.C. and hot water. In the major industrial towns developed in the nineteenth century and in the multi-occupied London terraces, the housing problem remains acute. Nor can many of the former be improved; they are too small, too crowded together, flimsily built and poorly maintained. They were built as cheap houses for an urban proletariat, to minimal standards and in shoddy construction. Many of these have long outlived their usefulness as shelter, and ought to be cleared as their population demands.

The inescapable problem is the residence in these areas of old, decayed houses, of the very groups of the population, the elderly and the poor, who are least able to articulate their own interests or cope with removal. For many of the elderly the house has become the essential haven, and has conferred a certain status in the local social order. Current administrative procedures do not permit consideration of individual needs and circumstances. In this respect the slum mythology lives on, as the more modern house is deemed to justify all other hardship. This is the task for community development, to create a political environment in which the views of individuals as to their future environment are given as much weight as economic circumstances permit, for household income and housing subsidy are the parameters which set the limits of individual choice. But the overriding responsibility is to plan for the future and not to attempt to salvage the past. Houses age, and in any society must be replaced.

In the century since the assumption of responsibility by the state for the slums, the urban problem has altered. Originally the problem was of acute overcrowding which worsened as cities expanded, and the slums were both the home of an urban social residue as well as the reception centre for new urban populations. As the rate of urban growth slackened, the problem became one of the senescence of the housing stock, a problem aggravated by the diversion of investment away from the inner city districts to the suburbs. British slums are blighted slums in which investment, and change, has been checked by rent controls, their designation as slums, and also the availability of alternative subsidized local authority housing. Generally, in Britain, slum living conditions are not so much problems of overcrowding as of neglect of the housing stock and its generally low profitability. In this, Britain differs markedly from America where it can be argued that the slum-ghetto is a 'high-price use', a highly profitable use of inner city land.[45]

The history of urban renewal in Britain bears no comparison with that of the American cities, or that of Paris.[46] In the latter the Nanterre sociologists claim that the current redevelopment programme has been more concerned with upgrading the population in the inner core, than with upgrading dwellings. They conclude that urban renewal in Paris had nothing to do with saving the environment, but everything to do with a reconquest of central Paris by the middle classes associated with the expansion of Paris as business and financial centre.[47] In similar vein, Gans argued that the clearance of the West End of Boston was designed to bring back upper income groups to central Boston, so checking the loss of rateable value to the city and improving its image for commercial investment, and that the residents of the area saw the scheme as 'a politically motivated plot to take the West End for private profit with government help'.[48]

With the possible exception of Covent Garden, and other scattered sites in West End London, there are no comparable instances of the use of clearance programmes by propertied or moneyed interests in Britain now. In the overcrowded rooming house districts, the contemporary reception areas, the present use is the most profitable one, and intervention means the expropriation of the property of the slum landlord. In the decayed slums of the provinces, as in inner London, the clearance programmes are a heavy charge on local and government funds, and offer no direct benefit to any group other than the construction industry, and those who move to a more modern house. The direction of housing resources by the state through the agency of the municipalities has meant a privileged position in the housing market for considerable sections of the working class, which, it is claimed, sets them apart as a 'housing class'.[49] Thorough-going support for the principles of comprehensive replanning of the cities comes from the socialists: the old cities are seen as a scandal from which municipal intervention can rescue the population, and the hardships of compulsory removal without consultation are overlooked.

But initiative for the direction of community resources to housing came from government. Working-class pressure against the state throughout the nineteenth century, and into the twentieth, was directed towards the improvement of working conditions, the right to vote, and the right to mobilize in trade unions, thereby securing independence and health for working-class families. The impulse for the departure in 1919 from the previous supine attitude of government to the housing of the working classes came not from the housing reformers, but from the growth in numbers of working-class voters, the militancy of the large unions, the growing unrest in the workforce previous to, and during, the First World War and the grudging acknowledgment that a population that has withstood the horrors of that war was not going to suffer the same conditions of existence as hitherto. It is therefore arguable that 'Government answered parliamentary anxiety about the condition of the people with legislation about their housing not because that was pressed for, but as a sop to quieten agitation about other matters'.[50] The reformers' arguments had greater impact on their own stratum in society, as it was the middle classes that had to be convinced of the necessity of this policy if they were to pay for the new housing programmes, in taxes and rates, and lose their monopoly over the secluded existence of the suburbs.

The problematic of the twilight areas

The new definition of the 'urban problem' reaches further afield than that of the slum. As the large cities have undergone a more rapid rate of change post-war, both in the reconstruction of their physical infrastructure, and the movement of population and activities within and between the city-regional complexes, there was a slow realization that the urban flashpoints were going to lie outside the areas defined as slums, in the so-called twilight areas, areas which had hitherto lain outside the reach of official policy.[51] Attention therefore was switched from relatively slow processes of physical decay, decay which might reasonably be attributed to physical causes of poor original structure, and old age, to the 'dynamic processes of obsolescence, exodus, displacement and social change'.[52] New demands were being placed on existing buildings, new pressures on existing communities; not only was the situation in the inner districts of the metropolitan cities one that was changing rapidly, but it also involved large-scale transfers of population relating to substantial changes in the national economy. The urban problem could no longer be isolated and treated as a limited issue of slum clearance and the rebuilding of selected neighbourhoods, but had to take into account wider issues of housing policy and social planning for the entire urban complex. Environmental reconstruction might still be necessary, but the basic problems related to economic planning, and the allocation of public resources.

Changing economic conditions in British society and the cumulative impact of existing housing policies had forced this change in definition of the issues of urban policy. Acute labour shortages in the major metropolitan centre had stimulated interregional movement and immigration, but with the closure of the areas officially defined as slums to newcomers to the urban housing market, areas outside the 'slum' districts became the reception centres for households new to the urban area. Housing was created in these areas through the subdivision of existing units of accommodation and the now familiar problems of multi-occupancy and overcrowding were experienced. The historic function of the central city slum as the home for successive generations of new recruits to the urban labour market was displaced to the once sedate and still structurally substantial areas outside the inner core. It is in these areas, designated by Rex as 'zones of transition',[53] that there has been the greatest housing hardship and the most intense pressure on community facilities. It is here that the overall competition for space and housing in the city between groups of different prestige, purchasing ability and political power has been felt most acutely. Into the twilight areas moved the newcomers to the metropolitan centres, who, from the mid-1950s onwards, comprised workers from former colonial dependencies as well as the outlying provinces—Scotland, Ireland, Northern England. The debate over the twilight areas was provoked by issues of race; underlying though unadmitted fears of incipient ghettos in the twilight areas pushed on policy discussions as to their future with an urgency that the slum issue had never generated in the previous decades.

These areas of 'housing stress' attracted special attention in the debates over policy, but alongside there was a growing consciousness of the conditions of multiple deprivation experienced by the poor resident in the local authority estates, the slum clearance districts, as well as the rooming houses of the twilight zone. The future of the children in such areas was held to be at risk: not only did they suffer from poor housing conditions, inadequate facilities, and an inhospitable environment, but their educational opportunities were being limited by the turnover and remoteness of their teachers and the poor equipment of the schools. The recognition that there were areas of educational priority, 'areas of special need',[54] was a first step in evolving selective policies for the inner cities. Community development programmes were yet another strategy in the 'modernization' of the underdeveloped central city districts. There was growing recognition that only co-ordinated social planning could prevent the long-term economic stagnation of these areas and their exclusion from the anticipated affluent society.

Palmer argues that the moves towards the adoption of comprehensive social planning have as their main objective 'the creation of a consumer society in which all members participate through their own unaided efforts'.[55] Growth in the economy has come to depend on increased rates of consumption, and so the presence of residual pockets of poverty (low consumption) is a brake to continued prosperity. Everyone must be brought into the mass markets if capitalism, as an economy always searching for new outlets, is to survive. He writes,[56]

> Pockets of poverty and underdevelopment represent imperfections for the market which must be removed if the techno-structure is to move away from war-oriented economic policies as the mainstay of further growth. Thus social and welfare planning will be a device increasingly used to ease out and finally eradicate imperfections of consumption.

In this interpretation the very prosperity of the capitalist system depends on the incorporation of the residents of the inner city, twilight areas and slums alike, into the mass society of high consumption. The suburbs cannot turn their backs on the inner cities, for their continued comfort depends on the transformation of the neglected inner districts into replicas of suburbia.

Conditions of existence in the inner city are therefore subject to a wide range of policy initiatives. The statement of the urban 'problem' now explicitly acknowledges the reality of poverty and deprivation, as well as the inadequacy of existing housing and planning policies. But, however broad the redefinition of the problem, it is still restricted in its terms of reference. The issue is phrased as one of inadequacies in living conditions, and the remedies investigated relate to the allocation of community facilities and resources alone. They do not call into question the use and rewards of labour in the economy. Implicit is the assumption that selective policies for certain areas, directed at community revitalization, can offset, if not eradicate, these imperfections of consumption.

It must be admitted that the overcrowding, neglect and poverty that characterize the metropolitan twilight areas derive from the position of their residents in the metropolitan labour market. Just as the reserve army of labour of the Victorian metropolis, its 'residuum', was housed in its rookeries and tenements, so the new arrivals to the ever hungry labour market of the metropolis are to be found in the overcrowded rooming houses of the twilight areas. While the metropolitan economy depends on a reserve of cheap unskilled and semi-skilled labour to fill the low wage jobs in transport, retail, catering, hotels, building, hospitals, cleaning, all the services on which the specialized activities of the metropolis depend, then there will remain these pockets of 'poverty and underdevelopment'. The problem of the twilight areas is conspicuous because of its environmental failings but it is not an environmental problem; its existence reveals the underlying weakness in the metropolitan economy which growth and diversification do nothing to mitigate. While the industrial worker prospers, and migrates to the suburbs, the families of unskilled service workers are progressively pauperized, and their living conditions in the inner city deteriorate, as available housing decays. The growth of the metropolitan economy not only increases the demand for such labour, but further exacerbates the housing problem by the increase in land values which it generates. In addition, the supply of housing available to the low wage service worker is reduced by the pressure of commercial development, as well as the competition from highly paid professional workers for inner city housing. Just as the American ghetto was brought into being by the demand for cheap labour in the northern cities, so the British twilight area derives from the nature of the metropolitan economy. The grim twilight areas are the obverse to the glitter of the West End, the glossy tower blocks of the City, the gleaming paintwork of the upper-class residential quarters.

The urban 'problem' therefore reveals itself as wider than commonly set out. The issue is not merely one of devising appropriate housing and planning policies for the twilight areas themselves, urgent though these might seem, but of controlling the process of metropolitan growth. It is for this reason that the issue of the twilight areas is termed a 'problematic'.[57] In phrasing the problem as that of intervention into a limited sector of the city's housing market there has been a distortion of the 'real problems' posed by urban growth. That is, the 'problem' as posed is ideological.

The features of a 'problematic' are, first, its limitation of the issues for discussion; it pre-selects certain areas of investigation and debate, and excludes others. So, for instance, the issue of the slums was phrased in such a way as to exclude the economic circumstances of their residents. Similarly that of the twilight area has localized the urban problem, and also brushed under the carpet the issue of race. In this manner the questions are posed so as to command the answers to the problem. The answer to the issue of the slums was given in their very definition as slums, clearance. As yet the answers to the twilight area 'problem' are less clearly enunciated, but seem to devolve on the redirection of housing policy. The lead taken by the Milner Holland Committee in calling for 'a common approach to the problem and for a fully considered development of policy based on an understanding of the whole housing situation'[58] has been accepted. One suggested policy has been that of the municipalization of all privately rented housing in the major cities; another, the institution of a special task force acting on behalf of the local authorities, directed to the progressive renovation of the twilight neighbourhoods.[59] In short, it has been assumed that more systematic, better considered intervention in the housing market could solve the problem of the twilight areas.

If the issues of urbanization are phrased in a partial, limited fashion, the inference is that the 'solutions' devised will not only be ineffectual, but the cause of further difficulties for city populations. Take the case of the slums: the remedies put into practice were twofold, that of their clearance, and their replacement by new suburban housing. The effect of clearance was to displace the demand for cheap, freely available and accessible housing to the zones outside the designated slum districts. The diversion of investment to the suburban estates and the encouragement of migration out of existing neighbourhoods, was to accelerate their physical decay and the attenuation of their communities. Nor will the remedies advocated for the twilight areas solve their problems. All the policies suggested, improvement or reconstruction, would have the effect of pushing up the cost of housing, and also of reducing the freely available supply.[60] One generation of newcomers to the metropolis, one group of its poor, would be rehoused, at a cost, but neither their children nor their successors as recruits to the metropolitan economy would be provided for. A selective policy in which only the areas of housing stress were controlled would have the effect of creating new crisis areas elsewhere; a comprehensive policy of control through municipalization would effectively close the housing market to newcomers. Hidden multi-occupancy and an increase in long-distance commuting for low paid workers, as occurs in the cities of Eastern Europe, or the growth of peripheral slums, or shanty towns, as in France, would be the result. In either case the problem of housing low paid recruits to the metropolitan economy would remain.

Under present circumstances of housing provision it is tempting to argue that no action be taken to remedy living conditions in the inner districts of the metropolitan cities. If redevelopment is undertaken, the resultant housing is too expensive for the present occupants, and its high density further accentuates the existing differences between inner city and suburban ring; rehabilitation is similarly costly, and will only lengthen the life of originally poor buildings, and perpetuate environments that are fitting memorials to the priorities of the speculative builders of the nineteenth century, and parsimony of public expenditure since. Either policy would only substantiate Engels's century-old contention that 'the bourgeoisie has only one method of settling the housing question . . . that is to say of settling it in such a way that the solution continually poses the question anew'.[61]

In at least two respects the solutions open to the State in the metropolis have narrowed in the intervening century: the very extent of decentralization has foreclosed the option of suburbanization of the labour force, and the accentuation of traits already apparent in its role as centre of nation-state and empire have intensified the housing problem for its poor. Continued dominance as controlling centre has meant a perennial demand for a supporting labour force without the political power, or potential mobility in the city-region of the industrial proletariat, who have yet to compete for living space with the more privileged sections of the national population—the professional, managerial, highly specialized labour force of central London. And the intensity of this competition for living space has been aggravated by the very planning policies intended to mitigate the effects of suburbanization (greenbelt policies) and urban congestion (density control). The problems of reconstruction on any scale are therefore rendered virtually insuperable. This urban problem not only raises the question of the differential in living conditions between classes, but the specific issue of the specialization within the national economy between London, as centre of power, and all other regions.[62] The de-industrialization of the core to the metropolis, allied with its maintained hegemony, nationally and internationally, places it closer to the 'modern big city' of *fin de siècle* Europe, than the centres of commodity production it commands.

3 The housing question

1 The house as home

The housing question is one that few in a developed society can ignore. Our housing situation is directly determinant of our comfort and well being, our status and autonomy, and the house as home is an integral part of our life experience. The dwelling is the background to our childhood, the scene of our primary socialization; for many women their life's work revolves round the home they create for their family; for all it holds strong associations. It is easy to disregard these non-utilitarian sentiments attached to the physical artefact of the dwelling and to treat housing either as shelter, or as property, a massive costly artefact. So many households are handicapped by the inadequacy of their shelter that the housing problem is seen as a numbers game. And yet housing attracts such attention because of the depth and emotionality of feeling attached to the home.

The house in itself is a bare shell—a shelter against the elements: the home is a social unit of space articulated round the family. In industrial societies, despite gloomy predictions of a breakdown in customary social life, the position of the family in its place of interaction, the home, has been maintained and even strengthened. Accordingly both housing and planning policies have been judged by their success, or failure, in providing environments for happy family life. In the massive disruption and social changes of the nineteenth century the ideal of the home as offering 'the last vestiges of human feeling, the only haven of communal emotion'[1] achieved great importance. The dwelling was presented as a private area free from the strains of urban industrial society in which the laws of the market prevailed. The family is *gemeinschaft*, the seat of relationships of 'mutual affirmation', the city is *gesellschaft*, with its calculative exactitude and personal debasement.[2] Tönnies in this typology articulated the gist of much nineteenth-century social literature, in which the family, and the home, are placed against the turmoil of the city. There was disquiet about the nature and future of the city, a man-made entity which had once been a source of pride, and now seemed an environment of social debasement. One response was to focus on the home as an island of peace and certainty, a known world in an incomprehensible and precarious existence.

Dickens, a self-made man, celebrated the Victorian ideal of 'hearth and home', one widely held by the liberal middle classes of Western Europe.[3] The house was already seen as the place where privacy was assured under the umbrella of the head of the household: 'the citadel of the private self'.[4] In this Dickens anticipated what was to be the solution to the city as problem: clean and moral homes were to be the means by which the poor and outcast were to be drawn into urban, or bourgeois, society. Those whom urbanization had unhoused even to the extent of sleeping on streets or under railway arches, were to be rehoused on terms set by those who provided the houses. Women, who came to the city to achieve independence and found the alternatives of sweat-shops, outwork or prostitution, were to be provided with an environment in which they could be brought closer to Dickens's image of the middle-class 'angel', safe in her home, nurturing sound social values in her children, and supporting her husband in his role in the world. In fact one answer to the perennial unemployment crisis of London was to create conditions in which women were to withdraw from the labour force.

So the family grouped round the hearth, the hearth that was to symbolize home even in *avant-garde* architecture in the twentieth century,[5] was to become the model social existence. In face of the hostility of the city this is the most valued social grouping, closed, introverted, adopting a defensive posture to the world outside the front door. The uncertainties and unpleasantness of urban experience and the harshness of the economic order are shut out. This enclave of intimacy depended on the compliance of women in the role assigned to them and their acquisition of home-making skills, of which in early experience of urbanization and industrial working conditions many had been divested. Women must acquire the skills and temperament suited to the creation of tranquillity and peace for the men returning from the realities of urban working life. In empirical terms, for the affluent middle-class family,[6]

> To the father, the house as home may mean not the material and visible sign of his success as a breadwinner, but the one place he may be himself, relieved of pressing responsibilities, free of competition, sure of warmth and companionship. For

> the mother, the house as home tends to represent her major task; the creation for the husband and children of an environment in which security and understanding are paramount.

Little sociological evidence exists on the expectations placed on the dwelling unit by families in this, or other modern societies. The world of women and children has been largely ignored; the attitudes of those who work without wage in the house, in the sustenance and maintenance of their families, have not had the importance of those of the wage earner on the shop floor to whose productivity industrial sociology has been directed. Indications only exist in scattered surveys and government reports.[7] Access to housing in the same society differs so greatly between social classes that it is scarcely surprising that expectations of housing differ in the same degree. But in one requirement, privacy, there would seem to be very little difference between classes. Although the lowest income households, unlike the more affluent, cannot achieve the zoning of rooms for display or privacy, for frontstage or backstage use,[8] both in America and Britain studies of working-class communities have indicated that neighbourly life is restricted to the public, or semi-public, domains of street, backyard or garden, and that it is rare for those outside the family circle to cross the threshold.[9] Where neighbours are unavoidable, or public life uncertain, the personal domain is defended against intrusion.

In fact the dwelling of the lowest income household offers few of the satisfactions and gratifications of that occupied by the secure middle-class household. There are, in addition to the homeless, those forced to double up with relations, those in the squalid rooming houses, where even the room may be invaded by the landlord or his agents, and those in the cramped conditions of the cottage housing of the nineteenth century, now categorized as slums. In these circumstances shelter from the weather is not assured, there are no status gains, no pretensions of property, and little privacy for members of the family unit. More seriously, perhaps, where shelter is provided in publicly subsidized housing, these basic conditions for the home often go unanswered. In modern mass housing there is insufficient protection from noise, from overlooking, from undesirable scrutiny by neighbours. The definition of the housing problem needs to be broadened out in acknowledgment that much new housing does not meet deeply ingrained requirements for the home.

In the circumstances of the modern big city, a stronger case might be made for planning and designing for privacy rather than community. Given the realities of working-class life, shift work, monotonous labour, inadequate space for household tasks and inability to choose one's neighbours, and in realization that there is little inclination for a more communal life among those for whom architects design houses, and planners lay out residential areas, then to anticipate a change in orientation to the family circle and the home, in the built environment, is to invite disharmony between family units, and within the family cell, so adding to the burdens of the housewife or home-maker. One of the requirements of the home in modern society is certainly the need to retreat, from neighbours, noisy children, movement, from all the pressures of a world that lies outside personal control.

The feminists' counter-argument can be succinctly expressed—already for the mother the house as home is her life's work, the home is no refuge. The woman both creates that refuge and is part of it, she is bound by her own social relations to that which she works to achieve. If conditions of 'privacy' were to be effectively created, then she is isolated in her work, and left without escape from her own web. In answer one can point to the fearful withdrawal in the mass housing projects where lack of definition of space round the home, lack of articulation of public, semi-private and private domains, or even too tight a delineation of the neighbourhood boundaries,[10] may impel total separation rather than a participation in common activities, such as the oversight of children. To plan for privacy should be to plan for an extended living space in which women can relate without feeling trapped by prying eyes, in which children have room to play safely, without surveillance from their parents. A sense of autonomy should not have to depend on the maintenance of tight property lines, and on a cast-iron demarcation of private space within and without the dwelling. Status should not have to depend on withdrawal. Utopian and idealist, perhaps, but a more testing challenge to the architect or planner in the disposition of the meagre resources allocated for housing, than enforced togetherness in a new 'community'.

Not only is there little privacy in modern housing, but little autonomy. Although the home is deemed to provide an impregnable enclave of intimacy and security, the sense of personal determination of one's way of life is illusory, in that the household is, with few exceptions, dependent on the judgment of others in the design, the layout, fittings and finish, of the house. Housing provision in a developed society has become a matter of matching statistical aggregates of population—income groups, age groups, household types—to certain calculable categories of housing space. The whole process of housing provision is depersonalized, and the householder ceases to be a client, and becomes a consumer at third, fourth or even twentieth hand. Housing availability is a function of national investment policies and priorities for allocation of shelter, and these in turn are controlled by the vagaries of international finance, and switches in political fortune. The security of the private realm of the household is exposed to issues that lie outside its competence, knowledge, or control.

In this respect there is no difference between capitalist and socialist orthodoxy. While the problems of housing shortage and slum living conditions were questions integral to the socialist critique of capitalist society, the question of personal autonomy in the selection of an environment for living was not raised. Engels, for instance, saw the housing question as a physical problem of overcrowding which would be resolved in the first instance by the redistribution of the housing stock after the socialist revolution, but which in the long-term would depend on the dissolution of the antagonism between town and country. The housing problem would be solved by the planned reallocation of housing and the redesign of the capitalist city. Possession of a house, security in occupation of one's home, any measure of control over the form of that environment, were conditions that belonged to the world of pre-industrial society, which would involve putting back the clock of world history by a hundred years, and 'the present day workers into just such narrow minded, crawling, sneaking slaves as their great-grandfathers were'.[11] He saw the separation of occupancy of housing and its ownership and construction as a means of liberation for the working class.

How then had this separation come about? Engels would have looked to the division of labour which had broken up a world in which each person turned out a separate and independent product. Housing, like any other product, is now a social product, 'the joint work of many individuals through whose hands it has to pass'[12] and only through a rational division of labour could housing be produced in abundance. Housing was not considered to be any different from any other commodity in this respect; in an age when even the urban middle classes expected to rent their accommodation, housing had no special claim to unique conditions of production in which the individual specified his requirements of the product. The task of socialism was to release the potential of the division of labour for all; only mass production could solve the housing problem.

The architects of the Modern Movement endorsed this position. Their basic premise was that technology, harnessed by rational design, was to be the basis for urban living in the twentieth century. The economics of mass production, standardization and repetition of housing units, the construction of new housing on a large scale, on at least the module of the neighbourhood unit, was the only way to provide a sufficient standard of housing for all. The very form of the house was to be transformed, as was the form of the city. New methods of construction enabled intensive uses of land: double-decker highways, and multiple use of the same plot of land in the tall building. There was to be a new concept of architecture to accord with the transformation worked in the social order by the machine age; traditional values associated with the house and the city were to be rethought. 'The twentieth century is called upon to build a whole new civilization. From efficiency to efficiency, from rationalization to rationalization, it must raise itself that it reaches total efficiency and total rationalization.'[13] It was a basic tenet of the movement that rational planning must replace the anarchy of the older cities; autonomy in the selection of an environment for living had no part in the modern urban world.

Mass urbanization, in sheer pressure of demand for shelter, would certainly seem to create conditions for specialization in housing production, but even in the cities of the Western world, mass markets have neither ensured the mass production of housing, nor the full use of modern technology. House-building is the most localized of all industries and, in Britain at least, still heavily reliant on building practices that pre-date the large cities. And in the Third World, where the rate of movement to the towns has exceeded the capacity of urban economies, urbanization has not created mass markets in housing. The urban resident has to provide his own accommodation, and the shanty town results. More important in ensuring the separation between housing ownership and occupation in Britain was the consolidation of land-ownership in large estates, and the transferral of initiative in housing provision to the land-owner, in the countryside as in the towns.

But the tendency of housing costs to rise with urbanization is important. There is greater need to maximize on site costs—a denser and more costly form of building results; dense living sets in train requirements of sanitation, water supply and fire prevention; transport adds to the cost of raw materials; different and perhaps more expensive building materials may have to be used; and labour must be paid wages comparable with those in urban industries. Whereas the labour input into the construction of a house in a peasant economy is not costed in monetary terms, once labour acquires an exchange value, housing becomes a very expensive commodity. This last is perhaps most important: the housing problem is certainly exacerbated by the very conditions of urbanization; but it derives its special character from the new value imputed to labour in an industrialized, full employment economy.

The special feature of housing in the industrially developed societies is its cost. Not only do housing costs rise with the passage from a rural community to an industrialized urban economy, but they continue to rise thereafter. With pressure of demand from households, the push of public standards for housing, the level of housing provision for each member of society is forced upward as the economy expands.

> The size and quality of housing built in the most industrialized countries depends on the standard of living the economy will support, and the expectations of those living at that standard . . . housing is . . . a central feature of a nation's living standards and largely sets the demand for many other goods.[14]

So housing costs do not fall despite improvements to the technology of production, and housing maintains its special position as a pace-setter in consumption, and as such a key element in the modern economy.

The consequences of continually increasing costs in housing are to render the householder further dependent on others for his or her occupancy of a house. Housing has become a very expensive commodity, involving an expenditure quite disproportionate to average annual earnings. The use of housing must therefore be paid for on an incremental basis, i.e. in the form of rent, or requires long-term credit. Where the providers of housing must themselves raise funds, they too must bear the costs of credit which are then passed on to the consumer of housing.[15] The requirement of credit raises the cost of the house to the occupier, and worsens his condition of dependency. This is so whether he seeks to buy his house, rents from a public agency, or relies on the market in privately rented accommodation. The occupier of housing is reduced to a passive recipient of housing space, which he then has to adapt to his own purposes as best he may. The housing problem is then phrased as a problem of providing housing for a population, of ascertaining their needs and assessing their capacity to pay for the housing so provided, and not of creating conditions in which 'people can act on their own environmental needs'.[16]

If, however, the practice of the Third World were to be adopted for Britain, America or Europe, it would have drastic effects on the shape of the cities, and the use of national resources in land. Extensive development of cheap housing around cities such as Athens, Ankara or Lima has only been possible because the countryside has been under-used, and households have been willing to forego the benefits of public transport, schools, sewerage schemes, or even water supply. Housing costs have been kept at a low level by reason of the low density, and cheapness of land in the absence of urban services. The implication of the movement against professional control over housing design, linked with financial independence in home life, is the dissolution of the cities as relatively dense aggregates of population. The building and design skills of the common man have never had to contend with the complex requirements of the modern big city. 'Do-it-yourself' housing, without oversight by the professions, and without resort to the techniques of the specialized builder, implies semi-urban, or, at best, suburban housing. The cry of 'three acres and a cow' as a solution to the problem of urban housing finds a new validation in the reaction of adherents to the ecology movement against their dependency in the urban environment.

2 Architects and housing

Until the twentieth century architects had very little to do with housing. The landmarks in architecture had been the churches and the palaces of the Italian Renaissance, worthy successors both to the temples and amphitheatres of Greece and Rome, and the great Gothic cathedrals. The role of architects had been in the ordering and ornamentation of monuments. But from palaces they proceeded to villas—for which the model was to become Palladio's designs for the bourgeois of the Veneto—and the country houses of Georgian England in which colonial and urban wealth was translated into rural serenity. The first architect-designed mass housing is perhaps that built on the large estates of suburban London (or in leisure centres such as Bath or Brighton). Bloomsbury in London, and the earlier (1600–10) Place de Vosges in Paris, were designed by architects to the brief of wealthy clients concerned to ensure long-term profitability for their estates, and a congenial environment for their own town house. Traditionally therefore there has been a very close reciprocal relationship between architect and client, and the world to which architects looked for approbation was this élite influential in setting the style and the tone for buildings, as all else, for their society.

Through the early industrial era, the tasks for architects were diversified. New monuments were required—cotton mills, warehouses, town halls, schools, then office blocks, department stores and banks. Eventually the skyscrapers pioneered in Chicago by Sullivan. And always the churches. But involvement in housing remained minimal—country houses and suburban villas, and the rare brief for an estate of model houses (as at Saltaire). But in a very real sense housing, mass housing, was not architecture. It did not require innovation in design, nor supervision over new methods of construction; with the exception of the tenement flats it was built to traditional patterns, and with traditional materials. And it had to be cheap. There was no opportunity for fine finish, illusion or grandeur—it was shelter built to minimal requirements if it was to be occupied at all. Where architects were drawn into the problem of housing for the working classes by upper-class patrons, their designs were expensive, and the houses inconvenient, as in the cottage ornée. 'Deep eaves which excluded almost all the light from tiny ornate windows, ornamental patterned slate roofs which could only be repaired at great cost . . . it became the country house of the . . . reasonably well-to-do professional gentleman with romantic tastes.'[1] Architecture was too expensive, an unwanted luxury for the working population.

The circumstances of the twentieth century, in the assumption of responsibility by State and municipality for the provision of housing, have brought the design of mass housing into the forefront of professional consciousness. Politicization of housing meant professional administration in that there had to be an alternative to the speculator/builder/designer who had put together the Victorian city. The provision of housing on the scale envisaged after 1919 called for planning, in layout and design, and the organization of building projects in a regular, orderly and predictable fashion. These were not necessarily architectural responsibilities. In Britain it was not until the renewed onslaught on the housing problems of the inner cities after 1955 that architects were finally able to establish autonomy of office within local government. Hitherto the task of providing mass housing had been shared with other professions such as the surveyors, and architects might be subordinate to housing manager or city engineer. Inner city housing, however, called for new skills in the organization of the environment, which were out of reach of the layman. The squeeze on inner city land, high costs, the obligation of rehousing, new requirements for schools, recreational spaces and roads, pushed house-building into ever more complex design problems, of circulation, and space, i.e. into architecture.

There has therefore been a steady increase not only in the number of salaried architects, but also in the number of private architects dependent on public sector commissions (and approximately half the capital expenditure of local authorities is on housing). In 1964 about 40 per cent of all architects were employed by local or central government (or the boards of nationalized industry), and the 50 per cent in private practice found, by the same date, that rather more than half their work came from the public sector. The remaining 10 per cent were in research or teaching.[2] In a profession numbering some 26,000, it can be assumed that a very high proportion will be involved at some stage in their career in mass-housing projects in the public sector.

There is therefore an extraordinary situation. Until the twentieth century architecture was a matter of private patronage, in which individual buildings were commissioned by clients with whom, in general, there was great affinity in tastes and culture. Now, not only does housing constitute a major proportion of the building construction for which architects are responsible, but the profession is directly responsible for providing housing for the groups in the population furthest removed from themselves in life-styles and aspirations. It is as if Mary McCarthy were to become features editor of *Woman*, or the music critics for Radio 3 to assume responsibility for Radio 1. But both these analogies break down in that the consumer has no equivalent sanctions. The only safeguards are the interest and vigilance of the committees to whom the architect is accountable, and the standards set by national government, itself advised and influenced by other architects.

New housing falls into three categories.

1 Upper class: 'tailored' housing

In this area of private patronage the architect has an opportunity (within the limits set by the client, and planning controls) to express that artistic autonomy traditionally associated with architecture. This very small area of activity is highly regarded by the profession, and in this manner influences the design of mass housing in a process of 'filtering down'. The reciprocity between architect and *client* (in itself a transferral of competence and aesthetic judgment from the earlier relationship of *patron* and architect exemplified by the Baroque of the Vatican, St Petersburg, and the country houses of Georgian England) is seen at its most fruitful in the work of F. L. Wright, and that of the English Romantics—Voysey, Webb or Shaw—in their design of substantial country villas for the middle class of Victorian and Edwardian England. The former's open hearth and free planning is an influence still felt in more pretentious houses, and the latter, in their use of brick, vernacular design or timbered gabling, mullioned windows and coloured glass, were the models for the inter-war semi-detached house.

2 Middle class: 'ready-made' private housing

While the direct client-architect interaction is not operative here, shared values and the selective processes of market forces and the greater choice of the users provide a house form of greater suitability in the buyer's terms than that provided in the public sector. Architects have hitherto had little direct responsibility for the mass of 'spec.' built housing, but if employed must meet the requirements of the market in providing a desirable commodity for sale. The scope for innovation is further narrowed by the hidden institutional controls. Building societies have been notoriously chary of innovation in housing—flat roofs, timber construction, non-standard layouts, have all met disapproval in their turn. To these checks must be added the conservatism of planning authorities, and the cautiousness of the small builder who has hitherto provided the greater part of housing in this category. The presence of these controls, plus building regulations, mean that the possibilities for 'user choice' are residual and eventually unreal.

3 Working class: mass housing

It is in public housing (whose share in the total housing construction each year since the early 1950s has varied between 40 and 60 per cent of total provided) that there is the greatest gulf between the providers and users of housing. Architects, recruited largely from the middle classes, are granted, by default, an autonomy in the design of this housing virtually unique in architecture. As professional officers in bureaucratic hierarchies, implementing state policies, they have had a direct influence on the form and quality of working-class housing. It is in public housing, potentially the largest area for progressive and creative planning, that the worst abuses of the power to determine a 'setting for life' have occurred. To a considerable degree this stems from the professionalism of architecture. The very code that was to ensure competence, and disinterested public service, has prevented the architect from ever reaching out to those for whom he has to design homes.

The profession of architecture

There had been a national professional institute for architecture since 1837 (RIBA—Royal Institute for British Architecture), but not until 1931 with the passage of the Architects' Registration Act, and the establishment of ARCUK (Architects' Registration Council for the United Kingdom), was it necessary to establish standards of professional competence.[3] The institute had the responsibility of offering protection to the public (by safeguarding standards of design and imbuing strongly held professional codes of conduct) in exchange for the protection of architects from competition. Contending claims to competence in the built environment had been felt most strongly from the engineers and the surveyors among the professions, and from the commercial builders. But the institute was never able to fulfil its responsibilities as a learned institute, and instead maintained the 'professional defensive' role.[4] And in protecting the profession from these competitors (not altogether successfully), architects have built up a language and a methodology which is remote from laymen, and, further, have steadfastly refused to introduce the public to the 'mysteries' of architecture. In excluding the public from all knowledge, from any feeling of competence in judging the buildings they use, they have created a gulf which is most apparent and most crucial in the field of housing.

Guidance in design criteria was in fact most pressing in housing. Architects were stripped of their previous assurance in dealing with clients whose tastes were familiar, and which they probably shared, and were now faced with a situation in which the users of the buildings were not the clients—those were the committees set up to safeguard public expenditure, and further, they were users of whose way of life they had no experience. Architects were forced to turn inwards for guidance, to other architects, to the profession. Good architecture, in housing as all other spheres, now emerges out of a professional consensus. Professionalism therefore has two aspects—one is professional closure, in the attempt to establish a monopoly of skill and an authority over the administrator,[5] and the other is the search for a creed which will set the guidelines for design in situations where this interaction between architect and client has been broken.

Architecturally (this century and internationally), the dominant beliefs governing professional activity have been those of rationality ('New Realism'), a belief in the potential of technology, a lurking sense of there being a 'pure' architecture—a geometry of space realized in three dimensions—and, in the vanguard of the profession at least, a strong sense of social commitment. One typology of the dominant definitions of architectural practice and responsibility might be:

(i) classical, or pure architecture. In this view, architecture has an enduring quality which outlives all generations; it is timeless. It is not to be assessed by its social impact, or judged by its particular structure or circumstances;[6]

(ii) functional, or rational architecture, in which rational analysis of space requirements, activity patterns, social needs, would produce buildings that would effectively express their purpose;[7]

(iii) social architecture.

Like all classifications it is imperfect, in that, for instance, Le Corbusier demonstrates elements of all three strands in his written work, as does F. L. Wright. Further it ignores the hidden dimension in architectural activity (i.e. not admitted to the layman), its essentially irrational and intuitive elements. This last was expressed by Le Corbusier in his discussion of sensibility,[8] and is writ large in his Ronchamp Chapel. But there is always a hidden tension in architecture, between the need to solve the run-of-the-mill problems of circulation and services, in which no questions can be asked of the brief, or the budget, and the Utopian air with which many schemes are conceived. Architects, by very reason of the constraints with which they are confronted, have designed in the belief that it was possible to create new worlds; 'worlds within worlds'.[9]

In the construction of mass housing in the public sector in Britain in the past twenty years, a main influence over architectural design has been that of the Modern Movement. The ideas and attitudes towards housing spearheaded in pre-1914 Vienna, represented by the Bauhaus in Weimar Germany, the young architects of the Soviet Union, and the French, personified by Le Corbusier, in which a reconciliation was attempted between the inherited canons of architecture as art, as classical form, to the functional requirements of twentieth-century building activity, is known as the 'Modern Movement'. Architecture was brought into the vocabulary of science and technology; and it was a constructive activity in society. The Russians expressed this hope for architecture, in terming it a 'social condenser'. It was 'a reflection of the new society and a mould in which the society was to be cast—a device for creating, transforming, improving man'.[10]

This new attitude or code for architecture had to be assimilated with the British tradition, in which the stereotypes of the Beaux Arts school coexisted uneasily with the tradition of vernacular and romantic design, with its roots in the writings of Ruskin and Morris, and the practice of Voysey, Webb or Unwin. From the 'arts and crafts' definition of building activity, architecture began to be seen as a technical product; from questions of 'style', architecture moved into techniques of industrialization; from fixed preconceptions of 'design' into what Martin terms a 'new matter-of-factness', in which the end product was more pragmatically based.[11] The crucial years were the 1930s in which the two traditions encountered, without a reconciliation, and the ideas of the Modern Movement began to be absorbed into the consciousness of the young architect in Britain.

In this Le Corbusier was particularly influential. He was a prolific writer who presented a grand analysis of the urban scene in dramatic and pungent style. The two most influential publications of the period were *The City of Tomorrow* (1924) and *The Radiant City* (1931).[12] These had tremendous appeal: their style, incisive, programmatic, even prophetic, inspired a sense of a 'new Jerusalem'. After his death one who had been a student in the 1930s reviewed his impact on that generation as 'the prophetic voice of the poet speaking with the certainty of art, and telling us what to do—the certain vision of prophecy giving us what every new movement needs—a system and a method'.[13] More than anyone else he was the inspirer of a new vision for urban living, and of housing design. These students were men whose careers were interrupted by the Second World War, and who in the circumstances of post-war Britain had been unable to carry out architecture in housing. Not until financial restrictions were lifted, and urban renewal programmes could proceed on a large scale, could any of these ideas be implemented. The mammoth and unprecedented task of rebuilding and replacing damaged and obsolete housing, with the new responsibility of architects in the public sector, found them unprepared. For inspiration they resorted to the ideas of the generation previous, and so Le Corbusier's models for housing projects in the inter-war decades are realized *en bloc* in industrial England in the 1950s. The movement had become a style in that its natural development had been inhibited, by fascism in Germany or Austria, by Stalinism in Russia, and by post-war shortages across Europe. The formulae had been frozen, in that there had been no opportunity to learn from experience.[14]

The Le Corbusian models were the inspiration behind the short-lived but massively costly departure from the tradition of low rise, and comparatively low density, housing in England, between 1950 and 1970. Records of Ministry of Housing approvals of tenders for local authority housing indicate the efflorescence of the high rise (see Table 1). 'Too many architects were captivated by the aesthetic of the tall block, and its apparent ability to achieve higher densities . . . and satisfy the politicians.'[15]

Table 1 *Number of tenders approved 1953–69*

	Over 20 storeys	*10–19 storeys*
1953–9	376	27,000
1960–4	9,436	78,000
1965–9	27,000	71,000
	(peak year 1968: 4,233; 1970: 637)	

But the Le Corbusian vision was interpreted in the narrowest sense by the planners and architects entrenched in the government policy-making bodies. His functional buildings were stripped of their expensive supporting facilities such as centralized food storage, nurseries, recreational space, extensive care with landscaping, and careful detailing and finish. The sculptural qualities of Unité d'Habitation, which led Picasso to inquire further into how 'one makes architectural plans',[16] were transferred and transformed into the flat façades of the prefabricated tower blocks in Glasgow; the sinuous, organic modelling of the Plan for Algiers in 1932 became the stark forms of Park Hill, Sheffield, isolated above the industrial valley; the classically sculpted *pilotis* of the Swiss Pavilion became the crude structural underpinnings to an industrialized building system.[17] His rationality of purpose, and idealism of intent, were lost in circumstances where the necessary resources were not available, and in the hands of designers trapped in the ideas of their youth.

Le Corbusier had had four guiding principles—rationality, order, sensibility and technology. He started from the premise that man was a creature of reason and intelligence; this was the very basis of civilization. Further, that man is seeking for order and perfection, which in architectural terms could be realized in geometry and the play of forms. Architecture had to be fitting for a culture, a society, so that the architect, in the search for order, had to make use of his 'sensitivity' or 'intuition', 'the sum of acquired knowledge',[18] to assess the spirit of an age. And for Le Corbusier the modern spirit was technology. Through technological advances man was pushing out the frontiers of his experience, and therefore his architecture should both take advantage of this technological liberation, and push it forward in freeing man from previous constraints and unnecessary use of time and energy. The machine-made environment was to free women from traditional tasks, just as it was to provide access for all to recreation and leisure.

This was undoubtedly a radical philosophy—the bourgeois clutter is to be left behind, and men are to be freed from 'the inhibiting social arrangements in which the glue of polite society had stuck them',[19] but it is frightening in its arrogation of responsibility to the architect. The architect declares, this is the machine age, this is the age of reason, this is the age when one asks oneself as designer of houses, not what people are accustomed to, nor even what they think they would expect of a new dwelling, but what they, in the opinion of this man, actually require to carry out the business of living. The task of the architect therefore becomes that of providing objectively determined requirements of space, light, sun, privacy, movement and services. The assumption is that the ordinary household is in no position to assess what would serve its needs best. Just as for the *avant-garde* of Soviet architecture in the 1920s 'architecture was a means, a lever to be employed in achieving the highest goal that man can set himself. For them architecture was above all a tool for transforming mankind',[20] so for Le Corbusier the architect is 'the superman struggling amongst men', who must admit 'the necessity that he destroy conventional wisdom before he can realize his revolutionary ideas'.[21]

It is also a humanist conception—the starting point and objective was man and woman (not children). He wrote 'the family is badly housed in the city. Garden cities house these needs better'.[22] But his conception of man in the city is dominated by the sensory, most particularly the eye. He speaks constantly of sensation, of comfort, and discomfort, resulting from the perceptions of 'the eye, the ear, the lungs and our legs'. The mind lies behind the eye, and 'When the line is continuous and regular, and the forms are full and founded without a break then the senses are solaced, the mind is ravished, liberated, lifted out of chaos, and flooded with light'. The eye mediates, it is the contact between the environment and man, it looks outward and the mind judges—'The eye is capable both of being battered into submission or of being stimulated. The mind too may be sunk deep or lifted high.' He states 'Our true basis' as being is a 'physiological and irrefutable one'.[23]

From this physiological conception derives the tower block in the park. Man's needs can be met if his dwelling opens out prospects of '*Soleil: Espace: Verdure*'. His design specifications for the dwelling in the city are (i) 'a vessel to hold each family, absolutely separate from other families'; (ii) a flow of light; (iii) pure air; (iv) organization of communal services to save time in the upkeep of the dwellings; (v) provision of facilities for the recuperation of physical and nervous energy in recreation; and (vi) sentimental needs. Of the latter he says, 'I express them in so far as they concern us architects, in two words: visual drama and architecture'.[24] It is a lyrical vision of rational man, mass man, not reliant on others outside the family cell. Man, in the Radiant City, stands alone: 'The keystone of the theory behind this city is the liberty of the individual. Its aim is to create respect for that liberty, to bring it to an authentic fruition, to destroy our present slavery.'[25] As Jacobs comments succinctly, 'he seems to have meant not liberty to do anything very much, but liberty from ordinary responsibility'.[26] The vision of Nietzsche, of man alone, is translated into architecture as isolation and solitude.[27]

Was the attraction of his schemes to his architectural disciples this very omission of the 'social', as Mumford suggests? Jacobs terms his schemes 'vertical filing cabinets', 'his towers in the park were a celebration in art, of the potency of statistics, and the triumph of the mathematical average'.[28] The logic of the commercial skyscraper which allowed for higher densities and higher rents, and a greater sense of space for the office worker, was translated into the residential building. It was a pattern for development which could be 'reduced to a mere mechanical formula and repeated anywhere precisely because it paid so little attention to a variety of human needs and the complexities of human association'.[29] The very conditions of the home and the neighbourhood which were not amenable to rational analysis were omitted from the specification for the design of the dwelling.

The consequences for the working-class families forced by homelessness to take tenancies in these unfamiliar structures are well known. The 'rationality' of design became the 'irrationality' of vandalism; order was reversed to create social disorder as 'each family, absolutely separate from other vessels containing families' was isolated from the community at large. The social effects of rehousing people in unfamiliar dwelling forms (the flat in the tower block), in a culture where the ideal is considered as a separate house on the ground, were unconsidered. And the problems of the women, isolated with their children, remote from play space and associates, is well publicized:[30]

> Who knows most about what it's like to live in the houses we've got? Women and children. Who better to define alternatives? To work effectively we should use our own experiences, and if we think about them our experiences can be pretty bad. Give me the child till the age of seven and I'll give you the man. If that child was reared by a lonely mother in a flat on the fourteenth floor and played in the gales among the stilts of the concrete giant he lived in, what kind of a man will he be?

In this as in all else the Le Corbusian vision has been distorted and brutalized. His own schemes had full provision for safe areas for children's play, and he expected the provision of crèches and nurseries, on the Russian model.

Why were these brutalized versions of Le Corbusian schemes adopted? Was his widespread influence simply due to his own sensibility, his power of assessing the spirit of an age: is it that 'his thought has run so closely along the grain of our age that fragments of it are scattered everywhere'?[31]

Architects have to operate within the political economy of a given society and its specific controls over the distribution and financing of construction. In this housing is especially vulnerable to political directive as to the nature, scale and impetus of construction. The heyday of the Modern Movement came in the grand reconstruction of Europe after the devastations of the First World War, but it was promptly checked by the economic crises of the 1930s, and the rise to power of fascist governments. In Germany the urban housing projects, for urban workers, were promptly abandoned by Hitler in the Reich, and in their stead the unemployed were to be housed in minimal dwellings attached to allotments large enough for partial independence.[32] And post-war housing policies in Britain have shown something of the same swings. The immediate need was to make up for wartime losses, and the deficit caused by a break of six years in the building programme. The most expeditious method was the rapid construction of first the single-storey prefabs, and then the standard two-storey house of the suburban cottage estates. They used traditional forms of building construction, simple layouts, were cheap, could draw on the largest pool of labour in the building industry, and the minimum of supervision. After 1955 there was a change in direction of policy—small-scale suburban housing was to be provided by the private developer, and the main responsibility of the local authorities was no longer the provision of 'houses for general need', but 'slum clearance'. This new policy was backed up by directive and subsidy.

The Housing Subsidies Act of 1956 lowered the subsidy to local authorities for houses for general need to £10, maintained the existing allowance for slum clearance at £12 per dwelling, and made provision for new grants for the construction of flats. Each flat at four storeys had a subsidy of £20 per annum, at five storeys £26, at six storeys £38, and above this an additional £1 15*s*. for each floor above six storeys. The subsidy for a twenty-storey tower block of eighty flats would be *c*. £6,000 a year. That for the equivalent number of houses would be £800. Under the 1961 Act subsidies were raised and, subsequently, additional allowances were made for the acquisition of expensive land.[33] There was therefore a direct incentive for local authorities to build high, although this involved them in high construction costs, and new problems in the use of materials, 'so that simple problems that had been solved for generations reappeared with the advent of "modern construction" ',[34] i.e. loose roofs, leaking window frames, condensation, poor sound insulation and high fire risk.

The arguments in favour of the use of the new technology of concrete, steel and plastic in its high rise form are various. In areas of clearance, the point-block could minimize upheaval of the population as its initial land requirements were small—a rolling programme of decanting from street to tower block could therefore be initiated. Second was its speed of construction—marginal, but an important advantage over traditional methods. This had two advantages: it accelerated the process of rehousing from other areas in the clearance programme, and it minimized interest charges—an important consideration as interest rates began to rise from the low level of the early 1950s. Third, there was a minimization of labour input. Building labour was increasingly expensive, and in some trades scarce. If optimal use was to be made of the existing labour force to boost the overall housing programme labour was to be released to the private sector, and capital intensive methods of construction were to be used in the public sector. An opportunity was opened for the large contractor in the one sector of the housing market that allowed for forward planning, and industrialization of production. The adoption of Le Corbusian models by architects in the public sector has therefore to be seen in the light of the economic and political pressures with which they were faced. Once the housing programme was defined as a political issue, it had to be seen to be done—speed of construction, and conspicuousness in the local landscape. 'Punctuation of the skyline' became an important criterion of a successful housing programme. The future occupants, dispossessed from their familiar neighbourhoods, accepted the new dwelling, thankful for the only opportunity they would ever have of a 'decent house'.[35]

The architects of the Modern Movement were radicals and saw their programme for the reconstruction of cities as complementary to a wider reorganization of social and economic priorities in society. But the transmission of their principles of architecture, and their idea for new environments into the architectural profession, has meant the mechanical application of these formulae for habitation in situations where there is no such reallocation of resources. The consequences are seen in the poverty of design, the poorness of finish, and the lack of the necessary back-up services for the residential unit. Le Corbusier's images of a new world have become bureaucratized into something repellently arid and inhuman. Although the Le Corbusian vision was narrow, the sheer quality of its realization by him prevented the monotony and the blankness, and the inhumanity, of which he is accused. Subsequently the creative activity has been routinized, architecture itself industrialized, and its practitioners co-opted into the hierarchies of the 'urban-industrial complex'.

Le Corbusier, as the prophet for the Modern Movement, was a key agent in instilling the sense of professionalism in which architectural activity is emptied of political content or bearing. The design process is seen as that sketched out by Le Corbusier, the application of rational thinking to a finite problem of the environment, and its realization with the full armoury of modern technology. This professionalism, in which the few are to exercise 'disinterested judgment' over the lives of the many, intent on keeping their code of design esoteric, is strengthened with the incorporation of the architect in the welfare/public sector. Architects, unsure of their role in modern society, torn between idealistic self-images and the brute reality of cost yardsticks and floor space rental calculations, were able to experiment with an idiom, a style of building, sure in their own rationality. In this manner an environment for living was imposed on the very groups of the population for which it was least suited, and with least opportunity for protest.[36]

There is always a measure of discretion in the design of built form within the matrix of political economy, and bureaucratic restraints, and the adoption of certain technologies, and the direction of resources is influenced as much by values and aims as their availability. 'There is always a choice, no matter how rigid the economies, or how close the demands of use.'[37] Although the full potential of architecture can only be realized in rare instances, when patron or client can command resources equivalent to the requirements specified by the architect, and is responsive to the architect's judgment as to the realization of the building project, it should be possible for architects, even within the tight limits imposed on 'working-class' housing, to provide a simple, agreeable, even beautiful environment, in which families can make their homes.

> Of the man-made things, the works of engineering and architecture and town-plan, are the heaviest and biggest part of what we experience. They lie underneath, they loom around as the prepared place of our activity. . . . Against this background we do our work, and strive towards our ideals, or just live out our habits . . . as engineer and architect once drew, people have to walk and live.[38]

3 Housing finance

In Britain there is no tradition of 'self-build' housing in the cities. Even at the height of urbanization in the nineteenth century, there is no record of the shanty towns that characterize the new urbanization of the Eastern Mediterranean or Latin America. The slums described by the medical profession and the social crusaders at the peak of movement into the cities in Britain were either rooming houses, or the cheap housing thrown up by speculative builders for the newcomers to the urban area. There was nothing equivalent to the squatter settlements of cities like Lima, the self-build suburbs of Athens, or the *bidonvilles* in the Paris suburbs.[1] In only one limited period (1919–39) in this long epoch of mass urbanization were houses built independently of the professional builder, on the rural fringes to the major towns, backland areas in the outer suburbs, and in unused plots of farmland. But the exercise of planning controls subsequently has checked all similar ventures except to standards equivalent to the house built by the recognized agencies, and public policy has been directed to their removal both as health risks and eyesores. Possibilities opened up by a new affluence, and mobility, to achieve control over the home in the face of financial dependency enforced by other methods of housing provision were checked in the interests of the British countryside. With this exception, there has been acquiescence in the view that housing, urban and rural, is something that is provided by others on given terms.

Throughout the nineteenth century urban housing was privately financed in a partnership between owners of land and lenders of capital in which builders and landlords were intermediaries.[2] Housing investment was quite widely spread, in that many small investors could have a stake in a group of houses through a chain of leasehold agreements, and in general terms adequate to requirements. Although housing investment was highly vulnerable to the state of the overall capital market (in that building rates were depressed whenever interest rates rose, and *vice versa*) and to conditions where there was an alternative local outlet for small savings, as in Birmingham during the 1890s,[3] until the twentieth century there seemed no reason to depart from this fragmented nexus. Although at certain periods, and for certain cities, there had been acute overcrowding of the working-class quarters, and even at the same time a surplus of new housing in the suburbs, there had been only a slow growth of specialized financial institutions, and no intervention by government to direct capital into house building. In this respect Britain differed from all the major Western European countries, in which, following the example of Germany, tentative steps had been taken in the last half of the nineteenth century to channel capital accumulation into property development.[4]

With growing efficiency in domestic capital circulation, allied with a highly favourable situation for British investors abroad, the weaknesses of reliance on these highly parochial networks of entrepreneurs and petty capitalists became more apparent. The building boom of the late 1890s, most marked in the London suburbs, faded away, and was succeeded by a rapid decline in housing investment. Between 1890 and 1895 an average of 63,000 houses per annum had been constructed; 1895–9, 115,000; 1900–4, 131,000; 1905–9, 112,000; and 1910–14, 64,000.[5] Fall in demand, associated with falling wages and rising prices, gave no incentive to invest in housing, and foreign investment outlets, mainly government bonds and railway stock, were accessible, and attractive to the small investor. Whereas capital outflows from Britain 1894–1904 had averaged £24 million per annum, by 1910–14 they had risen to a level of £185 million per annum, an estimated half of total domestic capital accumulation.[6]

Hitherto state intervention had been limited to safeguarding standards in the interests of public health and amenity. Successive governments had moved throughout the nineteenth century to empower local authorities with controls over construction, layout, servicing, and occupancy of dwellings. The costs of housing therefore inevitably rose. Control over construction means more expensive materials, and greater care in building; land costs also rise, either because fewer units are allowed on a given site, or because insistence on public provision of water, sewerage, gas, and eventually electricity, places a premium on urban land. At the same time standards of space occupancy were defined for the household so that small and cheaper units were not provided. Despite the persuasions of Octavia Hill, and the policies of the housing trusts and associations, middle-class standards of decency and convenience were accepted as the basis for mass housing for the working classes of England by the twentieth century, so raising the size of a working-class dwelling far above its continental or Scottish counterpart. This regulation prevented the builder or developer from adapting his product to the market, and the construction of low cost housing ceased.

As appalling as had been living conditions in town and country in the century of growth, the housing reformers had been unsuccessful in pushing the argument that direct government action in compelling the municipalities to provide housing on the German model was a necessary adjunct to legislation on working conditions. Government legislation in 1919 which established the double precedent of subsidy for the construction of 'working-class housing' and compulsion on the local authorities to do so, was taken in knowledge of the breakdown in the housing market. The war-time years had seen the cessation of housing construction; rent controls, introduced in 1915 in response to widespread protest about profiteering in key centres in the armaments industry, had checked any flicker of interest in renewed investment in cheap housing, and it was anticipated that post-war inflation would render impossible any further private initiative. In conditions of shortage of labour and materials, and a falling rate of return on investment, the necessary measures had to be taken to provide 'Homes for Heroes'.[7]

Although the prominent role played by local authorities in Britain in providing low cost housing can be interpreted as an extension of nineteenth-century policies for regulating urban growth, the concentration of capital and its centralization on the London market precipitated state action. The local authorities were substantial institutions which could compete on equal terms with other contenders for this capital accumulation. Further, the status of London as one of the world's best organized and long-established sources of private capital obviated governmental responsibility for the direction of cheap finance into housing.[8] Under the Addison Act of 1919 the local authorities were obliged to raise capital publicly before approaching the Public Works Loans Board, and with the exception of the period 1945–52 this has remained the rule. The rate, and the cost, of housing construction have therefore continued to be strongly influenced by the relative attractions of alternative investment, and the cost of money overall. Devolution of responsibility to the local authorities in this manner entailed minimal interference in housing finance, and considerable flexibility in policy locally, but left working-class housing vulnerable to local politics as well as national economics.

Housing provision in the public sector was not considered as a social utility. Local authorities were taking over the role of the private landlord, and their tenants were treated accordingly. Not only is occupancy hedged round with restrictions, but there is only grudging minimization of financial dependency. Unlike, for instance, the command economies of Eastern Europe and Russia, where rents on public housing are notional, in Britain, households occupying local authority housing have had to pay the capital costs of that housing (offset by government and rate subsidy), maintenance and running costs, and interest charges on the capital expenditure. Housing is not paid for out of capital, unlike road improvements or motorways.[9] Housing was seen as a similar investment to public works and civic improvements, since the 1880s financed by loans and not out of revenue,[10] except that the costs were to be borne primarily by the occupants and not the community.

The issues of housing finance stem from this conception of housing as either real estate investment or public utility, and not as the very basis of social existence. Housing is considered as a commodity, but one like no other. In the economists' phrase it is neither 'fish, fowl, or good red herring': it is immovable, fixed to the land, its exchange value a measure not of its intrinsic qualities as a product, but of its location; in situations of shortage, or surplus, it cannot be easily transferred from one section of the population to another: and it is a commodity of special character in its durability. It is, among all commodities, *par excellence*, property, and its security as a fixed capital investment must be assured. There is accordingly a consensus that housing, in design and construction, must have a long life, and professional standards and government intervention have been directed to this end. Accordingly, local authorities have only resorted to cheap temporary or mobile housing in extreme conditions of shortage, and efforts by the private sector to drop their market threshold by 'lowering standards' have been fruitless. Inputs of raw materials and labour therefore increase, and housing costs continue to rise. Most dwellings must be built or obtained with credit, which reinforces the consensus as to durability and long life.

Despite therefore rationalization in the circulation of housing finance, in that building society expansion has paralleled the growing responsibility of the local authorities, housing investment remains as vulnerable to movements in the capital market as it was in the nineteenth century. Although state intervention has offset the extreme variations of the era before 1914, regardless of political priorities housing investment has gathered pace, or been checked, in direct relationship to the state of the economy. Both in 1947 and 1967, Labour policies directed towards the public construction of housing with maximum subsidy and full provision of finance from government, were checked by the precarious status of sterling;[11] in the same manner maintenance of high interest rates in the London money market during the early 1970s checked the construction of housing in both public and private sectors. Present policies for the improvement rather than the demolition of older houses are in acknowledgment of the accuracy of the observation, thirty years ago,[12]

> if interest rates on the capital used in house-building are high, so as to raise the rents on new houses, the older houses will experience less competition from them, and will tend to have a longer life, and remain in occupation long after they ought to be scrapped.

High money costs enforce the long life of housing as a commodity, prevent the eradication of 'the slums', and debar elimination of the perennial housing shortage.

The housing situation in Great Britain now is vastly different from that in 1914. Then there were fewer than 9 million dwellings, now 18 million, of which two-thirds post-date 1919. Rather more than a third of the houses standing at the onset of the First World War have been demolished in slum clearance, redevelopment and war damage, and a substantial number of the remainder improved, so that only 15 per cent of households lack exclusive use of the basic sanitary facilities of hot and cold water, w.c., and bath or shower. Overcrowding has virtually been eliminated as household size has continued to fall: in 1961, 13 per cent of households were one person, 29 per cent two persons (in England and Wales), and by 1971 the proportions were 18 per cent and 32 per cent. Of the households in Great Britain 8.7 million owned their accommodation (48 per cent), 5.2 million rented from a local authority or new town corporation (29 per cent), and 3 million rented unfurnished accommodation privately. 800,000 were renting furnished accommodation. In 1914 it was estimated that only 10.6 per cent owned their dwelling, the vast remainder renting from private landlords.

In aggregate terms, therefore, there would seem to be no situation of hardship—a net surplus of dwellings, drastic inroads on obsolete housing (throughout the 1960s, replacement through urban renewal accounted for a quarter of gross housing production), a situation of housing affluence as the rate of household fission increases and occupancy rates decrease. On this analysis the relatively low proportion of Gross Domestic Product allocated to housing in the post-war decades (never above 4 per cent in the 1960s) was justified. High rates of investment in countries such as Western Germany and France (both averaging 6 per cent in the same period) reflect the more dramatic population movements from the rural areas into the towns, and the unsatisfied backlog of demand.

The aggregate statistics, however, disguise the shortages felt in all the major metropolitan centres (the London region actually has an inverse ratio of dwellings to household units), the concentrated pockets of damp, near-derelict housing in areas such as Broughton, Salford and Liverpool 8, the situation of elderly households concentrated in these unimproved old houses, and the plight of the young household unable to take out a mortgage as house prices spiral, without access to local authority housing, and increasingly restricted to high cost and inadequate furnished accommodation. There has been no increase in housing production to correspond to the increased rate of marriage in the 1960s, 348,000 in 1962 and 404,000 in 1971, the aftermath of the post-war baby boom. To these one must add migrants to the metropolitan centres, trapped in the 'zones in transition' by low wages and job demands. Incremental policy innovations, inflexibility in attitudes to the management, layout and design of housing, and overall failure to co-ordinate investment locally or nationally, have meant the dissipation of resources leaving the vulnerable sectors of the population as poorly housed as in 1914. Although housing investment has been maintained at a rate double that of the nineteenth century (as a proportion of GDP), decent housing remains out of the reach of successive generations of newcomers to the legacy of speculative chaos in nineteenth-century urbanization. Intervention has created dual markets in housing, that administered publicly, and that allocated on market principles, each with its own deficiencies, aggravated by the parallel contingencies of the other.

The public sector—local authority housing

In the first phase of experiment in housing policy, 1919–32, local authorities were handicapped both by their own inexperience and the vacillations of government policy, but in all 650,000 houses were constructed. In the second phase, 1933–9, they were without subsidy for all but slum clearance projects, and were unable to take full advantage of the cheap money policy which benefited the private builders; but in that period they constructed 355,000 houses. Only in the period after the war, 1945–52, were they able to benefit both from cheap money (the bank rate had been held at 2 per cent since 1932) and total government sponsorship. The pressing needs of post-war construction to make good war-time deficits justified the decision by the Labour government to license private construction, and to put full backing behind local authority housing programmes, though these were slow to gather momentum. The abandonment of the cheap money policy after 1952, the relinquishment of government responsibility for housing finance under Conservative governments, and in 1956 the direction of the local authorities once more into urban renewal (with the obligation to pay market price for the land required after 1959), marked the end of the era of cheap housing. In

England and Wales alone between 1945 and 1951, 661,000 homes, and from 1952–6 870,000 houses were built; by 1967, the peak year for local authority housing completions (159,000), a further 1.5 million houses had been built. And from 1968 to 1973 an additional 656,000. In that last year little over 79,000 houses were completed, compared with 203,000 in 1953 and 199,000 in 1954. From 1956 to 1973 1.4 million houses had been demolished by the local authorities, in slum clearance and for other purposes, and 3.7 million had been constructed.

It was a sizeable achievement, conspicuous in every urban area: suburban cottage estates from the 1920s to the 1950s, succeeded by inner area demolition and reconstruction accompanied by more peripheral estates, at ever higher densities, as land available to the local authorities ran out. But although backed by sizeable government subsidies, moving from an initial £9 per annum per house under the Wheatley Act of 1924, to a third of capital costs on reconstruction in the 1960s, it was increasingly subsidized by the practice of pooling rents on the housing revenue account. Rents on the older council housing stock were helping to keep the cost of the new housing within the reach of the households for which it was intended. But as costs continued to rise, expensive money, and ever more costly land, more complicated forms of construction in denser estates, rents were rising to embarrassing levels particularly in inner London.[13] Nevertheless until 1968 a mammoth programme of reconstruction was sustained, quite unprecedented elsewhere in peace time.[14]

Subsequently, however, despite a government subsidy on loan charges, more stringent cost yardsticks and the diversion of investment and labour into rehabilitation of older housing (including many inter-war council estates) have precipitated a drastic fall in public sector building. In the years 1968–73, in which local authority house building programmes halved, the take-up of improvement grants nearly trebled (114,000–319,000) and that of the increased discretionary grants moved from 46,000 in 1968 to 260,000 in 1972. Local authorities, of whatever political inclination, found they had to cut back their housing targets as cost yardsticks (on adherence to which Exchequer grant had been made conditional in 1968), and building costs moved further apart. Public expenditure on housing had to be cut, and the acute rise in government subsidies (which had risen from £60 million in 1963 to £175 million in 1971) could no longer be sustained.

The steady post-war increase in interest rates has exposed the weakness of the policy initiated in 1919. Not only were local authorities after 1956 virtually restricted to the most expensive part of the housing programme, urban renewal, but they were forced to compete for finance with all other interests. Although public housing is sponsored and managed by government agencies, it is not publicly financed, in that it draws from private savings rather than taxation. In 1971, for instance, local authorities borrowed more from the banks than from government, and throughout the 1960s, some 60 per cent of local authority capital came from the general finance market. Now the heavy indebtedness of the local authorities means that loan charges alone can exceed the revenue from rents on housing in their administration.

> Local authorities have reached the situation well known to under-developed countries in that what they borrow is used up in paying the interest on what they have borrowed already—hence loan charges represent 90 per cent of net (new) borrowing.[15]

There is an outstanding permanent debt which increases with each new item of capital expenditure (and housing accounts for some third of local authority capital investment). And it is this burden that is being transferred from the Exchequer to the local authority tenant under the Housing Finance Act of 1972. Just as consumers' expenditure on housing in the private sector was forced up in the building boom of the 1930s, so that of the tenant in the public sector is to be raised to defray the costs of building programmes which hitherto were assumed to be the community's responsibility. Rising incomes are deemed to relieve the state of more than vestigial responsibility, and market principles in the finance of housing are thus ratified for the public sector.

The private sector—sources of finance

In the inter-war decades, 4.3 million houses were constructed, 3 million after 1930, and of the latter 70 per cent were privately financed. Private builders constructed 2.65 million houses in rather less than twenty years. At 1939, one-third of the housing stock was less than twenty years old. The doubling in the rate of domestic construction between 1920 and 1938 was primed by government subsidy in the 1920s, but was accelerated by a flood of private investment channelled by the building societies into speculative suburbs. Before 1914 building societies had been relatively small, and as much concerned with the finance of rented housing as that for owner occupation. They had benefited from the reversal in direction of investment after 1918 in which foreign outlets were abandoned in favour of domestic needs. Government securities, public utilities, local authorities, insurance and the building societies all benefited. The latter, with their twin attractions of security and liquidity (and in the precarious financial conditions of the early 1930s, favourable interest rates), rapidly expanded to the position they have had since as the major source of private housing finance.[16]

Building society expansion inter-war, as post-war, was very much a reflection of rising incomes, rising expectations, and the absence of alternatives in housing other than renting from a local authority. Private construction to let did continue (between 1935 and 1939, a quarter of all privately constructed housing was for rental), but the growth of business was seen as the promotion of owner occupation where uncertain conditions in building costs, rents and government subsidies were of no account. And this was a demand that had to be created through extensive advertising and modifications to previous practice in safeguarding loans. In particular, means had to be found to clear the hurdle of the buyer's deposit which had hitherto restricted house ownership to the relatively wealthy. One device was securing collateral against insurance policies, another was to arrange with local authorities to guarantee the difference between the normal 60 per cent of 66.6 per cent advance and a 90 per cent mortgage, and the third, and by far the most widespread, was the 'builder's pool' in which the builder provided the collateral. By 1938, an estimated half of the business of some of the large societies was conducted in this way.[17]

Whole estates were financed on this basis. Speculative building was backed by the assurance of finance for would-be purchasers, and the builder could operate in the certainty of profit in conditions of cheap money, freely available land, and base costs for materials and labour, in a period of rising incomes and increased personal expenditure on housing. The interlocking of interests between the building societies, and the private developers, and government in its sponsorship of suburban housing in both public and private sectors, meant a transformation in the pattern of urbanization unparalleled in countries such as France, where cheap loans for housing are still difficult to obtain. All the major centres, even those with relatively high unemployment rates, demonstrated this suburban efflorescence into the countryside, but it was London's sprawl that generated the most protest, as it had at the turn of the century. There was a rising protest from groups of the population concerned both for the protection of the countryside and the orderly development of the towns. The London Green Belt Act of 1938 was the first indication that the investment outlet of speculative suburban housing was to be checked.

Post-war, private building was restricted and all available resources directed to the public sector. But under the direction of Conservative governments the contribution of private builders rose from a low point of 21,000 in 1951, to 210,000 in 1964 (peaking again in 1968, at 213,000 completed dwellings, and falling back to 174,000 in 1973). At no point have the construction levels of the 1930s been approached. In all, 3.5 million houses have been built by the private sector since 1945. In this, the building societies have been extremely influential in establishing the level of demand, and hence the rate of housing construction although finance for the builders comes from other sources, banks, property companies and share capital. Over the past two decades, building society assets have increased ten-fold and they now take second place only to insurance companies as a repository for private savings. Between 1966 and 1973, their share of the total finance for house purchase fluctuated between 80 per cent and 93 per cent of total. But despite their size (the five largest among 447 societies in 1973 controlled 60 per cent of assets), they are extremely vulnerable to changes in the level of savings and any pronounced movements in interest rates.[18]

Building societies have traditionally taken short-term savings, and lent over long terms—up to forty years. They must therefore maintain a high liquidity ratio, and ease of business has been dependent on stable interest rates. Between 1919 and 1931, the mortgage rate fluctuated between $5\frac{1}{2}$ per cent and 6 per cent; between 1932 and 1939 fell to $4\frac{1}{2}$ per cent; in the war years lay between $5\frac{1}{2}$ and $4\frac{1}{2}$ per cent; stabilized at the latter level until 1951; rose slowly to $6\frac{1}{2}$ per cent in 1961; then to $7\frac{1}{4}$ per cent in 1966; $8\frac{1}{2}$ per cent in 1972; and then in three moves in 1973, from $9\frac{1}{2}$ to 11 per cent.[19] Competition with the banks in a period when the minimum lending rate rose to 13 per cent, with local authorities forced to offer equivalent rates of interest both on the London market and abroad, and insurance companies exploiting to the full tax exemption loopholes to attract funds, meant a famine in capital available for mortgages, a hiatus in housing transactions, and a cutback in private building. And this impasse succeeded a frenzy of house buying in 1971–2, in which house prices had increased by 22 per cent in 1971, and a further 47 per cent in 1972. Violent fluctuations in interest rates and the volatility of consumers' expenditure on housing have exposed the weaknesses in the building societies' intermediary position in the housing industry.

The immediate cause of the house price spiral was a surge in demand after the rapid rise in incomes between 1969 and 1971, backed by a steady rise in the rate of household formation. The demand was fed by the societies flush with funds from those years, when the short-term yield from investment in other sectors of the economy had sagged, while the building industry was finding it more than usually difficult to respond to changed demand after a period of depression in the market. Difficulties of raising credit, shortages of materials, scarcities in the skilled labour force, and sharp rises in the price of land, all meant that response was slow. So although building society advances rose from £2,000 million to £3,600 million in the two years, the rate of increase in housing construction in the private sector was only 3 per cent per annum. Unlike the boom of the 1930s, these were conditions of land speculation rather than building speculation. Because of planning restrictions it is the landowner and not the builder who benefits from the appreciation of land values, and so there is no incentive to anticipate demand. Instead building society funds financed a house price spiral which added a further inflationary pressure to an already inflationary economy.

Land-use planning

Piecemeal accretion of legislation for the control of land use had paralleled the inter-war building boom, and was to culminate in the 1947 Town and Country Planning Act. In this the right of development was removed from the property owner, and a measure of control granted to the local authorities under the surveillance of central government. The initiative to develop still remains with the landowner or his agent, except where public agencies step in with compulsory purchase powers, but the prerogative of veto over all development, including changes in the use of land, as well as all new buildings, has been assumed by the state. The acceptance of such legislation in the post-war period was in large measure a reaction to the loss of the countryside to developments Bevan termed 'castrated communities'[20] in the inter-war decades; all this in addition to the war-weary drabness and squalor of the older areas, and the increasing inefficiency of layout of existing towns. Planning was intended, *inter alia*, to maximize the benefits of investment into the suburbs, so that the most efficient use was made of land, of public investment into roads, schools, as well as private savings. What was not apparent was the effect of planning, without actual control of the property market, on housing provision and finance.

Under the 1947 Act, and subsequent legislation until the Act of 1968, planning was conducted by reference to a mosaic of development plans in which land was zoned for purposes deemed appropriate to the area. For each site, also, there were to be well-defined standards of density and layout as well as use. The general effect was to remove the speculative impetus to urban development (while admittedly providing new ground rules for property appreciation). As so many development plans were largely a presentation of the existing situation, zoning acted to stabilize the market in land, and slow up changes in the use of property: it was a brake on movements in the map of land values. Zoning removed uncertainty as to the expected use of the land, both buyer and seller were better informed, and the role of an intermediary dealer-speculator was weakened, except in unusual conditions of site assembly, as in central London. Increasingly, therefore, development values accrued to the landowner, rather than the developer, unless the latter was able to hold land banks for long periods.

The removal of uncertainty as to the future use of land was not complete, as development plans were to be revised at five-yearly intervals and owners of property could appeal against restrictions on their development proposals, but there has been an almost complete separation out of markets in land, which has helped to prevent 'blight', i.e. depreciation in value of existing buildings in anticipation of future development, and also removed much of the impetus to suburban development. The urban land market was now in principle marked out from that in rural land by the delineation of the 'urban fence', in the same way that the market in residential property was protected from commercial speculation. Investment in farming and that in housing were therefore protected against short-term movements in the land market. But these safeguards on property investment meant that profit on new development increasingly accrued to the owner of the land, and not the developer. Possibilities of windfall gains from land speculation to offset possible sluggishness in demand for houses are now rare, and housing construction has to be tailored more closely to estimates of existing demand.

The shortages and costs of housing in Britain have therefore to be seen in the context of the exercise of planning controls over the location of housing.[21] As the allocation of land for new development has been largely restrictive, the potential supply of land has been fixed more firmly than in an open market situation, and the owners of land, particularly large landowners but also those with planning permission, have been placed in a quasi-monopoly situation, able to command their own price, either by hard bargaining, or by withdrawal of land from the market until such a time as the owner's estimation of value is met by the builder. Land for housing is therefore acquired at high prices, which either cuts the builder's profit margins, or puts up the cost of the finished product. The effect is the same in both cases: the supply of new housing is reduced. The shortfall in the production of new houses then pushes up the price of older houses, and the value of already developed land, the cost of as yet undeveloped land rises correspondingly, and so on. Particular conditions, such as the introduction of the Land Commission (1967–70) which slowed up the accumulation of land banks by developers, aggravate a cumulative situation.

In sum, the protection of the English countryside, the 'containment of urban England', without control over the release of land for development (except in the new towns), is not compatible with the provision of cheap housing. If restrictive planning policies are exercised for the major growth regions for a generation then housing costs must be pushed up. Financial conditions precipitated the rise of prices, but the rate of increase throughout the country mirrored the relative prosperity of the regions. That in London, the South and East, was approximately double that of the rest of the country.[22] If the base price of residential land in the urban areas is protected by planning, and new housing provision is slowed up by restrictions over development of rural land, then relatively expensive housing must result.

The British experiment in regulating urban development, and housing the mass of the population without the assumption of direct control over either the land or housing markets, and without full financial backing for planning or housing provision (except between 1945 and 1952), would therefore seem to have been checked. Houses started in 1974 by the private sector totalled only 110,000, the lowest figure since 1954. Builders have been hit by the rapid changes in supply in mortgage money, rising building costs, falling profit margins, and a general uncertainty as to future behaviour of the housing market in conditions of economic uncertainty. Inflation generalized in the world economy, allied with the weakness of Britain in international markets, has finally exposed the weaknesses in the regulation and finance of housing construction. Housing provision in Britain has depended on cheap money, stability of interest rates, steady demand, and also cheap land. All this has changed. The very institutions that fed the movement to the suburbs—municipalities, building societies and the building industry, all subsidized directly and indirectly by government—have been unable to sustain its impetus in the economic impasse of the 1970s. Improvements in living conditions initiated in the inter-war years, which considerably offset the continued inequalities in the distribution of wealth and income, can no longer continue.

In estimations of national housing expenditure, no realistic figure can be ascribed to the household's own input of labour. A recurrent spectacle is the reconstruction of older houses into passable facsimiles of a modern house, often by the occupiers themselves. In this way Edwardian villa terraces are converted into Georgian town houses, a 1930 semi-detached into open plan living, and small back yards become lavish patios. All this largely with private savings, domestic labour, with assistance from friends, neighbours, jobbing builders or moonlighting craftsmen. After 1918 home ownership may have been encouraged because 'pride of home is pride of country',[23] but nowadays the arguments for home ownership are economic in derivation. Do-it-yourself improvements, holidays set aside for 'self-build' housing within the shell of existing dwellings, all this obviates public expenditure on new housing. In this way, increased leisure, new affluence, untapped energies, are caught up in renovating the national housing stock in the name of home improvement. The ideology of house as home, citadel, security, refuge, status, means of personal expression and realization, is harnessed to the housing process. Urban or, perhaps more accurately, suburban man, becomes a peasant once more, but without the financial independence of that peasant, and his wife and children are co-opted partners in the enterprise of keeping house and home together.

4 Housing in urban sociology

The trademark of British urban sociology in recent years has been the emphasis on aspects of housing and planning. Neither of these were issues new to sociologists interested in urban questions, but the emphasis shifted significantly, from field research into the experience of groups of the population subject to planned upheaval in the furtherance of post-war housing policies, to study of the organizations implementing these policies, i.e. the local authorities and their professional staff. Subsequently much of the theoretical discussion has been directed towards those who allocate the community chest—public expenditure on basic services such as drainage and roads, public utilities such as housing and transport, and the social services, including education. There has been a greater emphasis on the urban managers than on the people who receive these Welfare State benefits and experience the effects of intervention into the urban environment.

In this reorientation the housing question has been central. Whereas the ecologists concentrated on the use of land in the city, in the British context it is impossible to disregard the intermediary structures of housing allocation. Where there are strongly developed policies for intervention into the markets in urban property, as in the British context, these, and not the market processes affecting the distribution of land uses in the city, become the analytical context in which specific studies have to be undertaken. The movements of the population through the housing stock, which the ecologists described in naturalistic terms as invasion, succession, and so on, are now considered as responses to the supply and demand equation in the local housing markets. The role of State agencies is therefore inescapable.

As illustration, consider the situation of a group of immigrants to the city. In the ecological model they found themselves at the centre of the city in the zone of transition blighted by the anticipated expansion of the central business district; that is, alterations in the pattern of demand for land could be considered to have initiated the movement of low status immigrants into a certain district of the city. In the situation of British cities during the 1960s, the segregation of immigrants into the twilight zones outside the established inner city slum districts was due to the restrictions on access to other types of housing. And these restrictions had to be related to the unanticipated consequences of established policy on housing provision, as well as to income and situation in the labour market of the immigrants themselves.[1] Conditions of living in the twilight area had to be related to the structures of housing allocation rather than the blind forces of the market.[2]

There is, however, a common feature to the two approaches which sets the work of the sociologists apart from that of the urban economists. Both models of the local community postulate that the location of any one group within the urban area is the outcome of a process of competition between all groups resident there. The segregation of the immigrant group is due neither to their own predilection for the kind of housing they occupy, nor even to the policy of any one section of the 'governance' controlling access to housing, or schools, but to the interaction between these, and all member groups of the urban system. The emphasis is on constraint, not choice; the unit of analysis is not the individual household making a personal decision as to the costs and benefits of relative locations within the urban area, but the urban system, however defined.[3] Any discussion of social conditions in the local neighbourhood must therefore be referred to a wider framework than that of family and community.

This emphasis on housing owes much to the work of Rex who sought to present the relatively familiar material regarding the operation of housing markets in the language of sociological theory. His approach to the sociology of the city derived from a conviction that there were 'particular processes at work in the concrete urban situation'[4] which structured the world of social relations in which individuals in the city encountered each other. The concept of 'housing class'[5] was introduced to articulate some of these processes, and to assist in a more precise delineation of the social structure and conflicts of the city. Housing is assumed to be an inherently scarce resource, which is so important for life chances (in that access to certain kinds of housing ensures advantages of location, education, and social services as well), that the city is defined for its citizens by the conflict over this one community resource. 'The class struggle over the use of housing is the central process of the city as a social unit.'[6] Instead of a social system based on awareness of common situation, and on individual identification with the urban community, as in Weber's analysis, the urban social system is defined by the exigencies of common residence, and its significance for its residents derives from its being a differentiated system of material resources.

In his substitution of class struggle for symbiotic competition, Rex comes closer to the ecological perspective on the city than does Weber, whose sociology of the urban community excludes class conflict; he is, however, in the tradition of Weber in his treatment of the city as a political community. The struggles over housing must necessarily be political in nature in that they engage agencies of the state. But this conflict is not derived directly from class position in relation to the mode of production. Intervention in housing has set up conditions of access into different sectors of the housing market, principally local authority housing, which differentiate the labour force on lines other than those of occupation and income. The resultant system of housing classes was considered by Rex to be 'part of the superstructure which demonstrably takes on a life of its own'.[7] That is, life in the community, even class struggles in the community, have a relative autonomy from life in the workplace.

Rex uses Weber's breakdown of 'class situation', in which the Marxist concept of class is relativized,[8] to justify his own distinction between class struggles in the sphere of production and the conflicts in the community, primarily over housing. Evident in Birmingham was a division between the established working population, who in the previous fifty years had obtained access to housing which was incompatible with their control over the means of production, and the immigrants, who were excluded from equivalent community resources. Newcomers to the local housing market were debarred entry into the owner-occupied suburbs by reason of income and job conditions, into local authority housing by residence qualifications, and slum clearance areas by removal of this housing from the open market; there was little alternative but the rooming houses of the twilight zones. The suggestion here is that urbanization itself creates certain relationships and problems related not to conditions of production, but conditions of reproduction of the labour force. For this reason alone an urban focus to sociological research is justified.

The approach taken here has been subject to considerable criticism, and Rex has subsequently modified his position substantially.[9] On empirical grounds, criticisms have been directed at the typology of housing classes[10] and at the presumed correlation between housing tenure and different districts within the urban areas.[11] The typology presented did not include key interests in housing—the absentee landlord, and the local authority were excluded—and was specific to the Birmingham situation. It would have no application outside Britain, and in its details, little outside the city of Birmingham, *c.* 1965. Theoretically, the most important criticism has come from Haddon.[12] In effect he argues that Rex has mistaken condition for cause, occupancy of housing as the basis for access to life chances, and not as a measure of the existing distribution of power and wealth. The conflict over housing is an extension of the major divisions in the society, and is only a class struggle in that the allocation of these resources is largely determined by the power each class has come to possess in a society with a long history of class struggles.

The debate cannot be resolved without taking into consideration the wider question of the status of the immigrant labour force of Western Europe, recruited from the Mediterranean countries as well as former colonial dependencies. Post-war economic growth required additions to the labour force, overwhelmingly concentrated in semi-skilled or unskilled occupations. For instance, in Britain in 1971 it was estimated that while 26 per cent of the total male labour force fell into these two categories, 50 per cent of West Indians, 42 per cent of the Irish and 65 per cent of Pakistanis were found in these low status, insecure, and often unpleasant jobs. Although therefore 'objectively the immigrant workers form part of the working class of Western Europe, within the class they must be regarded as a special stratum'.[13] Viewed in this light, the inadequate housing available to the immigrants—barracks and furnished lodgings in Germany and Switzerland, '*hôtels meublés*', hostels, *bidonvilles* in France, rooming houses or dilapidated slum housing in Britain—is an index of their insecurity in the labour force, their low status in the community, and their lack of political power. Indisputably the race question has come to the forefront of public consciousness in relation to housing and education, hence their importance for sociological analysis of immigrant–host relationships, but in no way can they be reduced to an urban framework.

The question of housing, and that of a labour aristocracy, have been interrelated for a century. The housing crisis was deemed an issue meriting departure from previous *laissez faire* attitudes precisely because shortage threatened the established differentiation in residence, as in conditions of employment, of the skilled craftsmen and the unskilled labourers. Just as the former began to benefit from national prosperity (in that wages rose steadily through the last quarter of the nineteenth century), and their political maturity seemed assured, rising rents and enforced overcrowding in the inner districts, particularly of London, threatened living standards and social stability. Enfranchisement itself depended on an adequate supply of housing. Engels had dismissed the state as 'the collective capitalist unable to remedy the inequalities of Capitalism',[14] and yet eventually the state was to intervene, so granting sections of the working class living conditions incompatible with their power in the market.

One suggestion has been that it was this housing question that provoked awareness of class differentials, in that the contradictions of capitalism were most evident, not in production, but in the sphere of collective consumption, i.e. in urban life. For instance, the growing turbulence of the London poor could be related to their worsening housing situation, as pressure for redevelopment of central area sites reduced the supply of housing in the inner districts. But worsening housing conditions only aggravated an economic crisis in the metropolitan labour market whose most salient features were 'the collapse or decay of staple industries, the expansion of causal and sweated trades, increased overcrowding, poor health, hard winters, prolonged unemployment, and chronic poverty'.[15] Despite the severity of the housing problem, unprecedented by provincial standards, it was not the metropolis but the provincial working-class communities that were to take the initiatives towards socialism as the divisions between classes became more acute in the era before 1914. The economic structure of the metropolis—the fragmentation into a mass of small-scale enterprises, its reliance on seasonal labour, and the sharp division between the skilled and the unskilled labour force—all prevented the emergence of a mass socialist party. The economic composition of the community, as related to its position in the national economy, was rather more important in provoking awareness of class issues than its living conditions.

It is difficult to sustain the argument that the central process of the city as a social unit is the class struggle over access to housing or any other community resources, thereby excluding the antagonisms and cleavages deriving from the scattered workplaces of the urban region. These are as much part of the urban scene as the council housing estates, municipal offices and community action groups on which the attention of urban sociologists has been focused. Urban sociology in Britain has subscribed to the fallacy of community analysis, namely the tendency to explain social relations in the community by conditions of living in that community. The analysis of the twilight areas in terms of the failure of housing and planning policies for the inner city is symptomatic of such a fallacy, as is the presentation of the situation of the colonial immigrants as that of a housing class, and not in terms of their status in the labour force. Urban sociology cannot restrict its field of study to the field of consumption, as hitherto, but must take into account the economic conditions governing the local class structure.

This approach certainly renders urban research more difficult, in that conditions of production in the local region will only rarely have the unity given to the consumption of community resources by their partial centralization in the hands of local government. Historically it is rare to find urban situations in which specialization or differentiation in economic activities does not prevent the emergence of a unified community class structure; such is the very nature of urbanization. The classic community studies relate to small, relatively isolated and slow-changing towns, or ones reliant on a single industry. The communities studied by Foster,[16] for instance, were then isolated, provincial towns, Oldham, Northampton, South Shields, at an early stage of development, but even there, interests in land-ownership, absentee financial interests, and diversity of economic enterprise, led to a diffusion and fragmentation of their local class structure. But if urban communities are to be considered as unique configurations of population, resources and institutions, which mediate in the individual's experience of the national society, it is precisely these conditions of diffusion and fragmentation that need to be examined. Cities are not systems, and yet the convergence of different interests in the same place gives the groupings of activity and population identity. It is these interests that need to be established.

Housing, alone with other aspects of the built environment, underlines the distinctiveness of each place in that it represents on the ground the past class structure of the town or city, its past and present wealth and influence, its situation in the international markets of capital and labour. So, for instance, London's housing market reflects its history as centre of nation and empire, with a large stock of housing built for middle-class occupancy, and its present housing crisis is a measure of its continued attractiveness as a world city, and its need for cheap labour. Its housing profile is thus that of a metropolitan centre, just as those of the towns of north Lancashire or North East England, with a huge store of working-class housing, and declining populations, reflect their provincial status. The long life and immobility of housing, and the strength of vested interests in property, underline differences between places in the one society that derive from their position in the national economy.

In any discussion of housing it is evident that the parameters determining investment, allocation and even design are national, if not international, and yet the markets in housing maintain their local character. And this is so now as in the nineteenth century. The standardization in design of housing is misleading; although mass housing has been brought closer to an international model, and the diffusion of styles for privately built housing is achieved more rapidly, local housing situations differ quite markedly. Compare Southwark with Kensington and Chelsea, the one with 60 per cent of its housing in municipal ownership, the other with 8 per cent. Or Tyneside, with the South West region, the former region with 43 per cent of dwellings in the public sector, the latter 29 per cent. State intervention has superimposed one rhythm of constraint over local building programmes on another, without diminishing the variation in local housing markets. Needs differ, there are differing rates of economic growth and labour requirements between regions, resources differ, and similarly the ability and willingness of the local housing agencies to match housing supply to housing demand. The centralization of power in Britain is blurred by the delegation of responsibilities to the local authorities who perpetuate the local community power structure in the built environment. All 'urban problems' have to be put into the context of this variability between places, to which housing is an index.

It is, however, far from certain that the existence of specific local markets in housing, and other resources such as education, is sufficient basis for delineating an 'urban' sociology. Marxists, and others, would claim that adoption of an urban perspective only confuses analysis of capitalism and diverts political energies to 'secondary questions'. It would not be agreed that urban social movements, defined by Castells as 'systems of social protest action which question the established order on the basis of specific issues within the urban problematic',[17] can have more than local impact. But the 'urbanists' would claim that state intervention in the provision and allocation of resources has blurred the inequalities deriving from the relations of production, and that confusion is there in the life experience of members of society in that they encounter the irrationalities of local bureaucracies, the arbitrary judgments of professional advisers to government, the particularism of local politicians, and the special features of the place where they live. On this score the orthodox Marxists and the urban sociologists are at loggerheads: the latter emphasize through their study precisely those circumstances of existence that nineteenth-century Marxism discounted.

At this point the debate is open to all comers. The housing question is an important one, and merits far more extensive investigation than hitherto, but it neither answers the particular problems of the relevance of urban sociology to the study of class in the societies of late capitalism, nor does it exhaust the scope of sociological investigation into urban communities. Without doubt it permits of a clear distinction between the local, or urban, and the national or societal, but the relationships between local modifications to the national pattern of resource allocation and variations in class conditions remain an empirical question. Variations in 'external conditions of life' there are, certainly, but then, 'the differentiation of classes on the basis of property alone is not dynamic; that is, it does not necessarily result in class struggle or class revolutions'. The connection has not been established between the everyday life and wishes of the population, in 'all the specificity of their situation in society', and 'the objective sociological process'.

4 The practice of planning

1 The town-planning movement

The town-planning movement may be dated from the last years of the nineteenth century; the Garden City Association, later the Town and Country Planning Association, was founded in 1899.[1] There was a sense that the old order called out for change, that technical prowess and commercial success had not brought prosperity and happiness for all, and that purposeful initiatives had to be taken by informed opinion to bring about an organized reconstruction of society. The town-planning movement was a small pressure group, drawing support from the professions (principally architects, surveyors and engineers, the three bodies who supported the foundation of the Town Planning Institute in 1913), with little mass support. Its closest connections were with other groups, such as the Fabians, the Sociological Society and the National Housing Reform Council (founded in 1900). All these sought to provide ideas for 'men of influence', and remained remote from the violent social upheavals of the period before and after the First World War.

Britain then was the most urbanized society in the world: only 10 per cent of the labour force remained on the land, and fully 80 per cent of the population was resident in areas defined as towns. And yet in comparison with other societies, powers of control over urban development remained minimal. The French had rebuilt Paris (although subsequent control was lax); the German municipalities not only owned much of the land in their boundaries but were able to regulate much new development by zoning and site design;[2] even America, the new society with escalating problems of urbanization, had remarkably capable city governments, who were prepared to take draconian steps to provide the necessary services of water supply, drainage, street paving, dock facilities and parks.[3] Systematic planning was accepted in principle long before the British were prepared to contemplate long-term interference with the markets in property: the Burnham Plan was presented in 1909, and the New York Regional Plan in 1922. It was the enterprise and grandiose vision of the American municipalities as well as those of private entrepreneurs that were to impress Unwin as they had Howard fifty years previously.[4]

Twentieth-century planning in Britain emerges out of nineteenth-century policies: there was no radical departure from the incremental legislation of the state on the urban question. Governmental powers over urban development had been grudgingly extended over the century, and the responsibilities of the municipalities as agencies for central government slowly confirmed. By the end of the century the local authorities had the organization, resources and influence to seek further powers for control over their domain, and could point to continental and American experience as precedent. The Housing and Town Planning, etc., Act of 1909 was promoted by the Association of Municipal Governments.[5]

The interest in town planning had different origins and expressions. Foremost in the literature at the turn of the century is the concern for housing. The 'town problem' resulted from the lack of 'scientific and well planned effort to supply effective housing accommodation',[6] with the result that jerry-building was condoned on the grounds that housing demand must be satisfied. In the new legislation, powers for compulsory acquisition and redevelopment of the slum districts, and by-law control over new housing construction, were to be extended to allow the municipalities to acquire land for working-class housing in the suburbs, and to permit them greater discretion in controlling the layout of new development. In isolation slum clearance had proved a weary exercise, exacerbating conditions of housing shortage, at considerable expense to the ratepayer, and the desolation of speculative suburban development was already apparent. Town-planning legislation must be seen as 'a logical extension of earlier legislation concerned with housing and public health'.[7]

The movement for town planning coincided with a major building boom in 1895–1905, most marked in the London suburbs. Burns, then President of the Local Government Board, when presenting the new legislation is quoted as referring to 'the extent of the damage that is being inflicted on rural England by the indiscriminate unorganized spreading of straggling suburbs'.[8] It was then estimated that half a million acres of agricultural land had been taken into urban development between 1893 and 1908. And yet the 'house famine' remained; this was middle-class housing to which the working class had no access. Although suburban development was accepted as the solution to the housing problem, there was realization of the failure to meet the needs of the mass of the population, and an active reaction against the dreariness of by-law housing, and the loss of the green fields of the Home Counties. The writings of Morris and his contemporaries expressed the widespread revulsion against the appearance of the new development, healthy and efficient, but lacking in beauty. The collapse of the tradition of management of the great estates, which had given London, Bloomsbury, Belgravia or Bayswater, and the breakdown of accepted canons of taste, now put the onus for civic amenity on the municipalities. They had to turn 'this land from a grimy backyard of a workshop into a garden'.[9]

Underlying the new movement was a longstanding concern for the relationship between town and country, brought to a head in the last decades of the nineteenth century by the plight of the urban poor, and rural depopulation. The poverty of the urban masses had roused the question of 'urban degeneration';[10] moral degeneration had long been assumed, but evidence was building up as to the decline in physique of second and third generation residents. Medical examination of recruits for the Boer War confirmed the differences between classes, and those between town and country: the rude health of the country lad was a necessary adjunct to Empire. Rosebery, in 1901, spoke as follows: 'An Empire such as ours requires as its first condition an imperial race—a race vigorous and industrious and intrepid . . . in the rookeries and slums which still survive, an imperial race cannot be served.'[11] The cramped housing, and lack of open space, sunlight and fresh air, the sheer urbanity of the cities, as well as their poverty, was considered to stunt the growth of city populations, while the absolute misery of the rural districts, low wages, bad housing, acute overcrowding, high rates of infantile mortality, precipitated yet further migration to the towns. It is in this context that Howard preached for garden cities:[12]

> the living stream of men, women and children flows from the countryside and leaves it more and more bare of active vigorous life . . . to enlarge our overgrown cities, to cause suburb to spread beyond suburb, to submerge more and more the beautiful fields and hilly slopes which used to live near the busy life of the people, to make the atmosphere more foul, and the task of the social reformer more and yet more difficult. . . . creating channels through which some of our population shall be attracted back to the fields . . .

A debate convergent with that on 'urban degeneration' was that on land reform. Much earlier in the century an essential part of the Chartists' programme and their theoretical analysis of the ills of capitalism had been that 'the land is the inalienable inheritance of all mankind',[13] and this aspect of their thought had passed into radical circles in the second half of the century. It is, as Saville comments, one of the paradoxes of British working-class history that during the 1860s and 1870s the country which 'was more proletarian than any other, and where decisive economic power was in the hands of those who controlled *la grande industrie*, political protest and social criticism were directed against the landlords and not the industrial capitalists'.[14]

Parallel to the anti-landlordism of the working classes was the dissatisfaction of the urban middle classes with the monopoly ensured to the landed aristocracy in the entail over their estates. Prominent figures such as Spencer and J. S. Mill were associated with the movements for land reform which provoked the House of Lords to initiate a survey into land-ownership nationally. The returns of 1874 and 1876 revealed that 1,200 persons owned a quarter of the land, and 7,400 owned 50 per cent.[15] The thought of Henry George who toured Britain in 1882–4, therefore, had a favourable reception. In his statements, the fruits of progress were alienated by the landlord, productivity was absorbed in rent, the capitalist penalized in that profits could not be reinvested, and the labourer was reduced to deeper poverty as wages were forced down to ensure further investment. He argued that 'Rent . . . is the price of monopoly, arising from the reduction to individual ownership of natural elements which human exertion can neither produce nor increase'.[16] His remedy was the 'single tax', to be levied on land values, and paid to the state; this would replace all other taxes.

Land reform was to become a controversial element in the radical social policies of the Liberal government of 1906–14. Taxation on development values, and on the capital appreciation of estates on the reversion of leases, was part of their strategy for the redistribution of incomes and the foundation of the social services, and provoked the constitutional crisis of 1910–11, but the revenue raised was small in comparison with that derived from direct taxation on wealth.[17] Public ownership of land was not advocated, and it was not until the 1920s that land nationalization became a live political issue. But the emphasis had shifted, and land reform was advocated as a palliative for unemployment, and not, as previously, 'one of the truest foundations of imperial greatness'.[18] Lloyd George, opening the campaign of the Land and Labour League in 1925, spoke movingly of 'an exchange of the green doors of the Labour Bureau for the green fields of Britain'.[19]

Garden cities of tomorrow

Howard, and the Garden City Association, in fact, make no explicit reference to either Henry George, or Liberal policies, but their proposals for the common ownership of land in the garden cities, and their belief that the benefits thereby accruing to the people would allow for a 'reconstruction of society', derive from this main stream of political thought. Retrospectively, Howard acknowledged the influence of the thought of Alfred Marshall, who during the 1880s had been an ardent advocate of the settlement of London's poor in the countryside; Wakefield, who had organized the settlement of Adelaide in the 1840s; Buckingham, who had produced a blueprint for a model city, 'Victoria', in 1849, again to solve unemployment;[20] and the writings of J. S. Mill and Spencer. The basis of his proposals for land reform he ascribes to Spencer, whose paper on land tenure had first been given in 1775, but whose writings had been republished during the 1880s by Hyndman.[21] His proposals therefore articulate the fabric of thought of radical, conservative and liberal elements of the middle classes of Victorian England. Eden concludes that 'Howard's great virtue was that he was able to create a picture that accorded with the aspirations of the class to which he belonged—the somewhat earnest, chapel-going, or chapel-emancipated lower middle class'.[22]

Howard's plan was very much more in the tradition of the 'bourgeois utopia' of Buckingham, than the socialist communities of Fourier and Owen. He was at pains to stress that it was not a socialistic or communistic city, though the realization of the project would depend on joint ownership of the land. Mumford terms his book 'a dry little prospectus';[23] Osborn stressed his realism and common sense. Nor did it turn away from the urban world as had earlier schemes; it was a blueprint for a new kind of settlement in which the advantage of town and country were integrated in a 'third alternative, in which all the advantages of the most energetic and active town life with all the beauty and delight of the country may be secured'.[24] He foresaw the development of urban regions as 'town clusters' in which each garden city was held separate from the next by a green ring, yet linked to the entire 'Social City' through rapid transport; it was 'a new open-work metropolis'. The manifesto of the Garden City Association is found in Howard's tract:[25]

> there should be an earnest attempt made to organise a migratory movement of population from our overcrowded centres and sparsely populated rural districts . . . a single movement . . . that the whole increase in land-values due to their migration might be secured to them . . . and that the golden opportunity afforded by the fact that the land to be settled on has few buildings or works upon it, shall be availed of in the fullest manner by so laying out a Garden City

> that as it grows the free gifts of nature, fresh air, sunlight, breathing room, and playing room, shall be still retained in all needed abundance. . . .

Howard saw planning as providing a middle way between the extremes of socialism and the existing social philosophies in that 'it encouraged individual initiative, and yet allowed the fullest co-operation'.[26] Howard and his contemporaries, Geddes and Unwin, claimed that planning would extend freedom of choice by offering an alternative to free enterprise. Unwin stated that it 'greatly adds to the sum of effective liberty by defining the limits and protecting the sphere within which each can move without being obstructed by others'.[27] They were concerned to win support for the practice of planning from the middle classes, and were at pains to demonstrate its inoffensive character. The Garden City Association acted as a lobby for garden cities (promoting two, Letchworth and Welwyn Garden City, as prototypes), and campaigned against the unrestrained suburban sprawl of the great conurbations, with a definite social philosophy. It aimed to 'make citizens as well as cities'.[28] Social planning was integral to their policies, they sought 'a new humanism' which would 'revitalise human, community, and civic values through a better ordering of spatial relationships'.[29] It never attracted a mass following, and in no sense was it a working-class movement. Howard and his associates believed that once the merits of their scheme were appreciated by the working classes, their demand for garden cities would overwhelm the vested interests of the propertied classes. Howard considered that the attractions of the new way of life he described so glowingly were self-evident. And yet it took forty years before leaders of the labour movement adopted the principles of planning outlined by Howard.

Although earlier in the nineteenth century working-class radicals had acquiesced in the policy of land reform as a solution to the failures of capitalism, by its end the labour movement had shifted its attack to the central fact of capitalist society, the exploitation of labour in the productive process. Trade unions had gathered strength, from a membership of 1.5 million in 1890 to 4 million in 1913 (rising to 6.5 million in 1919 and 8.3 million in 1920), and industrial unrest, in face of falling wages and rising prices, was mounting. Legislation over land-ownership, and control by planners over the layout of towns, was of trivial moment in the mounting tide of social protest: 'the Government was ceasing to govern and parliamentary institutions were falling into disrepute. The spirit of revolt was spreading from one section to another'.[30] A middle-class pressure group, restricting policies for collective ownership to the means of consumption, ignoring the means of production, had little working-class following in this political climate. Whereas Owen, a century before, had put forward plans for a social Utopia, designed to sweep away capitalism and the competitive spirit, the reformers promoting town planning and the garden city restricted their discussion to issues of environment.

Yet despite these inauspicious beginnings, the planning movement, and the first stumbling legislation, were to culminate in a surge of enthusiasm for 'town and country planning' after the Second World War. A system of planning controls was organized that for a period was hailed as the most successful in the world. Environmental planning, linked with progressive taxation over betterment of land values, was a central plank to the policies of the post-war Labour government, and the garden cities, as realized in the new towns, were to ensure British hegemony over the international planning world for a decade. Planning as a social movement reached its ascendancy in the 1940s, with the acceptance by government of the twin aims of the new towns and community planning. Strongly held views as to a new social world are to be found in the sober texts of the planning reports of the period,[31] and the idealism of a generation of young professionals had been harnessed to the planning process as an activity of government. Ideas that had had origins in eighteenth-century radical thought were to be realized by government as a means to the controlled reconstruction of the British economy in the second half of the twentieth century.[32]

2 The town-planning profession

Planning has not resolved the tension in aims evident from its earliest days as an activity of government and as a social movement. As the former it was concerned with land use, property interests and environmental planning, as the latter with community betterment. Intervention in the markets in land and property became in practice their regulation so as to protect previous investment, and to prevent the squandering of natural resources. In this sense planning was, and is, the protection of property interests, acting to stabilize the base price through limitations on further development, and preventing over-speedy obsolescence. But alongside this regulation of property interests there has been a continued interest in planning as a means to social reform. At all times there has been an interest in the welfare of the community, and a concern to redress the balance between groups in their access to urban resources. In this sense planning must be seen, along with activities of government in the provision of housing, education and welfare, as part of the shift in power at the beginning of the century away from the propertied interests to the mass of the population.

The consequences of this dual interest for the profession are seen in its ambivalence in definition of its own role. The debates over the last decade as to the proper stance of the profession with regard to the new urbanization with its contradictions of squalor and affluence have not succeeded in resolving the inconsistencies in ideology described by Foley;[1] if anything, they have worsened, and the lines of dissension are more firmly etched. The divisions are three-fold: between the environmentalists, who can be termed the conservatives of the profession; the methodologists, its progressives; and the social planners, its radical wing. Recent discussions have shown that the first group still predominates in the profession, a group convinced of the self-evident need for environmental planning in a society whose inherited resources are daily threatened by affluence and rapacity in the use of land.

In this attitude they are increasingly supported by the public who, too, have reacted with alarm to the depredations of development in town and country. The mushroom growth of the amenity societies[2] is only one aspect of this public concern for the conservation of the environment, preservation of historic buildings, and protection of traditional environments of town centre and village. In this respect there is more general support for the idea of environmental planning, though not acceptance of the authority of the planners. Planning decisions are more frequently than ever subject to criticism and hostility; a mood of apathy has been replaced by a critical interest by the more articulate in the planning process.

Meanwhile, the planners debate their future. Planning education, entry to the profession, the responsibilities of the planner, even his professionalism, are subject to deliberation. The sociologists' critical inspection has been only one element in a general uncertainty as to the role of planning in the corporate state. Town planners are having to face up to the exigencies of a professional role in a situation where they lack implementative powers and have only limited authority, and so the traditional assumptions as to the scope and role of planning have had to be reassessed. The debate of the last decade has ranged over 'non-plan', in which the planner ceases to impose his judgments on the market, permissive planning, incrementalism, planning as problem-solving, empiricism in which the aims of planning are set by the actors in the system, purposive planning, systems planning and corporate management. As yet there is no agreement in the profession as to whether planning is a field of activity, defined by its responsibility to the environment, or a distinct professional skill to be assessed by its management capabilities. The old certainties as to cities of the future have been replaced by a limited here-and-now pragmatism, so that planning theory has become a theory of methodology: 'Inasmuch as there is now any theory of planning it is a theory of response, to problems comprehensively perceived: it could be characterized as a theory of methodology.'[3]

The problem of planning is that few outsiders to the profession accept its claim to a unique competence out of the reach of the layman. As Davies writes, 'there is simply no such thing as an objective body of knowledge allied to a coherent theory which can be used as a basis of rational decision making',[4] and few planners would now make the claim that there is a science of land-use planning. The lack of authority accredited to the profession in part derives from its pragmatic tendencies. If planning becomes the practice of planning, then the planners cannot justify what they do except by what they do, i.e. their methodology. It is not possible to demonstrate that a proposal is the best or the truest one, only that at this moment in time it is the most useful for attaining certain objectives. So the way is open for the layman to draw his own, and different, conclusions from the analysis of the situation.

The planning profession would be in a stronger position if they could point to the beneficial effects of planning since the war. But spokesmen for the planners have found self-congratulation difficult. Buchanan, for instance, could point to little more than[5]

> the building up of the planning machinery itself and its corruption-free working; a massive if negative preventative effort to curtail the worst excesses of urban sprawl but insufficient to contain those etched out before 1947; extensive clearance of slums; a brilliant new schools programme . . .; the New Towns; the National Parks; the clean air action . . .

The two conspicuous results of planning have been the containment of sprawl through the designation of greenbelts, and the reconstruction of the inner city districts; neither is an advertisement for planning. Buchanan describes the outer London suburban belt as moving into 'social confusion, visual confusion, functional confusion, and transport confusion',[6] and the inner city districts are bleak wastelands, devoid of humanity and charm, 'social isolation wards' encased by motorways and commercial development. In the first instance planning has been conspicuously unsuccessful in controlling private initiatives in development; in the second it has collaborated with government in imposing an alien environment on those with least power of resistance. The Utopian vision which inspired the founders of the planning movement, and the zest and enthusiasm of the years immediately after the Second World War, have been dissipated in the bitter experience of bureaucratic organization, and loss of initiative for development from public to private sector.

The founding fathers

From the outset town planning contained contradictory elements, reflected in subsequent confusion in the aims of the profession. Although the Town Planning Institute was inaugurated (in 1913) with the support of the professions of architecture, surveying and civil engineering, it has never been able to establish an undisputed competence over issues of the environment, and the very diversity of recruitment reflects the different interests in its membership.[7] Even among the leading spokesmen for the new profession, Howard, Geddes and Unwin, there was confusion, even conflict, in ideas and ambitions. Notable was the muddle between Howard's garden city proposals and Unwin's statements on low density suburbs, which culminated in the garden suburb. There was also a clear contradiction between the methodology advocated by Geddes and that practised by Unwin, the first without final solution, the second searching for formal structures of urban design. The definition of professional aims had to be broad enough to encompass a diffuse commitment to environmental planning without disclosing the underlying dissensions in political interests. This confusion has increased as planning has been entrenched as an activity of government, and as its recruitment has diversified. Planning has increasingly become that which is legally practicable, rather than a strategy for the reconstruction of society, but it still retains the Utopian quality of its early days.

The early days of the planning profession were dominated by a few men, chief among whom were Howard, Geddes and Unwin. The former remained an amateur, disinterested in the new profession, but the latter two, in their different ways, left their mark. They are still important because the assumptions which governed their approach to town planning, and the arguments they entered into to justify its practice, have entered into the ideology of the profession. Their postulates form part of the mythology of planning to which all entrants to the profession are introduced.

Ideology has been defined as 'the language of the purposes of a social group'.[8] It defines the situation for the participants in an activity; ideas derived from previous experience are brought to bear on a situation, identifying the problems and the strategies that might prove effective. Experience is organized by previously acquired categories of thought: 'It is not men in general who think, or even isolated individuals who do the thinking but men in certain groups who have developed an endless series of responses to certain typical situations characterizing their common positions.'[9] A profession acts to codify the responses appropriate to certain domains of activity; it does this through formal education, meetings, journals, and the socialization of the younger members through contact with the elders of the profession. So the individual planner does not plan alone but acts with the full weight of the profession's collective experience behind him. The ideology of the profession constitutes the basic operating rationale by which he justifies his decisions, to others of the professional group, and to outsiders. 'The ideology provides an essential kind of consensus for the activity . . . a broad and attractive rationale for winning over and maintaining the allegiance of political leaders, appointed officials and citizens.'[10]

Patrick Geddes was the visionary, prophet, talker and teacher, a source of inspiration for the planning movement. He relied on the spoken rather than the written word and little of 'his restless physical energy, the boundless flow of ideas, the keen delight in living'[11] can now be realized. He was a major influence on both Mumford and Abercrombie, and through both on the mainstream of British planning. His was also an inspiring voice in the movement for the teaching of geography in schools and universities, and his writings for the Sociological Society on 'civics'[12] pre-date those of his contemporary, Park, by a decade. He therefore has claims to being the first 'urban' sociologist.

Like the other members of the planning movement, Geddes was antagonistic to the large cities; he called them 'sprawling man-reefs', or conurbations. He saw them as spreading like 'expanding ink-stains and grease-spots'[13] over the natural environment; city life was a matter of limitation of individual development, not of opportunity. He referred to 'the prejudices, the isolation from real life and the false self-sufficiency of the city dweller. London or any metropolis . . . in reality stunts the mind'.[14] He went further than most of his contemporaries in castigating the 'Paleotechnic order'; as an ecologist he slated the inefficiency and wastefulness of its industry; he condemned the instability of its financial system and the unfitness of its city life. The economy was based on primary and secondary poverty, and created nothing but 'Slum, semi-slum and super slum'.[15]

> The Paleotechnic order should, then, be faced, and shown at its very worst as dissipating resources and energies, as depressing life, under the rule of machine and mammon, and as working out according as its specific results, in unemployment, and misemployment, in disease and folly, in vice and apathy, in indolence and crime.[16]

For Geddes, the city lived, grew, had processes; it needed a new science, civics (the study of the life and working out of the city), with new practitioners, civic sociologists. For him the essential feature of this new science was its synoptic method which enabled the study of the totality of man's experience; this was the new philosophy. The science was to be the basis of the transformation of cities into the new social order. He advocated a method of studying[17]

> Each city . . . as an organic whole, whose life is subject to conditions laid down by the natural environment, and yet within these conditions exhibits its own characteristics and unvarying tendencies . . . the tendencies are seeking to clothe themselves in new forms suited to the age . . . the problem before the town planner is to assist this process, . . . he must start from concrete sociology as his base.

His commitment was to the city as a whole, not the different neighbourhoods, nor its social classes. He combined a classicist's view of the citizenry with the biologist's view of an organism in symbiotic relationship to its environment.

He persistently advocated the civic survey as the method of planning. He devised an empirical methodology which has been handed down to subsequent generations as 'survey, analysis, plan and implementation'. The new science was an empirical science, akin to biology, in which future action was to be derived from present analysis of the facts of the situation. In a lecture of 1923 he is quoted as saying:[18]

> As our surveys advance we become at home in our region, throughout its time and its place up to the present day. From thence, the past and the present cannot but open out to the future. For our survey of things as they are, that is, as they have become—must ever suggest ideas as to their further becoming—their further possibilities. In a word, the survey prepares for and points towards the plan.

Geddes was a visionary, with a belief in the possibilities of planning, and yet with a sense of the indispensability of the past. Despite its failings, the past would lead to the future with the intervention of the new social scientists, optimizing the benefits of the new technology. But there is a marked inconsistency in his teaching of the total interconnectedness of the present urban environment with the paleotechnic order, and his assumption that its failings could be remedied by the exercise of planning skills. He believed that through the application of civics as history, science and philosophy the idealism of the citizenry would be harnessed to the goal of the transformation of city and region. Like Howard and Unwin, he believed in the power of ideas to change the world; a reorientation of consciousness would effect the 'neotechnic reconstruction'. Men would transform the world, ideas had a strength which would overcome the limitations of the existing social order. Despite his advocacy of the 'economic interpretation of history' Geddes spent his life arguing for a 'Eutopia' which could only be planned and realized by its citizens.

However, although his ideas had an unmistakable impact on the planning vanguard, he was not successful in transmitting his vision of city planning as an organic science. His ideas were too diffuse, too rapidly spoken perhaps, to give the lead to a new profession. This was left to Unwin.

Unwin, unlike Howard or Geddes, was a professional planner, the leading figure of his generation. He was professional adviser to, for instance, the influential Tudor-Walters Committee of 1918 which set the initial standards for much inter-war housing; a senior civil servant in the ministries concerned with architecture and town planning from 1914 to 1930; an international planner, at home on both sides of the Atlantic, involved in the New York Regional Plan of 1922 and the New Deal legislation of 1935, as well as the London Regional Planning Committee of 1929. He, rather than Howard, had authority over development between the wars: 'It was town planning of the kind advocated by Unwin which established itself as town planning in practice.'[19] If we are to understand the directions taken by the planning profession subsequently, we must appreciate the reasons for his success.

The planning he practised and advocated was physical planning. Unwin had trained as an architect, and had a deep-seated belief in the need for control over use of the environment. In 1930 he concluded a paper on regional planning with a statement on the science and art of planning which would give 'the opportunity for a beautiful environment out of which a good human life may grow'.[20] Earlier in his career he had written persuasively of the need for beauty in urban life, over and above sanitary reforms and by-law control—'Not even the poor can live by bread alone'. His career has to be seen as that of the evangelist, preaching for 'the vivifying touch of art'[21] which was to transform modern towns from mere aggregations of people into expressions of a better common life.

Another aspect of his training as an architect was his belief in the simple and the obvious. The art of design, he argued, was to find 'simple solutions for seemingly complex problems', to search out 'the one simple and direct path to unity'.[22] Town planning was to impose order on haphazard urban development, and Unwin increasingly sought formal models of urban structure by which city growth would be controlled. Unlike Geddes's diagrams these were simple, easily understood, and remembered. The best known is his application of crystallographic principles in his diagram for a 'City with defined suburbs and satellite towns', which was to be a basis for the plans for the London region after the Second World War.[23]

Again as an architect, in the design of housing, he tended to discount the market, and sweep away 'the passing fashions or conventions established by the speculative builder'.[24] He argued for a common-sense functional approach to the housing situation, in which housing was to be designed according to needs for sun, light and space, and not for pretensions to middle-class living. In this respect there is a tempting comparison with Le Corbusier. Both rejected contemporary living patterns and bourgeois clutter; housing was to be redesigned for man's innate needs of sun, space and greenery; both sought a solution in the application of design methodology—the architect's analytical powers were to lead forward to a new solution; and both were concerned with the pure aesthetic values of beauty and order. The differences come in Unwin's nostalgia, Le Corbusier's extollation of technology, Unwin's reliance on common sense and craftsmanship, Le Corbusier's logic. Unwin was very much more in touch with the common man. Consequently plans for two very different cities were drawn up—the low density city clusters of Unwin, and the high density Radiant City of Le Corbusier.

Unwin had been strongly influenced by Morris, Ruskin and Lethaby, a group of men explicitly rejecting industrialism, the machine age, the large city, and the class divisions of modern society, and seeking the reintegration of a man's life in a work-place-folk synthesis. This movement of national romanticism influenced Unwin in two ways: in his design for new housing areas, and in his own early conversion to socialism. Unwin appears to have been the first to advocate neighbourhood planning (Mumford refers to a paper of 1921); he sought the redesign of cities as consciously organized communities and his housing design re-enacted the themes of cottage, village and village green. Access to the green fields and flowers of nature was a need common to all men, and housing was to be at a sufficiently low density to allow a subsistence garden for every family. Planning in providing for the whole nature of man was a means to override the existing class divisions in society, by setting a common standard of living. But through his career the emphasis shifted significantly; in 1909 he had argued that successful planning could only come from the expression of a truly common life, but by 1936 planning itself was to assist in the 'cooperation between men' on which a successful society depended. Unwin was by then articulating the reformer's myth that 'We have proved that if slum dwellers are given decent and adequate dwellings, the majority of them will adapt to the improved conditions'.[25]

His socialism did not prevent Unwin putting over his ideas in a way that was acceptable to government, and the middle classes generally, as for instance in his tract for the Garden City Association,

Nothing Gained from Overcrowding. He also believed in working within the structure of government, and was successful in putting the standard of twelve houses to the acre on the statute book. Creese describes him as[26]

> A deep-dyed individualist and rebel at heart [but] he never allowed his mind or imagination to wander off the reservation, to cut out on him, or his society. He learned to live and work within the actual environment and possibilities which surrounded him.

It is tempting to see Unwin as the prototype planner, in his willingness to work with the means at hand, within the limits of present society, and in his belief in the capacities of physical planning for effecting change in society. As a pioneer of planning it had been his task to convince other men of influence of the merits of a planned world.

The common denominator of these pioneer figures is their wholehearted rejection of industrial-urban society as it had developed. This rejection, coupled with their love of the countryside, led all of them to advocate a new form of urban living based on the house and the garden, the neighbourhood and the small town. They hoped to inculcate a new sense of civic responsibility in the context of these socially planned garden cities. Their attitudes, the careful conservatism of Geddes arguing for 'maximizing physical efficiency and economy throughout', their dislike of extremes, carefully choosing the middle way, finding common ground with those they sought to influence, their belief in the possibilities of progress in capitalist society, as well as their advocacy of suburban living, mark them as coming from a privileged stratum in society. They assumed that reform was within man's grasp, and that the individual could bring about change. They also shared a faith in man's innate good nature: once the blindness of his ways was pointed out, then there would be a change in attitudes.

Reissman argues that the town-planning movement had an organizing myth, that of 'reason'. The planners believed that 'man is rational. He can plan, and thereby create a better more harmonious, and more humane environment for himself'.[27] They worked on the postulate that 'reason was enough to change society from what it was, to what we would like to be'.[28] This generation of planners held that intelligent arguments would be accepted by all interests in society; these arguments were those of common sense and the conclusions self-evidently correct; there could not but be a consensus over the planner's conclusions. This belief in planning as such forms one of the major inarticulated premises of the profession, one which all its members must hold if they are to practise planning.

Equally important is the concern for the environment. From Geddes in particular there derives the premise that the environment must be protected against despoliation; from Unwin, the belief that the aesthetics of our environment are of vital importance in our everyday lives, that amenity and beauty must be the goals of the planning profession; from all three the premise that the basis of a future society can be provided for, through the restructuring of the environment. So is set up a major assumption of the profession—that its self-appointed task is to safeguard the environment. In this way the planning profession started to move away from town planning as a means to an end, and towards environmental planning as an end in itself, its merit self-evident to all rational men.

The third component in the planner's belief system is that cities have to be planned as whole entities; planning is by its very nature comprehensive, and as such overrides individual or sectional interests. So one finds Unwin at a lecture in 1937 describing city or regional planning in terms of implementing a previously derived design:[29]

> It is this capacity to work to a clear conception that makes a real town planner. He works rationally, but with his thoughts always for the realization in the future of the whole. . . . The right way is to start from a complete conception. . . . We have stressed the importance of thinking of the city as a whole community.

In both Howard's strategy for the garden city and in Geddes's methodology, there is this conception of the city as a complex and indivisible whole. The science of planning lies in the disposition of the component parts in their appropriate relationships. It is a science of forms rather than content.

Clearly it would be false to build up a profile of a profession from an analysis of its originating beliefs. All that is suggested here is that this is the inheritance of planning, this mix of enthusiasm, well-intentioned humanism and level-headed common sense, with which successive generations of planners have had to come to terms. Some ideas, such as those for the new towns, have carried on, changing in focus as defined needs alter; others have lost their initial sharpness of vision. In this way, Unwin's quest for beauty becomes the safeguarding of amenity, and Geddes's new science of civics a bald empiricist injunction. What is fascinating is the reappraisal of Geddes. His bio-social regionalism now finds favour with the current approach to cities as systems in their regional context, and his methodology is reinterpreted to accord with the emphasis on detailed survey and analysis in urban planning. The writings of an ecologist are seen as foretastes of contemporary concerns. The ideology of a profession does not stand still: it is being continually reconstituted as exigencies of circumstances change and the testimony of a pioneer sanctions the reorientation of the profession.

Professional reorientation

Writing in 1959, Foley identified the underlying rationale of British town planning as comprising three complementary ideologies. He described the components as follows:[30]

> 1) Town planning's main task is to reconcile competing aims for the use of limited land so as to provide a consistent, balanced and orderly arrangement of land-uses. . . .
> 2) Town planning's central role is to provide a good (or better) physical environment; a physical environment of such good quality is essential for the promotion of a healthy and civilised life. . . .
> 3) Town planning, as part of a broader social programme, is responsible for providing the physical basis for a better urban community life; the main ideals toward which town planning is to strive are a) the provision of low density residential areas, b) the fostering of local community life, and c) the control of conurban growth.

At the time he was writing, the profession had not found a basic operating rationale in its position as an activity of government. He was writing at the end of a long decline in enthusiasm for planning. The euphoria of the immediate post-war period when anything seemed possible, including the reconstruction of society, had been succeeded by the routinization and trivialization of planning activity in a sea of planning inks. Keeble's definition of planning as 'the art and science of ordering the use of land and the character and siting of buildings and communication routes so as to secure the maximum practicable degree of economy, convenience and beauty'[31] sums up the tedium of planning in the decade of the 1950s. It was seen that the aims of the planning movement were no longer realizable, that the profession lacked the support of government in the defence against the first post-war property boom, and there was little public support for their ambitions. A 'broad and attractive rationale for winning over the allegiance of political leaders, appointed officials and citizens' had not been established. The subsequent period has seen the readjustment of the profession to planning in a market economy, and the search for new legitimations for the exercise of authority over the environment.

By the end of the 1950s the post-war mood of optimism and confidence in the techniques of land-use planning had evaporated with experience of control over development once the initiative for change had shifted from government to the private sector. New entrants to the profession, and those concerned with reassessment of the scope and methodology of planning, found themselves confined by a legislative system formulated in the immediate post-war years, in very different economic and political conditions. The administrative machinery set up under the Act of 1947 had been planned in the expectancy of a slow-growing economy, stagnation in population growth, and public control over investment of industry and housing, backed by financial sanctions and incentives. Town-planning machinery had been created on the assumption that the market was subordinate to government, and a new world could be created. Hence the concept of a plan, a blueprint for a finite realizable world. But by the end of the 1950s there was an economic boom, a rising birth rate, increasing population mobility, rising standards of living, and a free speculative market in land had been reinstated.[32] The development plan had become a straitjacket as the range of planning issues and pace of development quickened.

The search for a more flexible instrument of development control, which was to culminate in the 1968 Town and Country Planning Act, has to be seen in this context. The very nature of development plans was felt to be illusory in that they gave an image of stability in circumstances of relatively rapid change; they claimed to have the blueprint for the future, and yet were unable to predict the nature of demands being placed on environmental resources. Different sorts of guidelines were clearly required for all those with an interest in investment in property. Structure planning, local government reorganization and public participation, as advocated in the reports of the Planning Advisory Group (1965),[33] the Redcliffe-Maud and Wheatley Commissions on local government (1969),[34] and the Skeffington Committee on participation in planning (1969),[35] gave planners a new organizational framework, different responsibilities, and a new conception of their role in local government.

The concept of the structure plan is central to the new planning. This is a statement of aims, of principles of development, rather than a blueprint for land allocation. It relies on a 'broad-brush' approach in which the physical structure of the town is indicated rather than designated—individual property holdings are not shown; it includes all aspects of the social and economic structure of the area that have bearing on the physical planning process; it is deliberately open-ended and flexible in that it is to be based on a continuous process of monitoring of the actual pattern of development, which may lead to a revision in the provisions of the structure planning documents at any time. In this approach,[36]

> Planning is less and less a matter of precise propositions committed to paper and more and more a matter of ideas and

> policies loosely assembled and under constant review, within which some project is seen to be as ready for execution as human judgement can pronounce. Once executed the project is fed into the process as an influence of all future decisions. This we see as planning for flexibility.

The plan is no more than a provisional hypothesis as to the direction of development in the area for which the local planning authority has responsibility.

The method of planning advocated is that of guidance or management of the urban system. The planner becomes 'a helmsman steering the city'[37] whose 'rudders' are public investment and development control procedures. At his side is a survey as to the present state of the system under management, and a document which points to intentions for its change, and the planner attempts to regulate development within the limits indicated. As decisions are taken, their consequences are fed back into the survey, and the plan, and both are subject to continuous monitoring and review. The plan is no more than a provisional solution, always subject to modification as the priorities of investors in the built environment alter, and policy objectives are switched.[38] The planning process is deliberately open-ended, here is no Utopia to be realized. Planning deals with the existing and predicted defects in the functioning of the system in a cyclical process of 'error-controlled regulation'.

The new planning makes great play of participation by the public in the planning process. Legislation (Town and Country Planning Act, 1971) now provides for consultation with the public in the preparation of the structure plans, as well as the local planning documents. The new process requires information as to what the public are seeking from the urban infrastructure, and in ascertaining new requirements, i.e. goal formulation. When, as at present, the mood of planning is strongly empirical, it becomes important to make contact with those directly affected by the plans. Participation by the public is therefore one method of establishing the parameters within which the urban system must perform; public opinion is monitored just as traffic movement, the demand for office space, or the condition of the housing stock.

There has also been a demand from outside the profession for participation, principally from middle-class amenity groups, who, like the profession, see 'intelligent planning as a non-political activity, which would bring benefit to all the community',[39] and from community action groups. But the activities of the pressure groups would seem to be symptomatic of a more general concern at the pace of change in the environment. Redevelopment proposals for town centres and inner city districts, the devastation of urban motorways, the use of agricultural land for housing and recreation, the submergence of villages in 'minor growth centres', the use of remote countryside for large-scale industry, are all seen as threats to the national heritage. Antagonism to these changes is projected on to the planners as the profession directly charged with control over the environment.

Town planning is a relatively recent activity of government, compared say with education or public health, and the planners find themselves assailed by criticisms in a way that must be unique in local government. They are accused by developers of obstruction, by conservationists of vandalism, the public of insensitivity, and the left-wing press of collaboration with property interests. Planners generally would acknowledge their lack of authority, as did Buchanan, speaking of the post-war experience: 'The planning effort has been maintained in the face of disbelief and ridicule by people who could by their patronage have greatly improved the results, and in the face of indifference from the public at large.'[40] In this context of at best indifference, and at worst hostility, public participation is a means for widening the basis of authority accredited to planning decisions. The legitimacy of the plan is no longer dependent on acceptance of the professional status of the planners, as the latter have appealed to the public and so widened the authority of the subsequent planning process. Unless this were done, then planning would be frustrated by objections and the antagonism of the community. The community has to be taken along with the plans and brought to acceptance of the increasing pace of change.

Other commentators have interpreted the town-planning profession's advocacy of participation in planning rather differently. They would see the profession seeking to achieve authority in the élite of decision-takers, by this process of consultation with the public. In effect the planners are appealing over the heads of the politicians and other professions in local government, to the public, to strengthen their own hand in the dialogue with the political representatives.

> Those who already have power to achieve their objectives do not usually offer to share this power with others. Where such an offer is made the explanations may lie in a need to recruit allies or to legitimate one's activities by demonstrating that they are supported by superior numbers . . . far from being in a position to offer a share in decision-making to its 'clients', it is still attempting to claim a share for itself.[41]

Structure planning has yet to become planning practice in that only a handful of planning authorities have prepared the new style policy documents, but already there are strong criticisms from within the profession, and a feeling that it 'was a conception of planning that was outdated even before it reached the statute book'.[42] Criticisms have been various, and can be listed as follows:

(i) The new system of planning has had to be carried out in a two-tier system of local government, in which the county authorities are responsible for strategic planning, and district councils for the preparation of local plans, as well as most issues of development control. The two levels of planning activity are therefore conducted in hierarchical rather than complementary terms.

(ii) The new emphasis in local government on corporate management is incongruent with a separation of function between regional agencies concerned with policy, and local agencies with powers of implementation. The vision of local authorities as integrated enterprises is shattered in the planning process.

(iii) The question of implementation is the subject of anxious discussion. Not only has local planning suffered as staff moved to the county authorities, but the few plans that have been prepared are considered to have been too narrowly interpreted in environmental terms, as exercises in 'convenience and conservation',[43] not as integral to social or community planning.

(iv) Participation in the planning process has been found to be lengthy, time-consuming, unrewarding and largely impracticable given the obligation of the local authorities to prepare policy documents for large areas in a relatively short space of time.[44]

(v) Flexibility has been found difficult to achieve, given the bureaucratic bottlenecks of local and national government. The process is considered to be 'over-sophisticated and over-elaborate'; both the débâcle of the Greater London Development Plan Inquiry, and that of the Roskill Commission, in which plans representing many man hours of anxious deliberation were first debated at considerable length and expense, and then summarily overturned on political and financial considerations, have strengthened the arguments for an 'incremental decision making process'.[45]

The doubts reflect the changes in organization of local government, and also alterations in the issues confronting the planners. The restructuring of local government has opened up new possibilities for the application of business management techniques to established procedures of local authorities, i.e. corporate planning. Its aim is stated to be 'to harmonise all the functions of an organization by reference to a statement of its general objectives, and to test the efficiency of particular activities against general criteria for the efficient use and allocation of available resources'.[46] Implied is a reappraisal of the efficiency of local government, and its responsiveness to the needs of the community. This adoption of a corporate attitude to the resources of the local authority in land, buildings, equipment, skills and information, meant that physical planning now had parity with strategic thinking about welfare services and educational needs, for instance, and the structure plan was just one document among many for the realization of the local authority's objectives for their constituency. In principle, therefore, it was open to professions other than that of town planning to join in the preparation of structure plans,[47] and equally, the planners could set up new empires over social planning, as well as their traditional domain of development control.

The shift in emphasis from the uncritical permissive attitude implicit in structure planning, to the more deliberate controlled emphases of corporate planning for community needs, reflects growing awareness through the late 1960s of the seemingly intractable social problems of many city districts, despite increasing local authority expenditure. The planning documents of the mid-1960s which provided the technical rationale for the new legislation had been prepared in the aftermath of the heady days of affluence, seemingly unlimited economic and demographic growth, and unceasing change in the use of the environment. The poverty and community deprivation of the inner city districts were then seen as isolated problems, 'pockets of under-consumption',[48] but now the problem is acknowledged to be more generalized: unemployment is as much a reality in suburban London as Wearside or Clydeside. Increasingly limited resources have to be more carefully budgeted, and more firmly directed to specific sectors of the population. The autonomy of the local authorities is whittled down in face of national stringency in public expenditure, and inability to raise local rates, and the powers of its professional officers are thereby reduced. The status and responsibilities of the planning profession are therefore in question: will the educational priority areas and community development programmes be extended to include 'an old area renewal policy',[49] instead of a new town policy, or will the rate of investment in all physical plant—roads, schools, airports, housing, commercial development—slow up so markedly that the planner is relegated to a back seat in the corporate structure of local government?

Community planning was one of the options put forward for the future orientation of the Town Planning Institute by its council in 1971. The changes in the context of planning, and new attitudes among its membership, were considered to warrant a reappraisal of the traditional role of the institute. One argument was for a much broader definition of professional interests in which the town planner would be identified by his skills in planning, as a process of analysis, for community needs. It was then stated that 'unless physical planning adapts to its changing environment it will become less and less appropriate as a tool for diagnosing and remedying community needs, and in certain local authority contexts its function and powers could diminish very rapidly indeed'.[50] Planning would cease to be defined by its sphere of interest, i.e. development of the environment, and would be marked out by its 'analytical methods and synoptic vision'. The rank and file of the profession, however, opted for a structure for the institute which would reinforce the traditional concern for the environment.

This debate within the profession is likely to recur: day-to-day work pushes the planner into ever closer involvement with issues of social policy, and yet physical planning as a means to achieve these ends is discredited. The conventional definition of responsibility within local authority organization as that of safeguarding the environment will either restrict the profession to 'town-tidying',[51] or force it back into the political arena from which it has struggled to separate its activities. The debate about planning methodology, and the attempt to achieve a new self-definition, a new ideology, have to be interpreted as a search for new bases to professionalism in these changed circumstances. The decade of the 1960s saw the profession reassessing its role in an affluent, educated society, not without misgivings from the older generation trained in the techniques of land-use planning. But now the cybernetic creed of the 'new' planning is itself under challenge from those intent to maintain the social commitment of the profession.

Professionalism

A 'professional' has been defined as 'one who possesses essential technical knowledge and skill of highly complex and esoteric nature'.[52] The definition of planning as a professional activity, handed down from the founding fathers, asserted first its scientific or rational quality, second, its specificity to a defined area of experience, and, third, its universality. Planning was delineated as physical or environmental in scope: the restructuring of society was literally its rebuilding; this, to the planners, was an issue beyond politics. In like manner, the present ecological crisis is presented as a non-political issue in which all must surely agree; just as the future of the world overrides existing international differentials in the use of resources, so the future of city-regional environments overrides immediate here-and-now distinctions, differences and hardships. The technical knowledge of planning as a profession related to the scientific analysis of the organization of activities and buildings in space, and it was to be distinguished from other professions by its holistic methodology, its synoptic vision. Architecture, or engineering, for instance, might have to do with aspects of the built environment, but only planning was concerned with the total environment, and the well being, 'economy, efficiency and beauty' of a whole organism, a city.

Over the years, with the incorporation of planning as an activity of government, professional attention was focused on the physical structure of the city, and the exercise of professional judgment was justified in bland neutral language with reference to concepts such as efficiency of circulation, conservation of resources, and protection of amenity. There was an assumption of a consensus over the public interest in the use of community resources.[53] In awareness that planning decisions had taken on an arbitrary quality, and that issues of environment were but symptoms of far-reaching shifts in economic organization and social structure, this definition of professionalism was challenged by the planning vanguard, who sought to redefine planning by its method, and broaden out its sphere of competence to include community planning. At the very moment in time when there was apparent a new interest by the public in issues of environmental planning, a more abstract definition of professional authority was sought.

In the debates about the future of planning, it was redefined as a distinctive method, a process of arriving at decisions, rather than as responsibility for a delimited field, or as a substantive body of theory. Planning was defined as an abstract process, which 'may be applied to many sets of circumstances, and is characterized by its analytical technique, its synoptic concern for the total environment, and its orientation towards problem-solving action'.[54] In this approach the British planners were following the American profession; Webber, for instance, had defined planning as a 'problem-solving method',[55] not as a body of substantive goals, and as such independent as an activity of the phenomena to be planned. Planning becomes a pragmatic problem-solving activity in which ideas are judged by their usefulness to achieve certain end-states, and in which 'science' becomes the practice of science. There is nothing sure, nothing firm; what answered yesterday as a solution is insufficient for today's problems. All the planner has to fall back on is a method, one couched in language inaccessible to the layman, and remote from that of other professionals.

Discussions of methodology therefore become of extreme importance, and it has been the methodologists who have formed the vanguard of the profession. In this debate systems planning, which attempts to bring together the synoptic vision of the traditional planner with the techniques of operational research, systems analysis and cybernetics, has been central. This has been considered to be of universal applicability: 'The fundamental principles of control in complex systems are universal, irrespective of the actual nature of the system—real or conceptual, animate or inanimate',[56] offered objectivity, and conferred academic respectability on the profession.

> The systems view of planning gives a coherent and philosophically satisfying core to a formerly incoherent discipline which was so concerned with development ends that it quite ignored its unique attributes, that of the process of thought and action required.[57]

In short, systems planning offered a new ideology to the profession in which the activity was defined by the exercise of planning skills. Or rather, more accurately, systems planning spelled out in new language the traditional conception of the planner as having a special comprehensive methodology.

The attraction of the 'new dispensation' lay in its separation of means and ends—the ends, or goals, of the planning process are set by the society to whom the planner acts as professional adviser: the planners' task is to make explicit evaluation through rational analysis of the means available to reach stated goals. In this respect it is fair to say that the planners 'sought methodological solutions to the essentially ideological difficulty'.[58] But there was another important aspect to the planning process—that of ascertainment of the needs and wishes, expressed and unexpressed, of the population. What was envisaged was the establishment of a consensus over the 'public interest'; the planners were to act as grand conciliators, seeking the points of common agreement as to the future of the city. If the planner were to act as mediator between the different interests, 'trying to propose new consensual values which permit some kind of truce between the different sides',[59] then the concept of 'disinterested service' inherent in conceptions of professionalism would take on a new complexion.

One of the major weaknesses of planning is in fact this assumption of consensus. In the debates over the reorientation of planning previous failures are interpreted as deriving from failures in communication between planners and planned, which could be remedied both by education of the public and greater consultation by the planners. A situation where different groups in society may have very different and irreconcilable requirements of the urban system is not envisaged. The major division in society is seen as that between controlling élites and the mass—'public opinion', whose sanction has to be obtained for the decisions taken by the élites, and whose long-term interests are identified with those of the élites. Only in this assumption of harmony in society can the planner justify his declared professional responsibility as that of maintenance of a system.

'Systemism', with all its technology, has joined 'environmentalism' as the professional ideology, so side-stepping questions of professional accountability. The planner is clearly responsible to the élite of politicians and professionals by whom he is hired as adviser, but is his competence to be judged by the optimal behaviour of the system, or the welfare of the community?

This dedication to the maintenance of a system has its roots in the view of planning promulgated by Geddes and Unwin: the primary responsibility of the planner is not to the citizens but to the city, this abstract but all too evident entity that has a life-span greater than that of any individual member. This is what Davies termed 'futurism',[60] i.e. the justification of actions by reference to the long-term interests of the city, and not by reference to the immediate needs of its people. In this way the concept of the slum changes into urban renewal: legitimation for redevelopment is no longer the welfare of the inhabitants, but the future fabric of the city, which, it is claimed, governs investment by business interests, and movement of population, and so determines the future prosperity of the area. Amenity, and not welfare, becomes the criterion for upheaval of the population. The planners' activity is justified by the efficiency and attractiveness of the urban community as an environment for growth, thereby ensuring the prosperity of all its members.

The systems approach still retains some of the Utopianism of earlier town planning, in that it seeks to move beyond the present consensus in establishing hitherto unexpressed wishes and needs, and in promoting an environment which provides a more satisfactory basis of existence for all. In this it differs from the more narrowly defined problem-solving or incremental approach, in which the planner is restricted to making 'an unacceptable condition acceptable'.[61] It is at least conceivable in the former approach that the professional may establish new problems, which extend common estimations of what is problematic in society; in this definition planning is more than a technical or remedial activity. It is still, however, seen as an a-political activity, as being the rational and sensible safeguarding of the public interest. In this self-definition it is still interpreted as an alternative to politics, rather than 'an extension of politics by other means'.[62]

Now, however, the very conception of professionalism is cast into doubt. The restriction of technical knowledge to a select body, the closure of the profession, is now interpreted as a monopoly power wielded against the interests of the mass of the population. As Goodman commented, 'The young shun professionalism, and for good reason . . . the very idea of autonomous professionalism has begun to fade: professionals are becoming nothing but the personnel of organizations.'[63] In that planners in Britain are largely the employees of the State (over 80 per cent in local and national government agencies), and that many are young, the move to a new definition of public service, in which the planner's responsibility is to be defined as that of furthering the interests of the people rather than serving those of the public bureaucracies, was predictable. The move to a pragmatic professionalism has been strongly challenged by a radical element concerned to direct planning to a political role in the allocation of scarce resources.

In this movement against professionalism they have been encouraged by the sociologists who have persistently criticized the profession for adopting language which obscures the political implications of its activities. It has been urged persistently that 'every stroke on a piece of paper or a map is a political statement of which they are a part'.[64] Faced with the frustrations of the cumbersome machinery of State-controlled planning, and the elephantine procedures of local government, younger planners in particular have whole-heartedly accepted the principles of advocacy planning, and its extension into the techniques of the 'bureaucratic guerrilla',[65] much to the dismay of their senior officers. In this way they hope to recapture the social commitment of the early days of the planning movement, and reorientate planning away from its fixation on systematic technology, to community planning.

'Planning for People' is therefore the new cry; planning skills must be placed at the service of the population at large—'the ability to hire and fire their own professionals means power to the people'.[66] One of the first spokesmen for this new approach was Gans. He argued that the planners should adopt a client- or user-oriented approach in which the starting point would be an empirical one in which the planner establishes how people live, what they want, and which of their problems need to be solved, before moving on to develop plans that achieve their goals, and solve their problems.[67] The weight of his argument was towards a new definition of professionalism in which the planner becomes the servant of the people. At first sight this argument seems attractive, but there are problems. The approach is straightforwardly conservative in basing planning on existing demands. Life-styles and cultures are developed in terms of present incomes and opportunities, present conditions of existence, and 'wants' are stated in terms of present experience. The argument reaches to the point of assuming that 'wants' and 'needs' are the same thing. By implication, needs can be realized within the present system of distribution of power and wealth. In addition there is an assumption of professional omnicompetence: the planner can deliver the goods. Gans here is sociologically naïve.

It is but a short step from this definition of professionalism to advocacy planning. In this the people are taught how to play the 'resource game'; the planner acts as their advocate in the seats of power, gives them information, instructs them on the technical means to realize their aims, acts as an intermediary between the world of the people and the world of the public bureaucracies. It is intended that the poor, those most vulnerable to change, should have access to the same range of professional skills as the agents of change. It is a means of democratization of community spending, and a removal of the barriers between the poor and the professions. Like social planning, it derives from the American experience of planning for the underprivileged populations of the big cities, i.e. their ghettoes. But, on the basis of this experience, it is already subject to a wide range of criticism.

The basis of criticism is the assumption that the welfare of the poor can be assured without disturbing the existing distribution of power and wealth: the only problem is an administrative one of ensuring that welfare provision actually reaches those in need of this unearned income. To make certain of this, the poor are to be taught to speak for their requirements. Goodman terms it 'allowing the poor to administer their own state of dependency'.[68] The professionals are only involved because the poor are deemed to be incompetent in the labyrinthine hierarchies of public bureaucracies: in the terms of black America it has been as yet another aspect of colonization of the ghetto.[69]

Further criticism is directed to the incrementalism of approach in that the total situation confronting the ghetto residents is broken down into its components of housing, schooling, employment, transport and recreation, and not treated as a syndrome of prolonged community neglect. Community needs are taken one by one, and in so far as they can be met, effective resistance to neglect is averted. The placatory effect of the localization of community spending is freely admitted by advocates of community control: 'There is no reason to believe that neighbourhood power would tend to increase black radicalism. Indeed precisely the reverse hypothesis is more plausible.'[70]

Both in America and Britain, experiments in advocacy planning have shown the fallacy of attempting social change on the basis of environmental improvements in one locality. McGonaghy, leader of the Shelter project in Liverpool 8, concluded that it was 'an heroic attempt to treat local sores without administering any systemic medicine',[71] and Goodman, on the basis of similar involvement in America, concluded that the principal effect was competition between groups of the poor for the hand-outs made by government to underprivileged neighbourhoods. Environmental professionalism was of little avail, it had to be phased out, after the planners had created a 'situation in which people can act out their own environmental needs'.[72] Instead he argued for a reorientation of consciousness in which the environmental professions were to participate, not as experts, but as community members. Change is to be effected 'through the creation of a culture which will not tolerate the repressive and competitive values which capitalism has induced us to accept . . . planning and architecture present an important opportunity for strengthening this process'.[73]

Goodman here articulates a widely held sentiment, not restricted to the town-planning profession—change must follow on from the building of a new culture—which challenges the very basis by which society is maintained. The hegemony of ideas by which the state holds power can only be broken down through purposeful orientation to action at the level of consciousness. The environmental professions, it is argued, as groups directly involved in construction and alteration of the situations in which people live out their lives, the immediate context of experience, could be central to this re-educative process by which people are brought to awareness of alternative ways of life. And yet the counter-culture will have no room for professionalism: direction by experts is to be superseded in a society where the people are to reassume control over their existence.

The theme of change through consciousness is a persistent thread running through the writings of the planning movement. Howard and Unwin clearly believed in the power of ideas to change the world, but it was left to Geddes to spell out the message of planning. The evils of urban life could be overcome if only the ideals of man could be made daily currency through citizenship, and the compelling arguments of the new science of town planning would sweep away the 'smoke-cloud of Paleotechnic industry', and the squalid environment it had created. The aims of civilized living could be attained if only the citizen worked with rational purpose towards common ends. Writers like Goodman are therefore representative of a long tradition, and as planning recovers some of its youthful zest there will be further restatements of 'a new, more human philosophy'.[74]

There is, however, a sharp division between those, like Goodman, who would contend that it is impossible to plan for people within the present structure of power and authority, and those, more representative of the profession, who would claim that, with planning, the complex economy of a post-industrial society can provide for all. Planning of resources in land, houses or community facilities can overcome the inequalities that would exist in an uncontrolled situation, and intervention can overcome the limitations of the market economy. In this view social planning can ensure that needs are met, and that the living conditions of the poorest are brought up to the standard expected by the mass of the population. The failures of advocacy planning are seen as deriving from its novel use of professional authority outside the existing structures of government, rather than a final failure in socially responsible planning. It is considered that a professional group firmly embedded in the structures of government, as are the town planners in Britain, can act to change the consensus by which priorities for spending are established, and in so doing effect a further redistribution of resources. The planner as 'the master-allocator of the scarcest resources' must see that 'he plays a much larger role in contemporary society than that for which his education has fitted him'.[75] Planning for people can be effected without a revolution in capitalist society in the systematic direction of resources in land and public expenditure on the built environment and community services to the underprivileged groups in society. The problem becomes one of administration, and the political economy is unquestioned.

3 The sociologists' critique

There has been a long-standing close association of sociology and planning: in both there was the same revolt against individualism, reassertion of tradition, and distaste for the cities of industrial capitalism. Mass urbanization into either the squalor of the metropolis or the 'sprawling man-reefs' of the industrial conurbations presaged social degradation, disorganization and anomie. Social solidarity seemed in doubt as the consensual ties of small-scale agrarian society were weakened, and a cleavage effected between work and domestic life for the mass of the population. But whereas sociologists such as Durkheim sought a reintegration of society on the basis of the division of labour, for town planners and urban sociologists the idea of community was central to their thinking. From the early years of the twentieth century the two groups have shared a common interest in the reconstruction of urban society on scientific principles of community organization.

However, the association between the profession and the academic group has been neither straightforward nor consistent. Urban sociology moved from social commitment to scientific neutrality, and back to action research and a concern for social welfare, and planning similarly has shown alternations between Utopian vision and hard-headed pragmatism in its transition from a social movement to an activity of government. But throughout, town planners have tended to define a narrow role for the sociologists in the planning enterprise. The first town planners were their own sociologists, sure of the merit of their approach, and their competence in ascertaining the direction of 'unexperienced futures'. Only subsequently, as planning devolved to less confident and less ideologically committed practitioners, have the sociologists been found a place: that of providing advice on community development and consumer preferences as to the shape of their environment.

The sociologists, however, are sceptical of the planners' claims to a unique sphere of professional competence, and far from convinced that planning is any more than 'sympathetic magic and public ritual'.[1] Where their interests converge, as in community planning, the sociologists consider they have good grounds for criticism in that planners have worked with vague objectives, over-simple analyses of social structure, and a persistent fundamentalist belief in the power of environmental planning. Consider the criticisms of community planning made by sociologists over two decades: Glass accused the planners of underestimating the complexity of neighbourhood social structure, and in dividing Middlesbrough into well-defined neighbourhood units of further dividing a community already segregated on class lines;[2] Dennis saw neighbourhood planning as having a 'limited reality component', and as misinformed in its attempt to substitute locality social control—'the mass of mediocre minds'[3]—for other sanctions on individual behaviour; Rex, in the light of the problems of County Durham, accused them of planning for the individual consumer, mass-market man, a suburban ideal of isolated nuclear households, and not for man in a community setting. He concluded that[4]

> the meta-sociological assumptions on which the planners . . . seem to draw are so trivial that they appear to be nothing more than rationalisations of what they are forced to do by their masters, or what they simply do for reasons of expediency.

There has therefore been an active rejection of the role assigned for them by the planning profession, and an angry criticism of planning practice and the orthodoxies of planning thought. Reade, for instance, in 1969, considered that 'the exposure of inconsistencies in planning thought and an analysis of the nature of planning and the assumptions on which it rests, would be the most useful contribution that sociologists could make to planning',[5] an attitude not dissimilar to the Weberian position adopted by Rex, in which the sociologist would seek to clarify the aims of planning, outline the choices confronting any group of decision-makers in a conflict situation, and make clear the implications of adopted policies. This might be considered to be the new sociology for planners, one which accords with the newly adopted professional stance of management and consultation, pragmatism and incrementalism, which can be contrasted with that advocated by Dennis: 'an applied sociology from below of community action', and 'the sociology of planners',[6] as a professional group embedded in the hierarchies of local and national government.

The main attack against the planners in recent years by the sociologists has been based on certain premises as to the role and responsibility of the profession in the structures of government. They were considered to be a particularly influential group of 'social gatekeepers',[7] having increasing responsibility for the allocation of the resources of land, buildings and community services. The statutory duties of the planners in control over land use which offered the possibility of veto over developments not in accord with the interests of certain sections of the community, and their responsibilities in the forward planning of the 'community chest', the basic infrastructure of the community—sewerage, roads, open spaces, health centres and publicly financed housing—seemed to place the planners in a commanding position as the 'urban managers'. It therefore became important to establish the attitudes of the planners —were they committed to the *status quo*, social reformers, or Utopians or radicals? Did they disguise their position to themselves and others in a cloud of technocratic terms: did the ideology of the profession into which they had been socialized lead them to make dispositive judgments that favoured certain groups at the expense of others? The planners' own definitions of responsibility, their domain assumptions, became of crucial significance in the study of systems of access to scarce urban resources.[8]

The thesis of 'managerialism' was premised on a definite model of the large-scale society as the corporate state. Society is thought of as a single vast corporation with every person an involuntary member and employee. The activities of the state ramify into every aspect of the life of society. Government is only part of the state, but government co-ordinates it and sets the pace. It is essentially rule by administration, rational, hierarchical and inclusive, and increasingly power rests in the hands of the executives—experts, professionals, managers—whose decisions are bound only by the rules of bureaucratic behaviour and the self-imposed discipline of professional codes of conduct. In this analysis, one initiated by Weber and his contemporaries in their observation both of political parties and the modern state, renewed in discussions of the organization of the large corporations, and taking on special force with increasing involvement of state administration in economic and social planning, the politicians are ciphers, easily manipulated by the experts, the men on the job, and power slips from the hands of the capitalists, as entrepreneurs and investors, to the managers. In the absence of, or apparent detachment from, local politics of an economic élite, and the increasing complexity of many local government decisions on capital investment, the argument 'in the end decisions depend on the presentations of policy options by the planner'[9] had validity in daily experience.

There is, however, a different interpretation of power in the corporate State: substantial power does rest in the hands of the experts, but their exercise of professional judgment is constrained by the need to satisfy the profitability and growth conditions of the large corporations.[10] Planners, along with other professional groups, have a certain responsibility as executives of government, but the options are implicitly foreclosed by the requirements of industrial investment and the demands of the corporations. Evidence for this view can be found in both regional and local planning policies. The vast programmes of modernization in areas such as the North East, involving not only the dissection of the region by motorways, the 'replacement of its outworn urban fabric', wholesale reconstruction of its city centres, and also the elimination of many of its former mining villages, were considered essential by local politicians and planners alike to attract investment to the region, and to raise the standard of living of its population. Planners were co-opted into a strategy in which the regional fabric—environment, local landmarks, local cultures and community groupings—was torn up, and the expressed interests of the local population overruled in the interests of growth. And many of these policies accorded with established planning doctrine: accessibility was clearly a prerequisite of regional growth, just as concentration of the population into fewer centres was necessary to provision of adequate shopping and welfare services. Comprehensive redevelopment seemed the only strategy to rectify a half-century of neglect. Fundamental changes in the way of life of the people were publicly presented as planning policies of self-evident reason and common sense.[11]

In like manner, the difficulties posed by the situation in the inner cities, their deprivation, unemployment, housing shortages, and lack of community services, were presented for public debate in the language of the planners. Comprehensive redevelopment, rehabilitation, cellular renewal, the provision of new towns for old, have been presented as the policy options; the crisis is an 'urban' crisis, one which can be remedied by better administration, more planning, a rationality in aims, and an efficiency in their realization. But in this area, as in the field of regional planning, the planners did not take the initiative. The interest in community planning for the inner cities dates from the middle years of the 1960s, a decade after the deterioration of living conditions and new social tensions consequent on the relaxation of control over management of private housing, and the arrival of the 'New Commonwealth' immigrants in the twilight areas. The planners followed on from the definition of the situation by the political élite, in whose terms the inner cities were now a major problematic.[12] The planner is bound by the consensus as to the problems of society: 'In order for him to act is is not enough for a problem to exist; its existence must be recognized by those power groups for whom he works.'[13] In both local and regional planning it would seem that the power of the planners is more apparent than real, their image as an autonomous profession a product of their role in preparing ostensibly non-political strategies for furthering ruling interests.

On a day-to-day basis, planning takes place in the context of a triangular relationship of public-professional-politician, in which there is a constant interchange of ideas and pressures, including those of community action groups, as well as longer established associations, for example, Chambers of Trade, trade councils and amenity societies. On a broader analysis, planning decisions rest on a consensus, a delicately shifting balance of contending forces, of property interests, public opinion, expert criticism, and governmental commitment and ability to intervene. In this situation the planner in everyday consideration of development proposals has to become an empiricist, sensing the 'mood of the moment', judging the likely implications of a given policy for investment decisions, the attitude of the government to further public investment, and the response of the local community to interference with accustomed practice, its tolerance towards change in familiar environments, and its receptivity towards innovation. One of the major weaknesses of 'master planning' was precisely its insensitivity to the differing interests in the community, and its unresponsiveness to changes in the consensus as to the scope and aims of development. This had beneficial effects, in that adherence to long-term strategy could avert the worst extremes of property speculation, and despoliation of the environment, but equally policies could be bulldozed through precisely because they had been official policy for two decades or more.[14]

In this 'consensus' public opinion is a vital ingredient, as the planners of Piccadilly know only too well, hence the importance of the Press and media in all environmental campaigns. And yet this force of public opinion is only operative in some issues, in some places. As a rough rule of thumb, public opinion has to be taken into account, the more nearly the central area of a metropolitan city is affected, or the closer the neighbourhood is to the established haunts of the television cameras or the journalists. Piccadilly or Covent Garden are hot news, as, with skilful manipulation of the media, may be Notting Hill; the fate of a back-street district of Blackburn, Batley or Sunderland, or a remote glen in Scotland, or a distant hamlet in Wales, is not, without violence, or 'illegal' activity. The consensus is not representative of all groups in society, for the poor and the inarticulate, in communities remote from the centres of control, are unable to command the mass media, the politicians or the professional groups. But even where a consensus is achieved between the planners, local politicians and the public over the aims of environmental planning, this is disregarded when decisions on, for example, the location of nuclear power stations, oil refineries, or even construction plants for oil platforms, are taken.

In this interpretation of the planning process, the planner has a conciliator, as a broker between interested parties, as one who sets the consensual values by which some planning decisions are taken. But this is not a position in which policies are initiated: 'Thus the man in the chic office turns out to be a broker, a decider between limited alternatives, a mediator, and arbitrator, but not an originator. And such a position tends to be inconsistent with originality.'[15] Experience of structure planning to date would confirm this analysis in that there has been little innovation, and a marked reluctance to depart from known trends in development.[16] The managerial stance of the progressive planner now makes sense; it is out of the question for him to direct the urban system; if nothing else, too many of the originating impulses for change come from outside the community, even from abroad; all he can do is to ensure that the demands placed on the structure of buildings, communication channels and spaces that make up the urban system are compatible.

If planning has so limited an impact, the autonomy of the planners localized and in general restricted to relatively mundane issues such as extensions to existing dwellings, the design and layout of new housing estates, the maintenance of conservation areas and promotion of pedestrian areas, then why should the sociology of planning occupy so important a place in urban sociology? Rather than undertaking detailed investigation of the latent content of planning theory, methodology and belief (as presented in the previous section), should not the quest for understanding of the direction of change in the urban environment, and the criteria governing resource allocation, be directed to the dynamics of modern capitalism as a mode of production? The 'urban' focus to urban sociology has misled analysts of urbanization into an over-emphasis on the environmental professions, in particular that of planning.

In general few sociologists would now agree that the planner is 'the master allocator of scarce resources',[17] and the thesis of managerialism in urban sociology is being discarded both by those who were its original proponents, and those who have conducted research on the premise that 'urban society is a managed society'.[18] In the interval between the original suggestions by Gans in the American context, and their elaboration by Pahl, and now, the ground-rules of urban sociology have shifted so that 'the problem now is to defend "mere" managerialism against the challenge of the more radical and elegant Marxist perspective'.[19]

Confused in the debate are two questions: that of the role of 'social gatekeepers' in different localities in the allocation of resources to different groups, and that of the responsibilities of the planning profession as one among many such social gatekeepers. On the first score, research possibilities are still wide open—housing managers, architects, city engineers, public health officials, education officers, within local government, are unexplored professional groups, as are social security officers and officials in employment exchanges. In the private sector, work has scarcely begun on groups such as building society managers, estate agents and private landlords;[20] the construction industry itself, the finance interests in property development, have not been penetrated. Among these various interests, the planning profession is inconsequential and its tortuous and reiterated debates of little import for community planning or for resource allocation in the Welfare State. A stronger case might be made out for a focus on housing managers, who already have in their charge a third of the national housing stock, or architects, who have so immediate an impact on the 'daily round, the common task'. Why then should the planning movement and the planning profession be of so great concern to urban sociologists?

In Britain through this century there has been great faith in the planning process as a 'means to revitalize human, community, and civic values through a better ordering of spatial relationships'.[21] In this respect the teachings of the planning movement have been only too successful. The British experiment in providing a comfortable environment for the mass of the labour force depended on the combined use of powers over housing provision and land-use allocation. The planning profession, as heirs to this social gospel, as designers of new worlds, creators of an efficient, convenient and amenable urban environment, has had to act as if it had autonomy of decision-making, and was capable of arranging homes, health and happiness for all. The sociologists, and others, have taken the profession at its own estimation of authority. There has been a failure to consider planning as 'an activity of government mediated through a bureaucratic hierarchy'[22] with strictly limited implementative powers. Alone among the sectors of local government, town planning has no substantial capital budget, and its powers of control over private investment are restricted to those of cajolement, bargaining and, subject to approval by central government, veto over development proposals. The principles of control over development set out by the Uthwatt Commission, which depended on considerations of neighbourliness, community betterment and public interest,[23] have foundered in the administration of plans concerned solely with the rearrangement of physical artefacts on the ground. The calls for a positively discriminatory approach in the allocation of urban resources have to go unanswered by the planners, for there are only limited possibilities for their shifting this previously determined allocation in favour of one group rather than another, without control of the legal rights in land. Where land is vested in a state agency, as for instance the National Coal Board, or the Central Electricity Generating Board, the local planners have rights of consultation only; where the development initiative comes from the local authority in, for instance, school building programmes, highway projects or housing policies, the planner has no rights of veto or implementation, only of consultation and advice.

The planners have therefore become the scapegoats for the seemingly irreversible decline in the 'quality of life'. In fact few of the pressures for change in the environment originate from them: the outline decisions on investment are either taken by government in the light of national economic and social priorities, or they are taken by individuals, institutions and corporations acting in their own estimation of benefit. This the planners have realized: 'the willingness to listen will be more important than the ability to write a plan. . . . The propensity to adapt an existing plan to changing circumstances will be more in demand than the obstinacy required to carry through an approved project',[24] but not the sociologists. Sociology has resorted to criticism without empirical analysis of the planning process, and endorsed the popular image of the planners as those responsible for the increased pace of development in conditions of economic growth, and inflation. If planning had been examined as a day-to-day activity, then experience of its routine would have demonstrated its inertia, total lack of imagination or innovation, and, for much of the time, sheer triviality. Or, alternatively, if the planners' own move from doctrinalism to pragmatism as the scope for initiative and discretion narrowed had been explored, there would have been greater awareness of the great gulf that exists between the rank and file of the profession and its vanguard, and more importantly, the tensions in a profession with visionary ambitions, caught in the machinery of the state.

In nearly all areas of operation the planners can be discounted as agents of change, and the sociologists' critique is misdirected. Detailed attention to the planning process is worthwhile only if it illuminates the interests that underlie changes in policy, and locates the technical language of planning in the formation of the consensus by which the public is brought to accept change. The public may be brought to 'an active and voluntary free consent'[25] in far-reaching changes in their daily lives, mystified by the language of professionalism and bemused by the ritual of planning machinery. The expensive, time-consuming, elaborate process of development control in Britain, locally detailed, centrally monitored and regulated, and now subject to sophisticated techniques of research, consultation and public inquiry, neutralizes articulate discontent by which the consensual might be shattered.

There are, however, circumstances in which professional groups can exercise a relative autonomy over the environment, that is, in conditions where business and financial interests are not directly affected by development, and where the public is unable to articulate its reactions and render its interests effective. The run-down of remote villages, deprived of services, has not caught public attention nor sociological criticism, but the policies of urban renewal in the inner cities have. Unlike the privately developed suburban housing estates which have to be laid out on criteria of marketability, public housing, whose occupants are there on a least choice basis, can be planned on principles of traffic segregation and closed circuit neighbourhoods, and the accommodation designed with closer reference to the tenets of the Modern Movement than the tenants' own perceived needs. In areas where the residents are poor, without prestige and no community organization, the consensus as to what is to be done is established by the professionals in association with the politicians. Professional authority is exercised without check or restraint, and becomes yet another element in the subordination of the poor and inarticulate. It is in areas such as these that the sociologists have called for the reorientation of planning, a surrender of authority. Anything less is taken as evidence of their total commitment to their status as associate advisers to government, and the final failure of socially responsible planning.

The sociologists' criticisms must go unanswered unless greater attention is given to the principles of local planning. The redefinition of planning as the practice of planning cannot offset present injustices in the allocation of urban resources, nor can it provide a rationale for development where the market presents no restraints. The pragmatic tendencies to planning which have been so heavily emphasized in the last decade have given it a formless relativism. Theories as to urban design are no more than hypotheses which help the planner to order his data, they are of no assistance in the organization of a new world. In James's terms, 'Ideas only become true in so far as they help us to get into satisfactory relationship with other parts of our experience';[26] ethical judgments are limited in scope to the immediate problem at hand. This means that there are no solutions, there is no ending to the process of planning; ideas, theories, policies are always provisional, and to be modified as circumstances require. In this pragmatic view, 'science' becomes the practice of science, and planning becomes the practice of planning. In the circumstances of the inner city such an attitude is more short-sighted even than that of the petty entrepreneurs who erected the buildings now replaced in comprehensive redevelopment. Given the lack of interest by planners in the very practical questions of layout and design of entire neighbourhoods responsibility goes by default to other professions, principally the architects. The high-sounding debates on the needs of the community and the public interest are emptied of meaning in their failure to provide guidelines for action in determinate situations.

The mutual awareness of planners and urban sociologists is indicative of their traditional concerns and common philosophical origins. The theories and methodology of the Chicago School of urban sociology were embedded in the pragmatism formulated by James and Dewey; their biological language of environment and organism, radical empiricism as a method of inquiry, and their pragmatic use of theory, in which 'theories become instruments, not answers to enigmas',[27] were all part of the world view put forward by social liberalism. Planning, too, derived its impetus from the same school of thought: the methodology advanced by Geddes had similarities with that of James, in that it depended on the exposure of planner as citizen and scientist to the totality of urban experience: ideas for the urban future could only be formulated in interaction with this 'flux of life'. The civic survey was a means to this end, and could do no more than 'prepare for and point towards the plan'.[28] Geddes harnessed the empirical methods of biology to a subjectivistic theory of knowledge.

Both planning and urban sociology have refused to allow of an intractable universe, unamenable to manipulation by man. Pragmatism conceived of man as experiencing indeterminate situations, what James terms the 'immediate flux of life', to which thought or inquiry gave shape. In this sense mind gave matter form; reality was a product of the human will; the universe did not exist save in beliefs about that universe. It is a philosophy admirably suited to the optimism of a rapidly developing society such as America in 1900, or the societies of post-war capitalism. To plan there must be a belief in man's power of control over his environment, to manipulate it according to his needs, and to shape the future. Planning refuses to admit the possibility of a logic to social experience which cannot be altered by the exercise of collective consciousness. The same attitude extends to the natural environment: over this, too, there is power of control, and its resources can be directed to the benefit of the community. The attention given by the urban sociologists to the planners is a measure of their agreement with this philosophy. The strong resentment by planners of sociologists, and the untempered criticisms by the latter of a profession embroiled in the development process, is a measure of the sensitivity of two groups who at heart share the same view of the world. The sociologists' critique, the internal disputes, dissensions and heart searchings of the planning profession itself, indicate the barrenness of liberal philosophy. The ideology which attempted to side-step ideology, to tread the middle ground of realism and rationality, has lost its credibility as British society moves out of the euphoria of post-war growth and stability, into an era in which class ideologies are reasserted. The question is now asked—why plan at all? To this neither group has an answer. The obvious response—the care and conservation of the environment—is illegitimate given the long-standing aim of 'the reconstruction of society'.

part two

Theories of urbanization

The theoretical offerings of urban sociology constitute a small part of the literature, and their pertinence to the study of urbanization is not easily seen. The 'classical' theories of urban sociology, in which are included the statements of the European sociologists Tönnies, Simmel and Weber (and by implication Durkheim), and those of the Americans, Park and Wirth, seem to have little more to offer now than the oracular conspectus of Spengler, the heartfelt concern of Masterman and his liberal contemporaries, or the dramatic vision of Zangwill in 'The Melting Pot'. The latter were contemporaries of the sociologists, and like them voiced the concerns of small groups in American and European society as the movement to the cities gathered pace in the last quarter of the nineteenth century.

It is with some sense of exasperation, therefore, that one speaks with colleagues interested in more specialized areas of urban research, and also with students in sociology. These theoretical statements cannot be superimposed on to research and yet students must be persuaded of the need to tackle demanding literature which seems to offer little return. The task is more difficult in the absence of concise appreciations of the sociological writings on urbanization. Presentations of the work of the Chicago School, for instance, abound, but all too often they formalize or codify their statements, so blanketing the contradictions and ambiguities for which they remain of interest. The statements are taken at face value, the question not asked of their questions; if the theories are criticized, their authors are impugned as at best old-fashioned if not intellectually incompetent. The problems to which they addressed their attention are ignored. The intention in writing this section was therefore straightforward: to present the principal theoretical statements of urban sociology in easily accessible form without belying their inconsistencies or inadequacies, so that attention was focused on the contemporary issues and problems, the structures of thought, within whose terms these men wrote.

Particular attention is given to the Chicago School for two reasons. First, although their exploration of the city in terms of ecology is now disregarded, their interests and concerns remain significant. They set the terms for a tradition of urban studies which remain of importance in geography and planning, as well as urban sociology. In this respect the break in composition of this book is misleading. Although the cleavage between the British experience of urbanization and the theories derived, and taught, by sociologists is most marked, evident over the past decade is a reworking of the intellectual traditions they endorsed, that is, positivism and liberalism.

Second, their work has bearing for any appraisal of academic sociology. It is easy to dismiss them as intellectual charlatans, their sociology peculiar to the mid-West intellectual frontier, but their failure to engage the issues presented in the large city of the modernizing world, their retreat into pseudo-scientific abstraction or ingenious empiricism, is indicative of the weaknesses of the sociology they transmitted from Europe into the American colleges. And it is this sociology which has been grafted onto British interests since the war. In their study of Chicago they expressed the domain assumptions of academic sociology in whose terms they are now criticized. To understand their theoretical dilemmas is to appreciate more fully the weaknesses of that sociology.

5 The critique of metropolis

1 Urbanism and Marxism

> the towns by their influence and importance achieve in the nation predominance over the rural organization. In consequence, country and village must use more of their own productive forces for the support and furtherance of the urban areas than they can spare for the purposes of reproduction. Therefore the rural organization is doomed to dissolution . . . urban life seems to exert by itself, producing and consuming, tending to dominate the whole country more and more, to draw from it its forces, and to lead to their destruction.[1]

First experience of a large city, or confrontation with the spectacle of mass urbanization, leaves few unaroused observers. For those resident in the urbanized societies of the 'developed' world, it is hard to recapture the incredulity, the incomprehension, the wondering amazement of the nineteenth-century observers of the great transformation of Western Europe. Only in confrontation with the new urbanization of the satellite countries in the underdeveloped world can any vestige of these reactions be recaptured. Urbanization has become a fact of our everyday lives; cities an environment to which we have become blasé. In Simmel's terms, only in a refusal to react to the life of the great city can we accommodate to the content and forms of metropolitan life. The intensity of reaction with which men like Tönnies or, much later, Wirth, observed cities and urbanization now escapes us.

This intensity of response is reflected in the sociologists' discussions of cities, which constitute savage critiques of the common experience of urbanization. The great city, metropolis, is the paradigm of an inhuman, debasing, social environment. For Tönnies as for Simmel, the money economy of the cities destroyed social life:

for Weber as for Wirth, mass urbanization nullified opportunities for political participation. Theirs was a sweeping condemnation of capitalist urbanization. The classical writings of urban sociology are more closely comparable with those of the poets, the novelists, the social commentators of the nineteenth century—a wide spectrum ranging from Wordsworth to Ruskin and Morris—than the clinical observations of their successors in the social sciences. It was left to men like Booth and Rowntree to commence the sociography of life in the cities. The rigid tenets of positivism were to provide the documentation of how men actually lived in cities, which ultimately was to invalidate the sweeping statements of the writers in the sociological tradition. But the restricted conceptual framework of positivism could neither encompass the city as a form of living, nor commence the analysis of the role of cities in the transformation of Europe from an agrarian household economy to an industrial exchange economy. This the classical sociologists attempted; with the economic historians they were the first writers to undertake the systematic analysis of urbanization as a force for change.

In their analysis they differed sharply from both the Marxists and liberal progressive thought. Marx and Engels had condemned the consequences of urbanization under capitalism, as seen in Manchester and elsewhere, but viewed the concentration and immiseration of the mass of workers in the new urban agglomerations as a necessary stage in the creation of a revolutionary force.[2] Pauperization and material degradation was one aspect of urbanization, but equally important was the destruction of the social nexus of the traditional community and its replacement by the utilitarian world of the city. Marx had described little communities as having 'subjugated man to external circumstances instead of elevating man into the sovereign of circumstances, they transformed a self-developing social state into never changing natural destiny'.[3] Man, as a rational being, was a product of the towns not of the countryside, the workers had to be freed from the confines of rural life, the conditions of rural idiocy. Bourgeois civilization, bourgeois philosophy, nurtured in the towns, was the necessary basis for the new 'higher relations of production'. Both for theory and for practice, therefore, communism depended on urbanism. Urbanization, at least ultimately, was liberative. There was nothing to salvage from the pre-urban past, for the social bonds of community were to be replaced first by the new groupings of class, and then by the 'development of human potentiality for its own sake and the true realm of freedom'.[4]

In this conspectus of urbanization there were certain similarities with the view of the new professional and business classes with which by birth Marx and Engels had the closest affiliations. There was tremendous contemporary pride in the new cities which were seen as engines of progress and the source of all fruitful change. Urbanization was progressive, providing the conditions not only for the transformation of mankind's productive capacities, but also for the realization of individuality. In Park's phrase,[5]

> The individual man finds in the diversity of interests and tasks, and in the vast unconscious cooperation of city life the opportunity to choose his own vocation and develop his own peculiar individual talents. The city offers a market for the special talents of individual men.

From the Marxists to the liberals, however, the analysis shifted from the alienation of labour in a productive system based on private property, to the realization of individuality through the sale of labour in a market economy. 'Freedom' for the Marxists could only exist beyond the city, for the liberals this was what the economic specialization of city life provided. Freedom, diversity and opportunity of choice were eventually to become the catch phrases of progressive urban planning.

Something of the same appreciation of urbanization is expressed by Mumford. He sees cities as enlarging all dimensions of life as the scattered activities of society are brought together, so releasing the energies of mankind in a tremendous explosion of creativity. The city has augmented capabilities for participation, and widened the basis of personal experience. He sums up his statement of the role of the city in history in terms of 'that magnification of all the dimensions of life, through emotional communication, technological mastery and above all, dramatic representation'.[6] But although he extols the immensity, the breadth of human experience, that the metropolis has to offer, it is the cities of classical Greece, and those of medieval Europe, that govern his assessment of the city as a social container. 'Freedom, corporate equality, democratic participation, autonomy, were never fully achieved in any medieval town, but . . . for a brief while "communitas" triumphed over "dominium".'[7] As in the case of the classical sociologists an image of the medieval guild city, the quasi-independent political community, governs discussion.

With the possible exception of Simmel, whose assessment of metropolitan culture owes nothing to medievalism, the European sociologists were strongly influenced by an image of the pre-industrial world of small cities embedded in a matrix of village and region. The medieval city had for Durkheim provided a basis for conditions of organic solidarity, for Weber, an existence in which public and private life fused, and for Tönnies, the highest form of community organization. Their critique of the modern metropolis was derived from a sense of a happier, more truly social past in these small towns. As Nisbet comments, 'It was medieval society that came more and more to provide a comparative offset to modernism for critics of the latter.'[8] Mass urbanization was neither progressive nor liberative, but signified a degeneration of social existence. From their work stems the perennial theme of the loss, the eclipse, of community, and the arrival of the mass society in which political life, culture and personality are in decay.

In this pessimistic assessment of the potentialities of urbanization the sociological tradition converges with the viewpoint of the neo-Marxists. In the writings of, for instance, Fromm, Marcuse, Mills and Fanon,[9] there is a consensus that conditions of capitalist urbanization are mutilative of the personality, inhibitive of community formation, destructive of social engagement or involvement, and conducive to apathy, alienation and anomie. Class consciousness is inhibited, and diverted in mass movements; unreason, not reason, typifies social response. The assumptions of orthodox Marxism are overturned, and it is to the groups least assimilated into urban society, students in the developed world, peasants and urban *lumpenproletariat* in the underdeveloped world, that the revolutionary writers look. Urbanization is no longer the *sine qua non* of a socialist transformation of society.

If urbanization is progressively bankrupt in human terms, cities themselves are seen as instruments of capitalist or imperialist domination. In the statements of Frank,[10] it is the cities, in their capacity as home of national, regional or even international bourgeoisies, that form the linkages along the chain of expropriation from satellite territory to dominant metropolis. The buildings of the cities are monuments to the labours of the rural, provincial and colonial masses, and not representative of the accumulated surplus of the honest toil of the citizens. The growth of cities depends on the effectiveness of their expropriation of the wealth of their satellites, i.e. their political control. Typically, the characteristic feature of the cities is seen not as their economic specialization, but their role as centres of dominance. They represent 'la sphere de domination sociale d'une classe spécifique, la bourgeoisie'.[11]

But cities are not only the controlling centres for their societies. They are the source of innovation and change, the source of new ideas for production, the leaders of taste, fashion, the pace-setters for consumption; they are also the guardians of culture, and the conservers of order in society. Consensus and continuity in a society are maintained from the city centres. Not only are economic enterprise and political power concentrated there, but so is social authority. To the concepts of expropriation and domination must be added that of hegemony.

The concept of hegemony was central to Gramsci's analysis of power and prestige. He specifically drew a distinction between leadership, or direction, and dictatorship, the former depending on the maintenance of a consensus, the latter on coercive force alone. Hegemony signified authority, the right to rule, intellectual or moral prestige, consensus. It has come to mean a social authority whose ultimate sanction is a profound cultural supremacy. In the context of Italy this authority had been a regional as well as a class prerogative. The north, focused on Milan and Turin, he termed the 'directive centre';[12] Piedmont, the leading state of the north, he argued, had had the function of the ruling class in the struggle for national unity; he considered that the entire region had exercised hegemony over the rest of the country and was in this respect comparable to a great city exercising control over its rural hinterland. There was a pervasive exercise of the authority derived from cultural and social prestige, from, and on behalf of, the metropolitan bourgeoisie. This hegemony is as important as dominance in understanding the control of the central city. The entire society is drawn into the way of life transmitted from the cities so that urbanism, the culture of the cities' ruling élites, is diffused throughout. Urban culture becomes the legitimation for control.

Superficially this general line of analysis comes very close to that of sociologists in the classical tradition. From Tönnies, Simmel and Wirth there is the same assessment of personality, community and culture in the urban centres; Tönnies and Weber, too, interpret cities as centres of dominance, creaming off the labour, capital and resources of the countryside, leaving the latter barren and underdeveloped. The longstanding antagonism between town and country had become an asymmetrical relationship in which the countryside was relegated to economic stagnation, poverty and decay of its culture through the loss and pauperization of its population. Equally, both these latter writers explicitly underscore the class basis of 'urban' culture. Tönnies counterposes the merchant of the educated classes

> without home, a traveller, a connoisseur of foreign customs and arts without love or piety for those of any one country, a linguist speaking several languages, flippant and double-tongued, adroit, adaptable, and one who always keeps his eye on the end or purpose he plans to attain

with the common folk belonging to 'a real people's *Gemeinschaft* of their own'. *Gesellschaft*, the urban culture, the culture of the educated classes, is solvent of the cultures of peasants and commoners:[13]

> To the extent that the common people, with its labour, is subject to trade or capitalism it discontinues being a people. . . . Science, which in reality distinguishes the educated classes,

> is offered to them in many forms and shapes as a medicine for their rudeness.

Weber's analysis of the Western city is equally specific. Both capitalism and the modern state had their origins in 'the medieval democracy of professional traders',[14] the politics, the legal forms, even the religion, of the new association of the urban community were oriented to the needs of this new class. The dissolution of the feudal order that was to result from the impact of the market economy and the new forms of social association, and the growing subservience of country to town, was to the advantage of this one group, the urban bourgeoisie.

These sociological writings have of course been fed into the mainstream of Marxist thought, and constitute one reason for the movement of the latter away from the orthodoxies of nineteenth-century Marxism, but it would be inaccurate to subsume the sociological critique of metropolis under a Marxist label. Not only were the sociologists, in different degrees, hostile to Marxist theories of history, viewing their own writings as modifications or refutations of popular Marxist doctrines, but the 'domain assumptions' with which they approached sociological analysis shifted their work decisively out of the Marxist tradition.

Above all else the differences in methodology are fundamental: the contrast is between the dialectics of historical materialism and the more cautious short-sighted empiricism of the sociologists narrowing down their overview of society to that which experience could encompass. Tönnies is explicit here:[15]

> In contradistinction to all historical theory deducing its findings from the past, we take as our actual, even necessary starting point that moment in history when the present spectator enjoys the inestimable advantage of observing the occurring events in the light of his own experience.

This changed view of the 'real' in society, in which the individual, the individual act, and the individual will, become the touchstone of sociological explanation, represents a rejection of the metaphysical terms in which the Marxists analysed social change, and their substitution by categories more readily reducible to empirical observation. So, for instance, instead of a definition of capitalism in objective terms, in terms of its 'relations of production', relations that are both indispensable and independent of men's will, there is introduced a definition in terms of the orientation of the individual towards his own and others' social action, that is by reference to the rationalistic conduct of life in general. The abstract categories to which Marx resolved capitalism—class, social formation, forces of production, alienation—are dismissed as metaphysical speculations. The ideal-typical concepts introduced by Tönnies and Weber are only means to give experience a more coherent and concise form: *gemeinschaft* and *gesellschaft*, the urban community, are no more than useful fictions.

The boundaries of 'experience' could be wide, as they were with Weber, able to draw on a since unmatched historical scholarship, but nevertheless evident was a tendency to limit sociological discussion to the immediate context of experience. Already apparent in Tönnies, it was to culminate in the extreme ethnocentricism and particularism of the Chicago School. This restriction of outlook is one of the most apparent differences between the sociologists and the neo-Marxists: whereas, for instance, Tönnies ascribes uneven development to the antagonism between town and country, later writers relate it to the terms of trade and exchange between national economies as well as regions, in which certain urban populations are the intermediaries. It is on the methodological postulate of nominalism that the 'urban' becomes the realm of observation and inquiry, and the answer to problems of development is couched in terms of urbanization.

Also apparent is a certain fatalism, an acceptance of the blind inevitability of the process of urbanization—'Rural life is doomed to dissolution', and also, a 'reification' of the city. The town, the city, the metropolis, for Tönnies, Weber and Simmel respectively, come to be viewed as entities independent of the populations they housed; their forms of economic organization, political association and administration gave them a corporate identity to which ultimately individuals were subordinate, 'the metropolis is the genuine arena of this culture which outgrows all personal life'.[16] The depersonalizing environment with which individuals are confronted in the modern city is taken as a datum line, not as something itself needing explanation. In a measure this attitude to urbanization must be related to the adoption of empiricist canons regarding scientific knowledge. Empiricism lends an inevitability to that which is observed in that it limits explanation to the circumstances of observation. Sociological analysis is therefore brought down to the study of 'the real life of society as we encounter it in our experience'[17] and the way is open for the study of cities in themselves and urbanization as a process independent of capitalism as a productive system.

As important as the contrasts in methodology is the opposition by the sociologists to the analysis of history and society in economic terms. The rediscovery of concepts such as 'community', 'association', 'status', 'authority', 'legitimacy', 'culture', marked a shift in focus away from the struggle with the material environment and the resultant relations of production, to the social bonds and forms of association as established between individuals. The resultant differences are very clearly seen in the discussions of community and alienation. To Marx, community could never exist under conditions of alienated labour. 'Species life, productive life, life creating life, turns into a mere means of sustaining the worker's individual existence, and man is alienated from his fellow men.'[18] All previous community formations had been for the dominated class 'not only a completely illusory community but also a new shackle. In a genuine community, individuals gain their freedom in and through association.'[19] Fragmentation of social being would be a condition of existence in all societies until capitalism had been superseded. Compare this view with the discussions of the sociologists: community had not only been the motif of existence in agrarian Europe, but was an integral aspect of all social life; the attenuation of such community relationships did not derive from the new conditions of production and productive activity but from the use of money as a medium of exchange, and the assertion of individuality in the metropolitan environment; alienation, as in the discussion of Simmel, is transposed from conditions of production to conditions of community life.

It is easiest to assess the contrast between the Marxists and the sociologists by comparing the writings of Tönnies or Simmel with those of Marx. The former moves away from Marx primarily in his breakdown of social existence to the forms of will from which action is derived. All social life is to be resolved into combinations of the two types of social association derived from natural and rational will, *gemeinschaft* and *gesellschaft*. This distinction forms the underpinning to the division of society between urban, bourgeois, exchange society, and rural, folk society based on a household economy. The major cleavage in society is not merely that between classes, but between two radically different forms of community existence. The money economy, trade, urbanization, in their substitution of rational for natural will, were effecting a transformation of society that no socialist revolution could ever match. Tönnies, by adopting a specifically social and individual frame of analysis, was able to argue that the major break with the past had already happened, so shifting attention from class formation and the need for a further transformation of society to the comparison between traditional and modern society.

With Simmel, 'the microscopist of society', we move yet further from the analysis of Marx. Discussion narrows to the immediate contexts of social association—the size of social group, the range of social contacts, even to the fragmentation of personal experience in the metropolis: 'the rapid crowding of changing images, the sharp discontinuity in the grasp of a single glance, and the unexpectedness of onrushing impressions'.[20] Simmel's sociology is immensely personal, assumes a unity to society, is totally uninterested in questions of class, and remote from historical concerns. The discussion of urbanization in 'Metropolis and mental life' accordingly has as its main concern the question of the achievement of individual independence, and the elaboration of individuality in the unparalleled social environment of the city. Simmel makes no distinction between the experience of social classes, as does Tönnies, and the national, even international, perspective shared by both Tönnies and Marx is dropped in favour of one restricted to the metropolitan environment itself. While Tönnies sets in train the redirection of attention from class to community, from industrialization to urbanization, it is Simmel who lays the foundation for the American ecological school which finally removed any discussion of economic conditions from urban sociology, and substituted ecology for the market economy accepted by both Tönnies and Simmel as the starting point for their discussions of urbanization.

The apparent similarities between writings in the two traditions are misleading. The sociologists, in a long succession from Tönnies to Wirth, were developing a counter-theory to Marxism for the explication of social change which led to acceptance of a fundamental cleavage between urban and rural, modernism and tradition, which was in stark opposition to any variant on Marxist theories of development. The urban–rural dichotomy which was to bedevil the study of development as it has urban sociology, derives from the substitution of a critique of metropolis for the analysis of capitalism. The 'urban' is accepted as a frame of reference, and 'urban society' as a specific mode of social organization becomes the object of scientific study. So is initiated the ideology of urbanism:[21]

> cette idéologie spécifique qui saisit les modes et les formes d'organisation sociale en tant que caractéristiques d'une phase de l'évolution de la société, étroitement liée aux conditions techniconaturelles de l'existence humane et, finalement, a son cadre de vie.

Such a reorientation of thought did not occur overnight: it took the half century from 1887 to 1938 to throw off the lingering after-effects of the ideological revolution occasioned by Marxism, and to explore the full implications of a sociological vocabulary for the study of urbanization. It is Wirth's essay, 'Urbanism as a way of life', that is now taken as the clearest statement of urban culture as a necessary stage in the evolution of society, but its theoretical legitimation can be traced through the mainstream of European and American sociology.

2 Tönnies: Community and Society (1887)

This compact, dense, intricate book constitutes the first of the classic sociological texts. In it Tönnies foreshadowed the later concerns of his contemporaries, Durkheim, Simmel and Weber: the impact of the market economy on traditional forms of social association, the implications of urbanization and the development of the state for the conduct of social life, the mechanisms of social solidarity in an individuating society. The distinction he draws between the two forms of human association, *gemeinschaft* and *gesellschaft*, community and society, has become the basis for a succession of typologies of which the best known are the pattern variables, formulated by Parsons,[1] and the folk-urban typology drawn up by Redfield and Wirth.[2]

Although Tönnies's interest throughout a long career was in building up an analytical sociology, in this early work scientific intentions are obscured in the vivid commentary on contemporary social change. The formal dichotomy which was later to interest the academic sociologists, such as Parsons, derived from a wide-ranging, empassioned critique of bourgeois capitalist society. Tönnies was 'with deep intuition, with profound imaginative grasp, reacting to the world around him, even as does the artist ... objectifying internal and only partly conscious states of mind'.[3] The book, despite his later disclaimers, must be read as a critique of capitalist society; the very language with which he describes the two types of social grouping debars any other appraisal of the direction taken by society. *Gemeinschaft* is 'the lasting and genuine form of living together';[4] *gesellschaft*, in contrast, is to be understood as: 'a multitude of natural and artificial individuals the wills and the spheres of whom are in many relations with and to one another, and remain nevertheless independent of one another and devoid of all mutual familiar relationships', and 'In gesellschaft, every person strives for that which is to his own advantage and he affirms the actions of others only in so far as and as long as they can further his interest'.[5]

The description is damning. The fundamental characteristic of *gesellschaft*, an exchange economy typified by the search for power through the accumulation of profits, is a negation of social humanity. Rational calculation of personal interest culminates in the sale of labour as any other commodity, its exploitation for the greatest profits, and an end to 'all creative, formative, and contributive activity of man'.[6]

However, for Tönnies it is neither the factory nor the large-scale bureaucracy, but the city or its higher form, the metropolis, that is the paradigm of bourgeois society. The metropolis represents the culmination of *gesellschaft* in community life as does the state in political life. In fact Tönnies presents a typology of community life related to the two forms of social will: the house, the village and the town representing *gemeinschaft* community formations derived from blood (the family), place (the neighbourhood) and mind (religion and friendship), whereas the city, the capital city and the metropolis are based on relations of the *gesellschaft*. The metropolis, the synthesis of city and capital city, of commerce, trade and political control, is described as follows:[7]

> In the metropolis money and capital are unlimited and almighty. It is able to produce and supply goods and science for the entire earth as well as laws and public opinion for all nations. It represents the world market and world traffic; in it world industries are concentrated. Its newspapers are world papers, its people come from all corners of the earth.

Not only does his distinction between the two types of social will form the basis of a community typology, it is also a conceptual framework for the analysis of social change. The historical development of European society is categorized as a movement from unions of *gemeinschaft*, such as the family or the village, to associations of *gemeinschaft*, such as churches, guilds, to associations of *gesellschaft*, which finally develop into unions of *gesellschaft*. The last phase represents the efforts of the state, the large corporations, and the trade unions to recover some of the communal securities of traditional society. Tönnies's formal distinction between the two kinds of society are ideal types with which he demonstrates the possibility of the scientific study of history; they are 'nails on which the facts of experience could be hung'.[8]

What Tönnies is describing is the collapse of traditional forms of community life under the impact of a fully developed exchange economy. The contrast he draws is between the household economy in which each region was largely self-sufficient and in which production was directed to the needs of the local people, and the exchange economy where production is for the widest market that can be reached. Production for sale, and remuneration of labour in money, strips labour of 'style, dignity and charm',[9] and the competition for markets leads to a growing impersonality and instrumentality of social life. The common culture disintegrates and the common people are drawn into *gesellschaft* as members of the proletariat; membership of trade unions and political parties signifies their full incorporation into the capitalist state. There is no doubt that in Tönnies's eyes the transition from agrarian, parochial Europe to commercial cosmopolitan society had been a bigger break in human experience, than any that a socialist revolution could effect. The consequences of *gesellschaft* forms of association could be mitigated, but they could not be diverted: 'the perfect unity of human wills'[10] found in the rural community could not be recovered. His fatalism is inescapable.[11]

The account of urbanization found in the treatise forms the basis of the conventional sociological stereotype later popularized by Wirth. The village and the town, in which 'the physical real soil, the permanent location, the visible land, create the strongest ties and relations', are compared with the city and the metropolis where the continuity and intimacy provided by neighbourhood is broken. The city represents 'the exaggeration of the principle of space'.[12] Localism is replaced by a restless cosmopolitanism; urban, national and even international interests replace those of home, village and town. At the same time there is a decline in family life and its substitution by 'special interest groups and conventional society', although here Tönnies draws a distinction between the 'common people' and the 'educated classes'. For the former, 'family life, along with neighbouring and friendship . . . is life in and for itself', whereas for the latter 'the family becomes an accidental form for the satisfaction of natural needs'.[13] Relationships of *gesellschaft* are attributes of class rather than those of urban living.

But it is the bourgeoisie who exercise hegemony in the cities: 'only the upper strata, the rich and the cultured are really active and alive. They set up the standards to which the lower strata have to conform',[14] and it is evident from the subsequent discussion that Tönnies envisages the extension of individual and family isolation to the mass of the population. A precondition of the rise of the 'proletariat' is the common people's divestment of *gemeinschaft* forms of association. 'City life and *gesellschaft* doom the common people to death and decay.'[15] Urbanization, as defined here, inexorably leads in the direction of individuation, towards a situation in which each person is his own master, free to move where he will, associate with whom he pleases, for whatever ends he cares to name. Relationships of rational will predominate in that action is undertaken for predicted ends, and every 'other' is dealt with on calculated and superficial terms. The 'self' of *gemeinschaft* is replaced by the 'person' of *gesellschaft*, and in the latter 'there exist human beings who conceive themselves as such, accept and play this "role", each one assuming the "character" of a person like a mask before his face'.[16] Instead of the organic unity of *gemeinschaft* is found an aggregate where independent individuals self-consciously maintain an artificial identity.

This disassociation of community life in the aggregate of population necessitates new mechanisms of social control and integration. Understanding, or consensus, had expressed the reality of life in the *gemeinschaft*. This was implicit in living, dwelling and working together, and by its very nature could neither be expressed nor comprehended. This tacit understanding of local groups of individuals culminated in the common culture of the people (*Volk*). In contrast, in the cities, the law courts and the police, 'the will of the state', formal legislation, convention and public opinion become the means for social control and integration. Convention is counterposed with consensus, but public opinion, scientific and enlightened opinion, prepared and offered by the newspapers, had no equivalent in the older form of social association. It acts in two ways—first, in the formulation of a new moral code, and, second, as part of the political process. Tönnies's presentation of public opinion has certain similarities with Durkheim's treatment of social facts in that it is considered to confront the individual with an extraneous power, but it is not society by which the individual is constrained, but class domination. 'Public opinion', writes Tönnies, 'passes easily from the demand for freedom (for the upper classes) to that of despotism (against the lower classes).'[17]

Tönnies is inconsistent in the use of 'class' in his analysis and seems to waver between the use of the formal typology to denote enduring social forms and its use to characterize specific historical class cultures. In fact the language of class is never used in reference to *gemeinschaft*, where, before trade for profit broke up the situation of reciprocal dependence, a situation of perfect unity is assumed. For the *gesellschaft*, at various points he denotes the 'trading class', the 'educated classes' and the 'upper classes' as the bearers of *gesellschaft* culture, and yet at other points in the discussion he writes of the change in community life in universalistic terms. In this respect he stands transitional between the economic historians and the new analytical sociology which sought to isolate the forms of social behaviour, and it is this intermediate standing that gives him interest now.

There is no doubt that aspects of his discussion would find favour with writers reworking the Marxist tradition in the light of the experience of the underdeveloped world. Outstanding, of course, is his sympathetic portrayal of peasant communities and those of the common people of the towns, worlds apart from Marx's dismissal of the non-urban, non-Western world as barbarian. For Tönnies, the extension of bourgeois society meant the disintegration of pre-existing cultures with the incorporation of the common people into the world labour market. And it is the trading class, the urban bourgeoisie, that initiates and effects the subordination of the whole country to this impersonal and inexorable regime. '. . . from their point of view the land and labour of the country like all other countries with which they deal are actual or possible objects for the investment and circulation of their capital. . . .'[18] As with, for instance, Frank, it is trade, rather than industry, that effects the extraction of surplus value, and the proletarianization of the people. Metropolis is primarily a commercial and financial centre, in which large-scale industrial production is incidental.

The lengthy analysis of the economic relations underlying the new urbanization, and the explicit references to class and class domination, set Tönnies apart from much subsequent sociological discussion of cities and urbanization. But his references to class, and the clear designation of *gesellschaft* as bourgeois society, receded in the minds of his readers and commentators, and in his own later writings, until forgotten, and *gemeinschaft* and *gesellschaft* were embalmed in the textbooks as the forerunners of sociological categorizations of human group life.[19] The direction in which academic sociology moved meant either the further abstraction of the concepts into the pattern variables, or their all too empirical grounding in the study of communities after the methods pioneered by the Chicago School.

The influence of Tönnies on the Chicago School is clear, particularly in the writings of Wirth,[20] but there is a significant shift of emphasis. Despite their concurrence as to the direction of social change with urbanization—the loss of neighbourhood affiliations, the break-up of the family, the growth of interest groups, the isolation of the individual, formal agencies of social control, and the use of mass media to establish public opinion—the one adopts the exchange economy of capitalism and the resultant class structure as his datum line, and the hypotheses of the others are drawn from assumptions as to the consequences of the ecological composition of the large city. Urbanization for Tönnies entails the acquiescence of the broad mass of the people in the life-styles of the dominant class, while for the Americans it was considered to bring about assimilation into society. The city's land market, not its exchange economy, was considered to be the great leveller.

3 Simmel: 'Metropolis and mental life' (1902–3)

Simmel shares with Tönnies the distinction of being an early proponent of analytical sociology, a method of inquiry which aims to present social interaction in terms of abstract categories. In Tönnies's hands these categories were mere clamps with which to grasp experienced reality, but for Simmel the study of society could only proceed by means of logical analysis of the forms of association which gave *a priori* grounds for its study. The forms are cognitive categories, and Simmel claimed that such a method provided a specifically sociological method of cognition. Tönnies had remained closer to Anglo-French positivism which sought to maintain knowledge at the level of experience, whereas Simmel belonged to the neo-Kantian tradition which frankly denies the possibility of the study of the natural or the social world without selection and ordering by the observer.[1] Knowledge, argues Simmel, was 'a process which we inject into reality';[2] human thought gave the observed world its seeming unity and the subject matter of the sciences 'have no counterpart in immediate reality'.[3] The definition of sociology as the study of forms of social interaction among individuals, not only delineated it clearly among human studies, but gave it grounding as a science, while avoiding the pitfalls of raw empiricism.

This apart, there is a common attitude to the task of sociology. It has to do with 'the actuality of concrete life', or 'reality in its total immediacy'.[4] Sociology was to turn away from the embracing theories of history and speculative philosophy, as well as the grand constructions of organicism, the schemes of Spencer and Comte. Specifically, the task of sociology was defined by Simmel as piecing together 'the real life of society as we encounter it in our experience'.[5] Accordingly, interest shifts away from the institutions of society to the relationships set up by man living in interaction with others. The terminal reference for sociology is this world of mutual influence and determination, by which, for Simmel, individuals are woven into society, so giving it unity. For Tönnies they are the basis of the common will on which the association of individuals depends. The sociological microscope is brought out, and the world beyond immediate experience is pushed out of focus.

Although Simmel wrote extensively on 'the origins, essences and destinies of cultural forms—music, painting, drama, science, philosophy, history, ethics and religion',[6] he showed little interest in the exploration of the major events of history, or in contemporary social and political issues. He maintained the same marginality and indifference to these as he did to academic life. His essay on 'Metropolis and mental life' represents a rare incursion into topical concerns, away from the introspective world of formal analysis and methodological clarification. As such it provided a vehicle for the exposition of ideas as to the nature of contemporary culture and individuality, and of theorems as to social interaction, which Simmel was exploring at more length elsewhere. It is therefore for Simmel unusual in that several strands of thought are being woven together in the course of a short lecture.

In chief, Simmel is concerned to expound on three themes: first, the consequences of a money economy for social relationships which he was exploring in the *Philosophie des Geldes*; second, the significance of numbers for social life which had already been the subject of an article in 1902, and was to form the basis of a lengthy section in *Soziologie* (1908); and lastly, the scope for the maintenance of independence and individuality 'against the sovereign powers of society, against the weight of historical heritage, and the external culture and technique of life',[7] a question with which he was preoccupied throughout his career and which forms a central theme in his engagement with sociological inquiry.[8]

In the *Philosophie des Geldes*, 'one of the neglected classics of sociology',[9] Simmel develops the proposition tentatively explored by Tönnies, namely that economic exchange must be studied as social interaction with certain consequences for the conduct of social life. The increasing use of money as a form of exchange promotes rational calculation in social relationships in that it renders these subject to precise assessment and manipulation. But here Simmel departs from Tönnies in declaring that it also permitted new forms of association, the voluntary association or interest group, in which individuals could participate without 'surrendering any of their personal freedom or reserve'. Money not only meant depersonalization of social life, but personal freedom. These changes in sociological form, he later suggested, are at least as important as economics in determining historical changes. The introduction of money had not only enabled trade, on which the Marxists admitted modern capitalism had depended, but a rationalistic world outlook which was as important in determining the shift to modern society as the economic transformation. In this way Simmel develops the sociological counter-argument to materialist theories of history, subsequently extended by Weber.[10]

Simmel's fascination with the significance of numbers for the structuring of social groups must be related to Kant's distinction between the form and content of knowledge, and the subsequent interest by the neo-Kantians in the derivation of mathematical concepts for the expression of scientific experience. Mathematics, the science of number, is to be considered the universal science of form; 'number' is an *a priori* form for the apprehension of reality phenomena. Simmel accordingly defines 'pure' or 'formal' sociology as the identification of the forms of sociation common to the most diverse social groups, and commenced the study of the geometry of social interaction, in, for instance, his studies of the dyad and the triad.

The essay that most closely relates to his exploration of the consequences of large cities for personal life is that entitled 'On the significance of numbers for social life'. This he commences with the statement:[11]

> It will immediately be conceded on the basis of everyday experience that a group upon reaching a certain size must develop forms and organs which serve its maintenance and

> promotion, but which a smaller group does not need. On the other hand, it will also be admitted that smaller groups have qualities, including types of social interaction among their members, which inevitably disappear when the groups grow larger.

This initial observation is developed in a discussion of socialism which he infers is impossible on the scale of the state by reason of the very numbers involved. He then turns to an argument later found in his essay on the large city, that of the estrangement of the individual from the society built up on his own social relationships. Simmel argues that this sense of estrangement is due to the size of society. The larger group gains its unity only at the price of a great distance between all the structures of social integration and the individual members of the group. In this argument alienation is an effect of the large-scale characteristics of modern society and as such is inevitable.

The possibilities of applying this argument to the city, the large-scale community having to provide conditions of order for an assorted population, are clear. Formal law, formal administrative bodies, impersonal symbols of identity, the separation of the citizen from the city as a political community, all can be related to the sheer size of the urban agglomerate. This, in fact, is what Simmel does in the essay 'Metropolis and mental life', an essay subsequently hailed by Wirth as 'the most important single article on the city from the sociological standpoint'.[12]

There are five main themes running through the discussion of individual life in the metropolitan city: the first of these is that of arousal in face of physiological and socio-psychological stimuli deriving from the turmoil of the urban environment. In consequence a blasé attitude is developed—a failure to react to surrounding events as the only means to preserve the inner self. The resultant strategy is that of reserve in face of the superficial and fleeting contacts of the crowd. Otherwise 'one would be completely atomised internally and come to an unimaginable nervous state'. In fact reserve is essential to social order in the metropolis, and the development of this aversion to others marks the individual's socialization into the city. The large city's heterogeneity and cosmopolitanism requires a certain kind of personality, thoroughly intellectual, unemotive, reserved and detached. This urban personality is, for Simmel, the basis of urban culture, or urbanism.

A second major theme in the essay is that of the money economy. This dominates the city, and is intimately associated with the intellectualism already attributed to the urban environment. Despite his own emphasis on the significance of monetary transactions for social interaction in *Philosophie des Geldes*, Simmel is prepared here to concede the causal primacy of the urban personality in establishing urban culture: 'no one can say whether the intellectualistic mentality first promoted the money economy, or whether the latter promoted the former'.[13] Money had also had an impact on the urban personality in establishing the matter-of-factness of urban social relations, and the effect of blunting discrimination and furthering the blasé attitude.

A third theme flows from the first two—that of punctuality and precision. Money has turned the world into a mathematical problem; there is a calculative exactness to life which corresponds to the ideal of natural science. But more than this, the aggregation of so many people with such differentiated interests itself forces precision of clock and schedule on urban man. Spontaneity is prohibited, the most intimate relationships regulated by an impersonal time schedule else 'the whole structure would break down into an inextricable chaos'.[14]

At a late stage in the essay, Simmel introduces a subsidiary theme: that of the city as the seat of the highest economic division of labour. Competition induces specialization, and hence differentiation and individuality. It is significant that this aspect of urban living comes last in Simmel's description of the city, and would seem to be derived not from technological conditions but from the scale of social environment in which the individual is placed. Unlike Weber, for whom freedom in the use of one's labour was the essential feature of urban living, for Simmel urban freedom rested in release from the 'pettiness and prejudices which hem in the small town man', and the division of labour, specialization, was only an incidental feature of city culture. The essential feature of urbanization was the freedom it provided for individuation.

Autonomy and individuality of existence in the metropolitan environment constitute the central concerns of the essay. Simmel argues that the metropolis offers a kind and an amount of freedom which has no analogy whatsoever under other conditions, for it has meant an enlargement of the circle which forms our social milieu. Metropolis is unique in that this circle is enlarged by more than the aggregate of the immediate social grouping. The city transcends this through its dominance over an ever expanding hinterland and its most important characteristic is 'this functional extension beyond its physical boundaries'.[15] The number of people in relation extends far beyond the city boundaries, and the effective society extends to national and even international horizons. In this characteristic rests the essential cosmopolitanism of the city, the possibility for every individual to work out his own incomparability and particularity.[16]

But, there is a price. In such an environment, the individual can enjoy independence, but the metropolis is independent of, external to, the most eminent individual personalities. The individual has become 'a mere cog in an enormous organization of things and powers which tear from his hands all progress, spirituality and value in order to transform them from their subjective form into the form of a merely objective life'. The very city itself—its buildings, monuments and institutions—crystallizes the culture which outgrows all individual life. The individual confronted with this immense, objective culture, 'has to exaggerate this personal element in order to remain audible to even himself'.[17] Instead, therefore, of genuine individuality, of man's self-realization, there is a contrived, exaggerated individualism in the face of an overwhelming crude and materialistic culture.

So Simmel, without any knowledge of the early writings of Marx, elaborates on the theme of alienation, from a neo-Kantian rather than a Hegelian position. His is a description of metropolis as a state of 'reification', a social environment created by man in his interaction with other men, which now opposes and threatens to overwhelm him. It only fails to overcome him through the development of this exaggerated particularism which draws him apart from other men. The cosmopolitan life of metropolis permits the working out of the incomparability which is every individual's endowment, but is a fragmented, self-centred lonely life; alienation is man's condition in the city, not community.[18]

Simmel refuses to condemn or evaluate this new existence. He points to the limitations of urban existence—in his terms, the necessary consequences of size, of numbers, of increase in scale, in a culture dominated by money—at the same time pointing out its advantages for those who are in a position to benefit from conditions of independence. And yet it is this refusal to evaluate the urban life he describes so vividly that makes Simmel's analysis so unsatisfactory. He sees this alienation, this estrangement of men from men, of men from their community; he describes the way in which the senses and the emotions are deadened and blunted; he even comments on the brutalization of personal culture, and yet he seems to accept it all as inevitable—a working out of great currents of life. He condones what others would condemn, and seeks in urban life the very antithesis of socialism—the development of individuality for the sake of the self.

Simmel's analysis of urban life was from the outset subject to criticism. A member of the Berlin faculty wrote of the lecture: 'It is hardly possible to treat of the mental life of the metropolis in a sparser and more biased way than he did in his lecture of that title. . . .'[19] All subsequent empirical observation of both the large cities of the western world with which Simmel was concerned, and those of pre-industrial or underdeveloped societies, have either invalidated his observations in total, or demonstrated their specificity of application to certain class cultures. His adoption of the individual in interaction as his problem for study, and the formality of his sociological method, led him into a timeless and universalistic treatment of disparate experiences and situations. All urban populations are treated as undifferentiated masses. The formality of method was itself misleading in that it permitted an analysis that ignored the class composition of urban society, and yet allowed for a certain moralism, in praise of individuality. Simmel's analysis, as much as that of Tönnies, is an outspoken commentary on contemporary culture as manifest in the metropolis. It is difficult to disagree with Sorokin's assessment of Simmel's work as 'the speculative generalizations of a talented man', as 'evincing pure speculation, metaphysics and a lack of scientific method'.[20]

Simmel's analysis has had, however, certain attractions for the urban sociologist, not least because he attempts to define the metropolis and metropolitan culture in sociological terms, i.e. by the forms of association of cities themselves. His explanation of urbanization is constructed in purely sociological terms. The chain of argument commences with the city as aggregation of individuals; this, in association with the money economy, itself considered as a series of social transactions, leads to reserve and a blasé attitude, rationality and calculability of social life, and the division of labour; these in turn are the basis of an urban personality from which the characteristic urban institutions, organizations and culture derive. Urban life is to be explained by the urban personality. But, as Martindale comments, 'it is one thing to recognise the development of a peculiar urban outlook, related to urban occupations, and the city environment, it is another to isolate this as the peculiar core of urban phenomena, making all explanations from it'.[21]

But Simmel, in the terms of his own sociology, could do no other. To him social change derived from changes in society, that is, in the forms of association. And these processes of association were no more than the individual in interaction; society does not exist outside of these individually derived processes. So when these relationships change, as he argues they do in the city, then the social order changes too. 'The sociological structure is the ultimate historical element which is bound to determine all other contents of life. . . .'[22] The individual is placed in a world of total subjectivity, without any constraints other than the social world created by his own processes of interaction. To Simmel, urbanization derives its direction from the social relationships set up in the cities; all other social change derives from urbanization.

4 Weber: The City (1911–13)

Weber's lecture papers, subsequently published as *The City*, are remarkable for the critical acclaim with which they have been received by other sociologists. Mumford speaks of the 'fullness of understanding of the normal processes the city furthers'[1] and Wirth, linking Weber's statement with that of Park (1915), claimed that they were the 'closest approximations to a systematic theory of urbanism'.[2] For all this praise, on casual inspection the relevance of Weber's comparative study of medieval and the ancient cities to the study of contemporary cities seems obscure. Not only does the content—detailed discussion of historical material—seem remote from contemporary concerns, but Weber's argument is elusive. He seems to be satisfied with the presentation of 'the materials of a human situation in such a way that certain conclusions inevitably follow, yet will not commit himself in print to drawing them'.[3] Weber's sociological arguments are embedded in a matrix of historical material which serves as a medium for a sustained critique of the direction taken by urbanization in the Western world.

Seemingly, his argument is simple enough: he commences by examining the existing definitions of the city, including that of Simmel, which he rejects in unequivocal terms.[4] From the economic definitions he turns to the political and administrative conceptions of the city as a corporate body with a given territory, and then to the considerations of military control. From this basis he proceeds to define the city:[5]

> To constitute a full urban community a settlement must display a relative predominance of trade-commercial relations with the settlement as a whole displaying the following features (i) a fortification (ii) a market (iii) a court of its own and at least partially autonomous law (iv) a related form of association and (v) at least partial autonomy and voting rights.

It is only when we come to the discussion of which settlements of which period and place would satisfy these criteria, that the standards of significance and completeness which Weber is setting for the city as a unique community become apparent. He eliminates all those cities where authority had rested on a charismatic or traditional rather than a rational basis; all where the law was enforced on a personal rather than a universalistic basis, for example where duelling, and not the application of statute law, was permitted; all those that retained groupings that were those of the family or the clan, for instance, cities of China, classical Greece, and Rome; those that were governed by religious groups; and finally he rejects cities whose strength derived from a military rather than an economic base. A military foundation implied overlordship, that is lack of autonomy, and the presence of an army which did not share in the freedom of the city.

The possibility of political and military autonomy—opportunities for an autonomous administration by authorities in the election of whom the citizens participated—and the ability to defend this new democracy against feudal lord, opposing cities, and the peasantry, were the critical conditions for the existence of an urban community. It is in fact a conception of the city as a social system, 'a plurality of actors interacting with each other in a situation that has at least a physical or environmental aspect . . . whose relation to each other is defined and mediated in terms of a system of culturally structured and shared symbols',[6] in which the fusion of interests and the unity of cultural life was symbolized by the walls of the city, on whose maintenance the continued autonomy of the community depended. The ideal-type city is therefore found at a watershed, the transition from the feudal order to the capitalist society of Western Europe. All previous developments had but prepared the way for the city in its perfect form as a community-association; subsequent developments had led to its disintegration as a social form.

Weber has been termed the 'philosopher of contradiction', and apparent in his sociology is a central tension between his views on the rationalization characteristic of capitalist society and the maintenance of individuality.[7] In his treatment of the city these two issues are central. Whereas, for instance, in his discussions of bureaucracy, he is forced to acknowledge an inexorable 'parcelling out of the soul' in circumstances of large-scale rational organization, in his discussion of the emergence of the Western cities he recounts a process in which the development of the rational-legal institutions that characterize the modern city enabled the individual to be free from traditional groups, and thereafter develop his individuality. No longer were the means of social control in the city myth and ritual, nor the immutable and unchallengeable 'cake of custom'. For the first time 'the oath of citizenship was taken by the individual. Personal membership, not that of kin-groups, or tribe, in the local association of the city supplied the legal guarantee of the individual's personal legal position as a burgher'.[8]

But Weber makes clear that the new political community, with its democratic forms of association, depended on the presence of a new class—the urban bourgeoisie. 'The peculiar political properties of the urban community appeared only with the presence of a new stratum.'[9] An association of citizens depended on the economic independence of the individual household, therefore the emergence of the new class of merchants and craftsmen. In principle at least all men had freedom in the disposal of their labour, and it was this circumstance, together with the identity of interests in defending the new association against the society which threatened to curtail its independence, that gave the city its character as a community.

Weber's study of the city is as central to his investigation of the development of capitalism as his work on religion. The importance of the medieval city for Weber lay in its providing conditions in which new forms of social association were to develop. The adoption of rationalistic standards of conduct, on which he argued capitalistic enterprise depended, had its roots in these urban forms of social association, the law courts, the guilds and their related churches, and municipal administration. Urbanization under specific economic and political circumstances was a necessary linkage in the chain of conditions leading to capitalism. So, for instance, part of his argument against designating the cities of antiquity as the ideal typical case was that their political structure prevented the emergence of a common law and the formation of joint city associations that might ultimately have led to 'unified state formation'. The medieval city was important in that it had been the basis of the new nation-states of Europe as well as modern capitalism. '. . . neither modern capitalism nor the modern state grew up on the basis of the ancient cities while medieval urban development, though not alone decisive, was carrier of both phenomena and an important factor in their origin.'[10] This then is the ambivalence of Weber's treatment of the city; it can only be defined in terms of historical circumstances in which it was destroyed as a unified social association.

Although Weber's treatment of the city evidently owes much to its previous examination by both Simmel and Tönnies, not least in this identification of the changed orientation to action engendered by trade, the money economy and urbanization, his methods of study place his analysis in an entirely different category. Conspicuous is the comprehensive quality of his investigation. Cities are treated in terms of their relationships to other cities, to other parts of their society, as integral parts of the social and political order. Although Weber emphasizes the closure, autonomy and separateness of the urban community, he is also careful to stress that the historical peculiarities of the medieval city were due to 'the location of the city within the total medieval political and social organization'.[11] The analysis underscores Weber's conception of the sociological enterprise as the understanding of the significance of the totality of a given configuration in all its universal relationships. The city therefore is studied from all aspects of social structure: economic, political, religious and legal institutions are all considered as integral aspects of the community. Weber's sociology, although grounded in the concept of social action, allows him to move outwards from discussion of the meaning or significance of a given situation for a hypothetical actor to a consideration of the structural determinants of that situation.[12] Therefore cities are treated as total social systems with many interrelated activities, and not, as in Simmel's analysis, merely in terms of individuals in association.

The effect of Weber's early training as an economic historian is very evident throughout the lectures. The use of historical materials in comparative analysis, the structuralist character of his definition of the city, his acceptance of economic arguments to explain the re-emergence of towns in medieval Europe, all derive from this academic background. Not least, is his acceptance of a materialist interpretation of historical configurations. This is very evident in the last section,[13] in which he clarifies his selection of the medieval city as the ideal-typical urban community in a comparison with the cities of classical Greece and Rome. This is undertaken in terms of what he calls 'structural differences'; the headings under which he groups his material are those of class, 'class oppositions in antiquity and the middle ages', economic and political organization, 'the ancient democracy of small peasants and the medieval democracy of professional traders', and military interests. Under this last heading he compares the 'military orientation of interests' of antiquity with 'the peaceful economic interests of the medieval city'. The invitation to further study of contemporary cities in the same structural terms is clear.

Weber's definition of the city in terms of its economic organization, and his designation of the medieval city as the limiting case, must have seemed outrageous to many of his contemporaries reared in the classical tradition in which the Greek city state of Plato represented the ideal. Equally outrageous is his exclusion by implication of the most distinctively 'urban' of all settlements, the metropolis of the nation-states. If the cities of antiquity could not be included in the definition of 'city', then neither could those which contain bitter class antagonisms and a plurality of interest groups, whose prosperity depends on the maintenance of trade and financial monopolies over large areas of the world, and where urban interests are subordinate to those of state or nation. Nationally-based and internationally-orientated bureaucracies have overwhelmed the localized and exclusive identity of urban communities. There can be no doubt that in his eyes, the urban settlements of his day had ceased to be cities—'there was a state of affairs less civilized than that to be found in the cities of the late Middle Ages'.[14] Conditions for democracy were no longer present. Much of his description of the medieval city, most specifically in his quasi-allegorical comparison with the cities of antiquity, must be seen as an indictment of the direction taken by urbanization in the West. Weber's definition of the city therefore becomes the measure by which we chart its subsequent disintegration as a political community.

There is in everyday language an image of 'city' widely held and understood; an image that has been derived 'out of what men have actually experienced'; it is 'built out of the language of history'.[15] What Weber would claim to do, through the analysis of the historical material, is to specify the essential features of what is understood in the idea of a city. He selects the perfect case where, he argues, all men valued the city so highly, because it represented their economic and personal freedom, that there could be no doubt as to its existence. But, although Weber would claim that his ideal-type definition is built up from the historical experience of men and their ideas as to that experience, it is clear that it is 'the one-sided accentuation' of one point of view—his own. Ostensibly a description of the city, this is in effect a critique of contemporary urbanization. Certainly Weber held no brief for the closed groupings of the pre-industrial world, but he attached tremendous importance on participation, on action, on behalf of others, through association. The fate of the Western world had been that of the pauperization of public life: 'The fate of our times is characterized by rationalization and intellectualization, and above all, by the "disenchantment of the world". Precisely the ultimate and the most sublime values have retreated from public life.'[16] His 'ideal city' is therefore the relatively closed system of the medieval guild city in which economic enterprise and religious activity, public and private life, are fused. From this standpoint he can only point to a progressive deterioration in community life with the development of capitalism.

Weber gives a cumulative definition of the city in this ideal-type construct which is very specific indeed; so much so that one wonders at its relevance for urban sociology. It is instructive to compare his methodology with that of Wirth. The latter attempts a minimal definition, stripped down to the seemingly universal demographic criteria of size, density and heterogeneity, so that it becomes suitable for all urban settlements at any time, in any place. It does not, however, give any indication as to what is entailed in the concepts of 'city', 'urban', 'civic', 'urbanity' and 'citizenship'. This Weber attempts, and succeeds in demonstrating how deeply rooted in the culture of capitalism they are.

Clearly, however, Weber's treatment of the city has tremendous limitations in application to contemporary society. It outlines too perfect a conception of urban existence to do more than demonstrate how imperfect, on this standard, are the present urban settlements. Despite Weber's declarations to the contrary, it is a Utopian conception and as such lends itself to critique rather than analysis. Its usefulness rests in, first, its methodology and, second, in its delineation of the essential features of social association in the cities of capitalism. He emphasized the city's rationality of organization, the indispensability of formal social controls such as the law and the police, its secularization, the predominance of impersonal contractual relationships, and its individualism. The city is a free association in which the individual participates in his personal right. And finally, there is Weber's designation of the city as dominated by the bourgeoisie. The secular, rational, individualist, market-oriented community formations of the medieval cities derived from and permitted the activities and interests of this new class. The medieval city therefore foreshadowed not only the social and cultural structure of the Western cities, but also their political character, i.e. the dominance of an élite, the rationality of organization which in itself renders null individual participation, and the apathy of the citizenry confronted with a community power structure with only residual autonomy.

Unsatisfactory and limiting as is Weber's approach, it is still possible to argue that it tells us more about the nature and derivation of urbanism than do the statements of either Simmel or Wirth. Both the latter commence their analysis from the datum line of the city itself, and derive propositions as to social structure and culture from that alone. They both intended to derive statements of universal applicability. Weber commences his discussion of cities from the analysis of a particular social and political structure based on a certain economic order. Urbanism therefore becomes inexplicable save in 'the real, i.e., concrete individually structured configuration of our cultural life' and through 'The analysis of the historically given individual configuration of those "factors" and their *significant* concrete interaction'.[17]

5 Wirth: 'Urbanism as a way of life' (1938)

The issues posed by the European sociologists, and the modes of inquiry adopted, lived on in the statements of Wirth and Redfield,[1] and through their reformulations came to govern the field of urban sociology. In particular, Wirth's essay, in which in easy evocative prose he sets out some of the major propositions of Tönnies and Simmel in the light of the experience of the Chicago School, came to hold a dominant position among statements on urban sociology. To many social scientists it was the sociological statement on urbanization, and for a generation it served as the point of departure for urban research. Now it is hard to understand its attraction. The ecological premises from which he derives a definition of the city and urbanism, the despairing tone of the account of the process of urbanization, the description of the city as *gesellschaft* or mass society, the self-consciously sociological frame of reference, all so evidently warrant investigation as problems in the sociology of theory. The problem becomes just how Wirth came to arrive at these particular propositions which neither represented the research of the Chicago School, nor did justice to his own attitudes to urbanization, and further, distorted the statements of the European sociologists. The attraction of the essay subsequently is another question.

Although the evidence of the European sociologists' influence is strong in the essay,[2] the immediate intellectual environment of the Chicago School of sociology was more influential than probably Wirth would have been prepared to admit. Not only had he been an undergraduate at Chicago, and one of the star post-graduate students, but he spent all but three years of his teaching career there. He was 'a charter member of urbanism incorporated'.[3] In that the Chicago School in its 'Golden Age' (*c.* 1915–30) dominated American sociology, Wirth had as his background the mainstream of academic sociology in the crucial formative years of the profession.

The Chicago School had moved out of a background of Social Darwinism and general speculative sociology into a 'period of investigation and research'[4] in the second decade of the century. It is therefore for its empiricism that the department became known. Fieldwork, investigation of the subcommunities of the city, such as that prepared by Wirth in *The Ghetto*,[5] and presentation of statistical data as to different facets of urban living, came to characterize sociological investigation. The philosophical underpinnings for the new methodology came from pragmatism. Knowledge became an interactive process between knower and known; in Wirth's terms, 'the object emerges for the subject when in the course of experience, the interest of the subject is focused upon that particular aspect of the world'.[6] First-hand experience is the basis of knowledge. At one point James had described the basis of knowing as 'pure experience' which he defined as 'the immediate flux of life which furnishes the material for our later reflection',[7] and in the same vein Park and Wirth continually reiterated the importance of direct experience and confrontation with social reality, as the basis of social scientific research. This entailed a rejection of the analytical approach of the European sociologists in favour of 'concrete empirical procedures'.

Although the leading members of the department, Small and Park, had looked to the European sociologists, in particular Simmel, for a lead in their redirection of sociology away from speculative discussions about large-scale social processes to a more careful investigation of a specifically 'social' field, their borrowings are scarcely recognizable in their American guise. Although individualism was 'a part of the very air men breathed',[8] in both psychology and sociology there was an early declaration of interest in social processes and the study of social groups. In 1905 Small had written:[9]

> While individuals are the real existences, and the groups are only relations of individuals, yet to all intents and purposes the groups which people form are just as efficient moulders of the lives of individuals as though they were entities that had existence entirely independent of the individuals.

The writings of Durkheim were therefore sympathetically viewed by Park although the influence of Simmel is more evident in his organization of sociological inquiry. Sociology came to be defined as 'the science of collective behaviour', and the 'central fact and central problem of society' was considered to be that of social control.[10] In Park's hands sociology became 'concrete, dynamic and orientated to social consensus, whereas Simmel's was abstract, structural and oriented to sociological dualism'.[11] Sociological inquiry was focused on the conditions for group cohesion and regulation at a time when traditional mechanisms of social solidarity seemed to be in rapid dissolution.

Equally important in establishing the innovatory status of the Chicago School in American sociology was the formulation of human ecology. This, a unique hybrid between biology and sociology, was derived from a distinction made by Park between the natural and the moral order, between a 'symbiotic society based on competition and a cultural society based on communication and consensus'.[12] Competition and conflict were relegated to a subsocial world whose behaviour was governed by similar laws to those operative in plant and animal communities, and which could be studied by methods derived from the natural sciences. In the form propounded by McKenzie and Park, it gave clear theoretical justification for the study of social phenomena which were not characterized by consensus, by statistical methods which both Park and Wirth rejected as inappropriate to sociology.

Wirth's definition of the city as a relatively large, dense and permanent settlement of socially heterogeneous individuals is in fact an ecological definition of the urban community, and the essay may be considered as an attempt to work out the implications of a given set of ecological conditions for the social organization and culture of an aggregate of population whose only initial common bonds are those of existential interdependence. His argument, summarized, is this. The aggregation of population, and to a lesser extent its density, leads to the multiplication of persons with whom any one person interacts. For this reason full contact of individual personalities is not possible and man does not meet other men as himself. Secondary relationships predominate and urban men meet each other in highly segmental roles. Contacts are impersonal, transitory, leading to reserve, indifference, a blasé attitude, and the immunization of self against others. Divested of family and neighbourhood ties, free to associate in a wide range of voluntary groupings, urban man is brought to a state of anomie. In consequence of this breakdown of traditional mechanisms of social control, accentuated by the division of labour, is the development of bureaucracy, communications through the mass media, and government by representation. All mention of the money economy, which in Simmel's essay played as important a part in creating an urban mentality as the ecological conditions of city life, is omitted. Wirth is unswerving in his argument that ecological conditions have created, and must create, urbanism as a culture.

Wirth's indebtedness to Simmel and Park is easily discernible, though his statement lacks any of their subtlety and breadth of imagination; less obvious is its relationship to Weber's examination of the city as a political community. The problem posed for both by the city derived from its territorial definition as a social grouping. Like the nation-state, membership of it was conferred not by ascribed categories of sex, race, family, class or religion, but by criteria of residence. That is, all are members of the city who live in the city. The shift to the latter criterion of community membership had meant a much greater degree of personal freedom as well as political individualism. All this is summarized in the concept of citizenship, in which individuality is overridden in the name of equality. The problem was that the modern city, or metropolis, was no longer the basis of citizenship, and it was hard for Wirth to envisage the circumstances under which collective action which would define it as a social entity might come about.

> . . . a human aggregation cannot be regarded as a society until it achieves this capacity for collective action, although it may manifest a high degree of symbiotic or functional interdependence between the individuals composing it. Such an aggregate may constitute a community without a society.[13]

But, whereas Weber had dismissed the modern city as a sociologically relevant entity, Wirth was prepared to concede that collective action could be achieved on the basis of the consensus generated by the manipulation of mass communications. It is of the city as a mass society that he writes. Although the city comprised a mass of unattached individuals, and lacked common customs, institutions and rules, mass communications enabled at least an elementary level of common understanding. If Simmel's argument had concerned the fate of individuality in a democracy, Wirth's concerns democracy in conditions of individuation.

Wirth's statement differs from those of his predecessors in several respects. Very evident is his despairing assessment of the consequences of mass urbanization:[14]

> If the individual would participate at all in the social, political, and economic life of the city, he must subordinate some of his individuality to the demands of the larger community and in that measure immerse himself in mass movements.

The progressive creed of Park, and the controlled ambivalence of Simmel, had been transformed into a cynicism and scepticism about the direction of social change in the cities. The essay in which Wirth had intended to advance the scientific study of urbanism through the construction of a theory of the middle range became a thoroughgoing statement of the degeneration of social and political existence in the metropolis. This is surprising because his intentions in setting out this statement were so explicitly professional: to further research into the city through the setting out of 'a comprehensive body of compendent hypotheses implicitly contained in a sociological definition of the city'. This concern for the professional advancement of urban sociology may help to explain his greater dogmatism. There was no place for equivocation or provisional statements in presenting 'an ordered and coherent body of theory upon which research might profitably proceed.'[15]

Different also is his use of an ecological framework. The focus of Park's ecology had been on the aggregate of population, not the environment; 'Human ecology . . . is . . . not man, but the community; not man's relation to the earth which he inhabits but his relations to other men that concern us most';[16] his intention had been the conceptualization of the behaviour of a system, not individual interaction. In following Simmel, and adopting the individual as his unit of analysis, Wirth thereby is forced into a more deterministic approach than that of Park, in that the aggregate of population opposes the individual. This is precisely the unsatisfactory feature of Simmel's discussion, in his case mitigated by a greater sensitivity to the non-ecological features of the metropolis.

The use of an ecological framework, the apocalyptic and dogmatic tone of the essay, even the attempt to set up a deductive theory for urbanism, are all unexpected, for Wirth, given his provisional and pragmatic views on theory, his sceptical if not hostile attitude towards ecology,[17] and his own previous research into the ghetto which contradicted the analysis presented here. Wirth was consistently sceptical about the preparedness of sociology for fully developed theoretical schemes, and particularly critical of deductive theory. Bendix summarizes his views as being that 'all efforts to build deductive systems merely lead to elaborate proofs of what had already been assumed',[18] and Wirth had dedicated much of his teaching of theory to the exposure of the assumptions on which other deductive theories had been built. Equally, from his knowledge of pragmatic philosophy, and the work of Mannheim in *Ideology and Utopia*, he was well aware of the 'positive and constructive significance of the evaluative elements in thought'.[19] And yet in 'Urbanism as a way of life' we have an empassioned critique of American urbanization in the guise of a deductive theory, open to all the criticisms that Wirth normally made of formal or theoretical sociology.

One possible explanation for his approach is his desire to turn sociology into service in the organization of society. His entire career as a sociologist was guided by a basic commitment to sociology as science, and throughout his many writings on aspects of social policy, town planning, housing, race relations and community organization, he writes as the scientist advising the layman. His interest in sociology depended on his belief that it could assist in answering the question he posed: 'What are the paths open to us in society for intelligent self direction?'[20] To him there was an integrality of theory, research and social action. The consequent air of depression with which he recounts the failure of sociology, despite the attention given to urban research, to arrive at a theory of urban existence which might guide policy for the cities, is unmistakable. He felt the need for a clear unambiguous statement of the corpus of sociological knowledge for those outside sociology.

A second explanation is that, like Park, he wished to provide a master statement on the city which would enable the student to see the city as a whole entity while engaged in smaller-scale investigations. Park's essay of 1915, 'The city', was by then outmoded, and Wirth evidently felt the need for a new statement embodying the best of sociology that would command the attention and respect of the young student and research worker. The work of the Chicago School in its 'Golden Age' had depended on Park's overall direction. Each individual study had contributed to the 'mosaic of the theory of the city and knowledge of Chicago that Park was building'.[21] Without such co-ordination the work of the Chicago School could only fragment into the abstracted empiricism and liberal practicality subsequently decried by Mills.[22] In 1938, Wirth viewed a research scene now dominated by the single study, and his statement may be construed as an attempt to regularize and systematize this diffuse research effort into a unified project which would advance the scientific study of urbanism.

For, unlike Park, Wirth was very much the total professional in his dedication to the advancement of sociology. Much of his time was spent as an active member of the various professional bodies, and in addition he was involved with any number of agencies, committees, groups and commissions interested in redirecting 'community life and social policy'. A man without background, his role and status as a sociologist became his identification, and his many activities outside the university affirmed this identity.

Wirth's professionalism explains much about this essay. His 'public' was first, the layman looking to the scientist for advice, and second, the graduate student requiring guidance from one of the elders of the profession. Both required an unambiguous and unqualified statement. So despite his considered and provisional orientation to theory, Wirth in his role as sociologist was impelled to take on the activity of theory construction to which he was by inclination unsuited. So, like a zealous convert, he constructs a specific theory which embodies the most deterministic and dogmatic aspects of the ecological and mass society theories. Ideas which Park would have termed 'suggestions' now become 'hypotheses'; mass society is no longer an eventuality but an inevitable occurrence; and environment is no longer mediated through other men but impinges unilaterally on the individual.

Wirth's depreciation of the social changes associated with urbanization, 'personal disorganization, mental breakdown, suicide, delinquency, crime, corruption and disorder',[23] and his full endorsement of the most pessimistic aspects of mass society analysis, are similarly surprising. Martindale describes him as 'a hard-headed urbanite'[24] and, unlike Tönnies, he never extolled life in the small community. Elsewhere he spoke of the 'intensely interesting and far-flung community life' of the city, 'its variety of cultural institutions . . . where men may find stimulation through association with each other', and wrote approvingly of the 'security and enrichment of life that can only come through participating in the wider world'.[25] This to him was preferable to the provincial and sectarian life of the small group—his assessment of ghetto culture. And yet he was frankly sceptical about the possibilities for creating a new basis for consensus in the large-scale society. The bondage of the small group had been exchanged for a subordination to social processes over which the individual had no control; neither form of social grouping allowed for individual freedom and participation in the consensus on which society depended. This scepticism came to a head in the 1938 essay, written at the height of the power of fascism, after the evacuation of his own family from Germany, and the flight of so many colleagues from Nazism.

But Wirth was also handicapped in his assessment of urban culture and social organization by the categories of sociological analysis derived from the European sociologists and transplanted into the setting of the American colleges and universities. Sociology had to do with the study of 'human group life', a field of investigation both specific to sociology and germane to the ongoing discussions as to the bases of social order in a society which had recently undergone both rapid urbanization and mass immigration. Wirth accordingly states: 'I regard the study of consensus as the central task of sociology, which is to understand the behaviour of men insofar as that behaviour is influenced by human group life.'[26]

There is 'an identification of sociality with consensus'[27] which led Wirth to disregard the possibility of social life on a basis other than the full participation of individuals in the 'established habit of intercommunication, of discussion, debate, negotiation and compromise'.[28] When, as in his survey of the city, it proved impossible to discern any basis for the generation of communication, and the creation of conditions for participation in these terms, the analysis of the city as a mass of disassociated individuals followed. Wirth's sociology, in which the key concepts were the 'individual', 'social interaction', 'communication' and 'consensus', prohibited a description of urban life in any other than pessimistic terms. Wirth was only able to reconcile the inherent tension between his own individualistic-liberal concerns with the conservative language of sociology by defining consensus in perfect terms as 'conditioning, and conditioned by, the participation of individuals in a common life'.[29] With this language he was unable to embark on a more dispassionate description of a social existence marked by disorder, conflict, remoteness of political institutions and apathy towards common issues.

Wirth was not a highly original thinker; the main thrust of his work was towards the discussion and transmission of the ideas of others for a wider audience. His statement on the metropolis must therefore be seen as the culmination of a line of analysis originating with the work of Tönnies, a half-century previously, gathering impetus with the statements of Weber and Simmel, and extended by Park in the American context. The redirection of sociology by these men to empirical study, their restriction of sociological discussion to a well-defined field of 'social' facts, their exclusion of the questions of scarcity and power in society, and their steadfast belief in both the need for, and the possibilities of, a science of society, are tacitly accepted by Wirth and form the basis of his exploration of urban living. What makes his statement particularly interesting is his capacity to 'develop idea[s], once germinated, into all of [their] logical possibilities'.[30] None of the European sociologists, Tönnies, Simmel, Weber or Durkheim, could have envisaged an analysis of city life in ecological terms yet Wirth's adoption of such a frame of reference is implicit in aspects of their work. Human ecology only became necessary to urban studies because of the limitations of classical sociology.

The ready acceptance of Wirth's statement as the basis for the sociological study of urbanization and cities never ceases to surprise. In its entirety it offers an open invitation to rebuttal and criticism. Empirically, few of Wirth's generalizations have stood the test of later research, either in the context which influenced his own discussion, that of American urbanization, or in circumstances with which he was wholly unfamiliar—urbanization in Europe or the underdeveloped world.[31] If the essay is subjected to logical dissection, inconsistencies and contradictions abound. Wirth so obviously mistakes congruence for cause, argument by analogy as full demonstration of an unchallengeable case. But above all it is his assimilation of a specific form of social organization to a general thesis of the growing complexity of the territorial organization of society, as an inevitable and natural outcome of environmental conditions, that confounds readers with a background in any one of a number of intellectual traditions. Weber, for instance, speaks for many in the mainstream of European historical scholarship in his peremptory dismissal of Simmel's basic premise; more recently, from the context of French Marxist sociology, there comes an even more cavalier dismissal:[32]

> Or, il suffit de réfléchir quelques instants pour découvrir l'absurdité d'une théorie du changement social fondée sur la complexification croissante des collectivités humaines à partir d'une simple accroissement démographique. . . . Toute évolution de la dimension et de la différenciation d'un groupe social est elle même le produit et l'expression d'une structure sociale et des lois de transformation.

In this view (and that of Weber) 'urban' culture is inextricably linked with capitalist society, and the literature of 'urbanism' only serves to confuse the different issues of culture and space in the one debate.

The thesis of urbanism, however, articulates a long-held view that environment, and in particular size of social grouping, is critical in determining social behaviour. Utopias are always small communities, the large city castigated as the despoiler of innocence, source of poverty and despair, creator of social disorder. Remedies for evident social evils are accordingly to be sought in the reorganization of the cities. Castells terms this a myth, in that it recounts in ideological terms the history of the human race; Williams agrees, suggesting that the interpretation of the transition from rural to urban society as some kind of fall is 'a main source for that last protecting illusion in the crisis of our time, that it is not capitalism which is injuring us, but the more insoluble, more evident system of urban industrialism'.[33]

Is there not some substance to the sociologists' claim that size of social grouping affects its mode of organization, and that the urban environment does foreclose some aspects of experience while opening up others? Even from a Marxist standpoint, steadfastly hostile to an 'urban' interpretation of the sources of social change, it would be hard to escape the conclusion that man's consciousness of social reality is in some way mediated by his experience of his community.

> If man draws all his knowledge, sensation, etc., from the world of the senses and the experience gained in it, the empirical world must be arranged so that in it man experiences and gets used to what is really human, and that he becomes aware of himself as man . . . if man is shaped by his surroundings, his surroundings must be made human.[34]

6 The Chicago School: urban experience

1 The ecological city

> It is the structure of the city which first impresses us by its visible vastness and complexity. But this structure has its basis nevertheless in human nature of which it is an expression. On the other hand this vast organization which has arisen in response to the needs of its inhabitants, once formed, imposes itself upon them as a crude external fact, and forms them, in turn, in accordance with the design and interests which it incorporates.[1]

The Chicago department of sociology was the first group to mount a sustained examination of change in a modern city from the standpoint of academic sociology. The 'Golden Age' of the Chicago School was initiated in the period immediately preceding the First World War when W. I. Thomas, then the leading member of the department, was joined first by Park and then by Burgess, and faded twenty years later. Under the guidance of its first Chairman, Small, the department had taken the lead in America in studying the 'real conditions of life', so redirecting sociology away from speculative generalization and the world of the library to increasing circumspect empirical investigation of the world of personal experience.[2] The Chicago School's reputation as sociologists was based on their close examination of the social fabric of their city, Chicago.[3]

The leading figure in this research activity, certainly after the removal of Thomas in 1918, was Park. Student of philosophy and psychology first with Dewey, subsequently with James; journalist; student under Simmel and Windelband in sociology and history; assistant to Booker Washington; and then at fifty, an academic;[4] Park took the lead in directing research inquiry towards life in the city in his paper, 'The city: suggestions for the investigation of human behaviour in the urban environment' (1915).[5] He assembled a novel synthesis of the writings of American and European sociologists, presented in the *Introduction to the Science of Society*,[6] and formulated what he anticipated would be the basic science of human existence, human ecology. As with so many of his ideas, its further examination and elaboration was relinquished to a former student, McKenzie.[7] But Park's influence was exerted not so much through formal publications and explicit statements, as through his teaching and dealings with students and colleagues. The research enterprise at Chicago was held together, for a time, by his personal influence whose stimulus and novelty it is hard to grasp at this remove. Park provided 'a vision of the character of cities and city life', and an individual student 'relied, implicitly and explicitly, on the knowledge that had already been gathered, as he contributed his own small piece to the mosaic of the theory of the city and knowledge of Chicago that Park was building'.[8]

The department had a then unrivalled opportuntiy for research. The university itself, founded in 1892 on novel lines, had rapidly established a reputation for innovation in teaching and research in both the natural and the social sciences: the intellectual environment was known to be stimulating.[9] It therefore attracted large numbers of graduate students. The work of the social science faculty was further encouraged by large donations for research from the Rockefeller Foundation, channelled through the Social Science Research Council set up in 1923,[10] and by gifts from the local community. Then, of course, there was the stimulus of Chicago itself, the 'shock city' of the new America.

> From a struggling village sunk in the mud of a prairie creek, Chicago arose within memory of living men to a great metropolis, ranking fifth in the roll of the world's greatest cities. It was inevitable that Chicago should assume this rank, for Chicago is a city of destiny.[11]

The department had the exhilarating experience of living and working in a city changing from a small provincial centre to an ebullient industrial, commercial and financial metropolis.[12] In it migrants from rural America jostled with those from underdeveloped Europe in the search for fortune, overwhelming the local population. Chicago was a teaming polyglot city whose growth and prosperity was symbolized both by its skyscrapers and its lawlessness. Chicago became known for 'illegal gambling, prostitution, gang warfare, intimidation of witnesses, serious defects in judicial administration, election frauds, and police and political corruption'.[13] The city called out for sociological investigation. The mobility of its population, the rapidity of change in community structure, its diversity of social groupings, the problems of personal disorientation and public violence, the very speed with which the physical structure of the city was changing, all challenged previously held views on community living and the social order. Small-town America, from which these sociologists derived, was being superseded by the world of property speculation, organized crime, mass communications and machine politics. And yet the city continued to function, law and order did not break down entirely.

Despite their opportunities, the work of the Chicago School is disappointing. The verve, subtlety and provocation of Park's earliest writings were matched by few of the studies undertaken under his direction—those of Wirth and Zorbaugh[14] are outstanding —and the research programme subsequently disintegrated into an aimless empiricism. The puzzle and paradox of the city presented by Park in the quotation heading this chapter was resolved by him in the designation of a field of study, 'human ecology' distinct from that of sociology, relating to a 'natural order' to human existence. Urban studies came to depend on explanations derived from plant and animal ecology, for it was assumed that 'the same processes of competition and accommodation are at work determining the size and ecological organization of the human community'.[15] The Chicago School tantalize in the pertinence of their observations and their insight into the social worlds of the metropolis, and infuriate with the diversion of urban studies down an intellectual blind alley. It is as though their insight was not matched by their knowledge; their sociology did not equip them with the language or skills to deal with the large metropolitan city. They were evidently bemused by Chicago, seemingly beyond individual control, and yet showing an ordered arrangement of population and activities. Accordingly,[16]

> The conditions of social change became to them the *facts* of social change. Thus their universe of discourse became limited to externalities, and the interpretation of social life hinged upon its most concrete aspects. Reducing social behaviour to a common denominator of the tangible and the measurable, congenial to the pragmatic intellectual temper of the country, human ecologists became the expounders of the socially 'given', the realists of the established order.

Park and Burgess identified competition as 'the elementary, universal and fundamental form' of interaction;[17] the struggle for existence identified by Darwin in the natural world had its counterpart in the world of men. Economic organization, as the effect of competition, was therefore ecological organization, a natural, inevitable substructure to social existence. The other forms of interaction identified—conflict, accommodation and assimilation—

belonged to the social, or moral order, which although itself accommodative to the natural order, in turn imposed control on these natural processes. In this way they were able to consider the unmitigated competition between classes in Chicago for space, community resources and political control, the profiteering activities of entrepreneurs, and the exploitation of labour, all as having inception in 'the traits of human nature and the needs of human beings'.[18] Capitalism and the capitalist city, as experienced here, were, if not inevitable, natural. The central problem of sociology was posed as social control: not control of the competitive processes by which resources were allocated but control of the activities of the individual in the interests of the social cohesion of the group as a whole.

One of the interests of Small had been in economic theory, and in a dissertation on Adam Smith he had attempted to demonstrate the interdependence of economic and moral, or social, activities.[19] This interest seems to have been transmitted to Park, who in his writings on the city seeks to explore the interplay between the world of competition and that of consensus in a territorial setting. But whereas Small had been insistent that 'economic relations must be expressed in terms of the whole moral process',[20] Park, while never able to clarify his views as to the balance between the 'double aspect to society', posed the problem in such a way as to commit the ecologists to a presumption of the independence of the processes of community organization from the social order. The city as an ecological community imposed an order to social existence which had to be ascertained before the strictly sociological concerns of interaction, association and communication could be explored.

Despite the bewildering rapidity of urbanization in Chicago, the sociologists took heart from the land economists' deductions of order and rationality in the use of urban land. Analyses of the market in land and property such as those of Hurd[21] must be considered as one starting point for their derivation of human ecology. Although the city grew on without plan, a frenzy of subdivision and 'platting' extending the urban grid of roads and buildings yet further from the city centre, nevertheless activities found a location within the city which acted to their economic advantage. Similarly immigrant groups all found a home, usually in the inner city, and the resultant changes in residential location seemed to follow a regular sequence. Also, despite the lack of co-ordination or planning, the city was evidently becoming the powerful centre for a new industrial economy, and many living in it were prospering. It seemed probable that if urban land use followed laws that determined the eventual shape of the city, then the city's social organization would also have an orderly pattern of development. The task of the new profession of sociologists was to elucidate this process of change in the community for a bewildered public, thereby encouraging further readjustments to the changes in circumstances brought about by urbanization.

Park's writings, never concise, became increasingly incoherent with age, and are marked by inconsistency and vagueness. Nor is his confused presentation of the ideas formulated in the course of the department's research in Chicago clarified by the writings of his associates, Burgess and McKenzie, who both stressed rather different aspects to human ecology and urban research.[22] They did, however, attempt to formulate a theory of urbanization which out-reaches any other in its comprehensiveness, and is still the most systematic theory available to urban students. For this reason it is necessary to set down their key ideas, and the stages in their argument.

The city as ecological community

The key statements in the urban sociology advanced by Park are as follows:

1 (a) The city is an organism whose structure and functioning is to be understood in terms of special laws that differ from those of governing society.
 (b) The city is characterized as an ecological community; as an aggregate, and through the mutual interdependence of its population, it creates its own future. It is 'an externally organized unit in space produced by laws of its own'.

2 There is a distinction between a double aspect to society, that signified by competition, the natural order, and that by consensus, the moral order. From the first derives the symbiotic community, from the second, cultural society as evidenced in collective aspirations, traditions and ideals.

3 Land values, 'themselves in large measure a product of population aggregation in the long run give the aggregate an orderly distribution'. They are related to the natural order.

4 The ordered arrangement of population and activity in the city is arrived at in a regular sequence equivalent to the processes of invasion and succession observable in a plant community. The end product of these processes are the natural areas.

5 The natural areas are the habitats of groups of people, who on the basis of existence in a common territory will develop distinctive cultures. They are 'natural' in that they are not planned—'not the result of design but rather a manifestation of tendencies inherent in the urban situation'.[23] These natural areas are important in Park's sociology of the city as

(a) Methodological tools
 (i) They define the association of space and social activity. Only in so far as these coincide can social relations be measured.

(ii) They are the 'frames of reference' within which the fieldworker can operate, and enable him to generalize from his own research to other areas of the same kind.

(b) The basis of community organization

(i) They are the basis for a theoretical model for the city: it becomes a 'functioning superorganism in which the natural areas are the cells'. They are functional in that they serve a need, allowing territorial differentiation for the different peoples of the city.

(ii) These local areas are to provide the basis for community organization in urban society. The end-state of the process of urbanization will be a constellation of subcultural groups founded in the neighbourhoods, held together through the interdependence of the division of labour.

1 *The ecological community*

The basic unit of analysis in human ecology is the population aggregate. The assumption is that it has attributes and character as a unit: the aggregate is more than a sum of the parts. The city as an aggregate 'is a thing with a characteristic organization and typical life history',[24] which comes to exist independently of its residents. It develops a momentum of its own to which the individual must be subordinated—an irresistible juggernaut sucking up the sands of humanity. Park's debt to Durkheim in the development of his own sociology is very marked, but there is an interesting shift in the level of thinking: 'Where Durkheim dealt with whole societies Park was interested in communities.'[25] For Park considered the 'social' to be that which may be experienced directly, that is, in the day-to-day existence of the urban locality, and on this basis 'social' can be no other than 'communal' and community is defined spatially as locality.

The city, like any other form of human settlement, is considered to be an ecological community. The ecologists used the term 'community', whether plant, animal or human, to refer to a habitat, and its population. An ecological community is defined as:

(i) a population territorially organized, i.e. in a bounded environment

(ii) more or less permanently rooted in the soil it occupies

(iii) its individual units living in relationship of mutual interdependence that is symbiotic rather than societal.[26]

The attention given to population and demographic variables by the ecologists in their definitions of the city, as, for instance, that of Wirth, derives from this ecological formulation. Park writes: 'the organization of the city, the character of the urban environment, and of the discipline it imposes, is finally determined by the size of the population, its concentration and distribution within the city area'.[27]

Order in the community is maintained through symbiosis, a relationship of mutual interdependence in which the struggle for survival is mitigated through the co-operation entailed through coexistence. Wirth is said to have defined symbiosis as 'a condition in which men live together by virtue of sheer existential dependence upon one another'.[28] No social interaction is entailed: men meet each other as impersonal categories with no recognition or communication. The ties which hold the community together are 'physical and vital rather than customary and moral'.[29] The ecologists' conception of community therefore stands in stark antithesis to that of the sociologists. Tönnies's conception of the *gemeinschaft* as a relationship of deep intimacy, and Weber's blunt declaration that 'communal action refers to that action which is oriented to the feeling of the actors that they belong together',[30] become statements about society. Park flatly refused to admit that impersonal utilitarian relationships were part of the social or moral order, and therefore relegated relationships that Weber designated as 'societal' and Tönnies as those of the *gesellschaft* to a natural order based on competition.

2 *The natural and the moral order*

In his introductory chapter to the 1921 textbook Park, in a discussion as to the nature of social control, argued that man and society had a double aspect in that they were products both of 'nature and design, of instinct and of reason'. Society therefore had to be considered as having its roots in 'nature and human nature'.[31] Subsequently 'nature' became the environment in which man is located and in which he competes for space (land and accessibility), and for sustenance (economic survival). The ecologists carried over into the urban environment the approach of the anthropologists investigating isolated, technically primitive and largely self-sufficient rural communities, or that of the human geographers studying the 'mutual relationships between man and his natural environment studied from the standpoint of man's adjustment to environment'.[32] Although the city's environment is man-made, it nevertheless possesses a natural order, 'a free and natural economy, based on a natural division of labour',[33] which provides conditions of interdependence and the basis for the cultural organization of the people, i.e. its moral order.

Park draws a distinction between 'community' and 'society' as double aspects of social groups. While competition characterized 'community' society was marked by 'solidarity, consensus and common purpose'. He acknowledged the force of Dewey's observation that 'Persons do not become a society by living in physical proximity any more than a man ceases to be socially influenced by being so many miles from others. . . . Consensus demands communication',[34] but nevertheless considered that every community must have some attributes of society. The differentiation of 'community' recognized that there could be relations that were not social, but which provided the 'necessary conditions in which societies are rooted'.[35]

Although Park acknowledged that the distinction could not be validated empirically in that there was no community that was not at the same time a society, the distinction is the keystone to his theory of the city as an ecological community. The conception of a natural order to community life legitimates his adoption of ecology as the language of urban exploration, and the necessary underpinning to urban sociology. The significance of this distinction finally led him to the conclusion that social existence should be studied by two quite different methods. The study of community as the natural resultant of the competitive process constituted the field of human ecology in which statistics were to be the normal method of research; and the study of society as the result of the cultural process was to be conducted with methods similar to those of social psychology or cultural anthropology.

Throughout his career Park remained uneasy about the purity of the distinction drawn between these two aspects of society, and the validity of the comparison with the plant and animal world. Although at times he sought to demonstrate that 'the cultural superstructure rests on the basis of the symbiotic substructure', he also admitted that 'in human society competition is limited by custom and culture'.[36] Subsequent to the exploratory statements in the 1921 textbook, he never set out his theoretical position with any coherence, and used his concepts loosely. Alihan, totally exasperated with his inconsistency and vagueness in the use of the key concepts, concluded that the difficulty derived from methodological naïvety, which might have been avoided by use of ideal types.[37] And yet Park was fully acquainted with this methodology in European sociology. The confusion is symptomatic of the tension between the analytical sociology of Tönnies or Simmel, and the holistic empiricism of James and Dewey in which total experience was the basis of knowledge.

3 *Land values*

These are associated with the natural order, and 'offer a new device by which we can characterize the ecological organization of the community, the social environment and the habitat of civilized man'.[38] The division of the city into separate neighbourhoods thus results from an economic process: rents mediate between incomes and land values. This explanation of the natural areas of the city must have presented Wirth with problems in his discussion of the Jewish ghetto,[39] but once more it is left to Park to discredit this version of the ecological argument: 'the population tends to segregate itself not merely in accordance with its interests, but in accordance with its tastes and temperaments. The resulting distribution of the population is quite different from that brought about by occupational interests or economic conditions.'[40]

Undoubtedly part of the fascination with land values lay with the possibilities they appeared to offer for accurate measurement of the urban scene. If they could be considered as indices to social organization, as a measure of social distance, then a greater precision to the science of society would be possible. However, the ecologists shied away from analysis of the property market, and turned instead to human ecology to encompass their interest in the influence of the land market over the transformations in the local communities of the city.

4 *Dominance, invasion and succession*

Familiarity with the city of Chicago had led to recognition of orderly sequences of change in the character of urban neighbourhoods. The extension of the commercial centre to the surrounding residential districts, or the movements of immigrant groups into the inner city, triggered off a predictable pattern of displacement and movement of population through the rest of the urban area. This sequence of population movement formed the basis of Burgess's zonal model of city structure,[41] and Park's concept of the natural area. Explicit use is made of ecological terminology. The area of 'dominance' in the city is its commercial centre; this is equivalent to the dominant species in a plant community, and sets the conditions for the existence of subordinate species, in this case the city's social groups. Populations or activities 'invade' territories, displacing previous occupants, until they 'succeed' to the control of that space. The resultant segregation of the population, although clearly related to land values, is therefore equivalent to the symbiotic balance maintained by plant and animal species in an ecological community, and is therefore natural in derivation. As Alihan comments, 'Actuality here merges with inevitability'.[42]

In choosing to explain segregation of the population in terms of ecological concepts, and in adopting the aggregate of population as the unit of study, so the Chicago School gave a mechanistic impersonal air to change in the urban environment. A blind inevitability is given to the most extreme subdivision of the city's territory. It is instructive to compare a social historian's account of the making of a ghetto with the ecologists' misleading formalism in description. Osofsky describes Harlem, built up as a middle-class suburb to New York between 1880 and 1904, in a series of speculative building cycles as transport to the northern end of Manhattan Island was improved.[43] In 1904 the property boom was over, the bubble burst, the market collapsed, and the speculators brought in black tenants to occupy unlet apartments as the only group other than the middle classes prepared to pay high rents for their accommodation. The turnover in population was hastened by 'blockbusting' tactics in which black tenants were brought in to force white owners into selling at low prices. The 'invasion' was promoted by speculators in housing, and the 'succession' was hastened by hard-headed businessmen after a quick profit. Segregation, and the 'dominance' of the black population in Harlem, was assured by the number of negroes arriving in New York, and debarred by restrictive covenants and discriminatory letting from tenancies elsewhere in the city. In fact the black ghetto of the American cities comes closest to Park's conception of the 'natural area'.

5 *The natural area*

Perhaps the most distinctive feature of the large city, any large city—pre-industrial Peking, medieval London, industrial Chicago, underdeveloped Calcutta—is its division into distinctive quarters or neighbourhoods. No large city is an undifferentiated mass but a recognizable clustering of neighbourhoods distinguishable not only by their physical appearance, but their population composition, and their reputation. This observation of the spatial separation of the city's population, the existence of separate worlds of experience, was the basis of Park's vision of urban life: the neighbourhoods, colonies and segregated areas of the city are seen to be 'a mosaic of little worlds which touch but which do not interpenetrate'.[44] It is these that give the city its diversity, novelty and excitement. An individual may move from one to the other, maintaining separate identities in each, the eccentric find a sympathetic circle, the uprooted peoples of the rural world maintain their self-esteem in confrontation with an industrial culture.

From initial observations as to the correspondence of the spatial divisions of the city with the 'racial, cultural and vocational interests' of its population, Park and McKenzie developed a theory as to their evolution. The separate communities of the city were the end product of the processes of invasion by which groups established control over a locality although they were in turn liable to be disrupted in successive phases of the city's growth. Apparently it was considered that 'biotic balance and social equilibrium' would always be reassumed. Phases of rapid change would be followed by the crystallization of the neighbourhood as a community with a distinctive array of institutions, attitudes and even personality structure. Consequently,[45]

> Every natural area has, or tends to have, its own peculiar traditions, customs and conventions, standards of propriety and decency and, if not a language of its own, at least a universe of discourse in which words and acts have a meaning which is appreciably different for each local community.

Park went on to argue that this statement held even in areas where the traditional marks of a community were absent, as in Bohemia, or the rooming house districts. Here too there is a convention of behaviour, even though that convention may be unconventionality.

Park had a seemingly unshakeable belief in the environmental or spatial basis to man's social existence. Despite a recognition of the privacy and peculiarity of personal experience acquired through an individual's mastery of space, he was still prepared to state: 'All forms of association among human beings rest finally upon locality and local association.'[46] Then, as later, he was reluctant to attach any credence to theories of social organization based on principles of occupation, class or association, and was unconvinced of the social viability of organization based on occupational or vocational interests.[47] The solidarity based on the interdependence of associational groups is fragile; the interdependence less real than the competition between groups for status and position within the city. It is a solidarity based not on sentiments but on interests, a 'sociological monstrosity'.[48]

In this way Park set urban sociology upon its subsequent course: the investigation of the local bases of social solidarity. The separation of life in the community from the world of productive activity is tacitly accepted, and the latter dismissed as inconsequential for the problems Park and his contemporaries considered crucial to sociology—social control, consensus, the maintenance of individuality, and the future of democracy. The major issue confronting the European sociologists, the relations between social classes, went unrecognized by the American academics.

The interest in the spatial delineation of human association was strengthened by methodological concerns. The pressures for the 'scientific', i.e. statistical, advancement of sociology were strong, and although Park never allowed of their use in sociological inquiry he was prepared to make a case for the use of statistics in ecological studies. Statistics could be used in accurate measurement of spatial distribution of the population which 'determine so irresistibly and fatally the place, the group, and the associates with which each one of us is bound to live'.[49] A correlation between space and social structure is a necessary precondition of statistical methodology.

There was, however, a stronger methodological argument for the adoption of natural areas as the unit for study. They are seen as a 'frame of reference' within which the sociologist can operate. Without some such model for the social and spatial structure of the urban area, work in the social-anthropological field tradition which Park advocated as the method for sociology cannot be undertaken in the large city. There, only in limited respects can the local community be isolated for study. Its origin, continued existence, and eventual decay are determined by changes in the pattern of investment in land and use of labour, and consequent movements in population over the city region. Community studies undertaken in the neighbourhood without reference to some conceptual framework for the urban community run the risk of misinterpreting the local situation.[50] The success of the Chicago School depended on the skeletal framework Park provided for the very diverse studies undertaken by his students.

Not only are the natural areas, in Park's view, the only basis for urban research, but they also 'establish a working hypothesis in regard to other areas of the same kind'. That is, Park assumes that having established the distribution and characteristics of the natural areas of one city, this model of the ecological community could be extended to other cities of the same general type, and so findings on one natural area in one city could be extended to similar areas elsewhere. To quote: 'It is assumed . . . that people living in natural areas of the same general type and subject to the same social conditions, will display, on the whole, the same characteristics.'[51]

This is precisely the assumption made by all urban sociologists in using local community studies. Suburbs, slums, twilight zones, Bohemias, and ghettos, all these are assumed to be generic categories. It is clear that Park hoped to accumulate a set of hypotheses as to the nature of urban existence generally, through case studies of individual areas that might be treated as types and used for comparative studies. There is a possible comparison here with Durkheim. Park is careful to draw attention to the natural areas as 'things' and not events; and his methodological arguments are reminiscent of Durkheim's treatment of the normal and the 'average' type. But whereas Durkheim is seeking to simplify the description of many instances, Park is trying to generalize from the one instance on a demonstration of its typicality.

These methodological difficulties broached by Park remain outstanding. Any sociologist exploring one of the social worlds of the city raises two problems: the relation of that particular social group to others in the same urban area, and the comparability with other similarly positioned groups elsewhere. Failure to consider these problems means either the accumulation of particularistic monographs as sociological curiosities, or their use for uncontrolled generalization. Park simplified his task by assuming that all cities were somehow similar, and that the social worlds of the city were in regular relationship. Now that sociologists are prepared to consider cities and not 'the city' as their subject matter, there is no longer a basis for comparison between places, and for generalization from local studies. Rejection of the urban–rural distinction as sociologically suspect, and at the same time modifications to the alternative class theories of society, has left an indeterminate void between environmental and social theories of urbanization.

The concept of the natural areas is the basis of Park's theory of urbanization. The end state of the process of urbanization is not a homogeneous mass, but a heterogeneity of life-styles each founded in control over a territory. Although Park uses the language of functionalism, and plainly sees the natural areas as essential to the city as a social unit, they are equally a response to the varied and fundamental needs of its population, and as such will be of enduring significance. Park answers to the problem posed by Simmel, that of maintenance of individuality in the metropolis, and that posed by his contemporaries, the prevention of breakdown in urban society, in a theory of community development which allows for self-esteem without prejudice to the way of life of others. Social control can be achieved without sacrificing the freedom of action opened up by urban living, in the development of distinctive cultures in the neighbourhoods. And this diversity of opportunity can be achieved without risk of breakdown of the city as a social unit, for, although communication between the different cultures may not be possible, they are inextricably bound together through the natural order founded on 'competitive co-operation'. Conceptualization of the city as an ecological community enabled Park to present the new metropolis as an ordered entity in which both community and freedom were realizable.

2 Community and freedom

A major issue posed by the middle classes of established America, including the new profession of sociologists, at this time was the maintenance of social order in the large cities. Rapid industrialization had entailed massive urbanization and this, linked with uncontrolled immigration from Europe, had brought together dispossessed, rootless populations among whom there was no 'language of discourse', and whose circumstances of existence challenged entrenched beliefs as to the preconditions for democracy. It was considered that in the city 'the problem of poverty, of medicancy, of unsanitary surroundings, and of debasing social influences, are met in the most acute form'.[1] Life in the cities was posed as a problem, a problem of social disorganization caused by urbanization, and the immediate task of sociology as a science, and as a practical enterprise, was the elucidation of this social problem. Sociology in its early days was an integral part of the 'social gospel', the 'ruling motivations of which were a passion for social reform and an adoration of science',[2] and its advocacy has to be seen in the same light as the concurrent enthusiasm for eugenics, which similarly seemed to offer possibilities for the control of 'the dependent, defective and delinquent classes'.[3]

One aspect of this concern with the failure of traditional mechanisms of social control in the new cities, was the initiation of a community settlement movement which was to 'remove hollowness by reconstructing localism, by building within the city little centres of neighbourly communication'.[4] Hull House, Chicago, had been set up in 1889 by Jane Addams with the express intention of recapturing a sense of community in the cities.[5] This concern for participation and communication is carried forward in the writings of Dewey, who throughout a long career examined the conditions necessary for a reinstatement of a communal reorientation for a democratic society. Mills cites a passage from a late book:[6]

> Democracy must begin at home, and its home is the neighborly community. It is outside the scope of our discussion to look into the prospects of the reconstruction of face-to-face communities. But there is something deep within human nature itself which pulls towards settled relationships.

Dewey in America, Durkheim in France, Weber in Germany, Morris, Howard and the Garden City Movement in Britain, all represent middle-class groups who, while liberal in their political attitudes, were nevertheless deeply disturbed by the direction taken by industrial society. For the American middle classes, themselves drawn largely from agrarian America, the change was doubly disturbing in that the environment of their own childhoods was being jettisoned.

The writings of the Chicago School are frequently interpreted as belonging to an anti-urban tradition which reveals 'a romantic attachment to the pre-urban community'.[7] A concern for social order, treatment of city life as pathological, and the search for bases for community living in the city comparable with those of the village, are seen as expressions of a small-town reluctance to envisage forms of social solidarity other than those previously experienced. This attitude was strengthened by evolutionary conceptions of social change imported from the European sociologists. 'In the writings of Park there is perpetuated that sociological romanticism which stressed the aloneness of the individual in an urban environment increasingly impersonal and indifferent to his hopes and desires.'[8] But this assessment is notably unfair to Thomas, Park or Wirth, all of whom evidently relished urban living and had no desire to direct the future to a rural past.

More careful interpretations of their treatment of urbanization do, however, present a more equivocal view of their position. Goist, for instance, views Park's work as representing a 'double-visioned' response to urbanization in which he tests the validity of older forms of association in altered circumstances, and assesses the potential of the new forms.[9] He argues that Park's theory does not belong to an anti-urban tradition but to a quest for community which has been a mainstream feature of American culture. Stein similarly makes clear Park's consciousness of 'the creative as well as the disorganizing potentialities of marginality'.[10] Potential for innovation would be unleashed in conditions of rapid change. And Sennett, drawing largely from the 1915 essay, presents Park's writings in terms of a concern for urban freedom. The city's division into the natural areas meant that it was impossible to compel conformity. 'Thus Park saw the city as the medium for the emergence of free men, whose personal development could transcend general societal standards, whose innovations could provide the basis for historical changes in society itself.'[11]

For all Park's alertness to the new forms of association and communication emerging in Chicago—voluntary associations, interest groups, newspapers—he did place his authority behind the older forms of association. Although he had a deep sense of the essential isolation of the individual, 'each in his own little world hugging his personal secret',[12] and for him the fascination of urbanization lay in the coexistence of these strangers, each of whom had to be considered as 'a unit of thought and action',[13] he defines the central problem of sociology as that of social control. He, along with the others at Chicago, stressed group cohesion, social order and consensus, in a thoroughly sociological way. Individualization and social disorganization merit 'a new parochialism': it is only the few who can take advantage of urban life and move out of 'the little world of the locality'.[14]

There is a consistent ambivalence in the writings of the Chicago School indicative of the individualistic tendencies in American intellectual life, and its conservative concerns. The very traditions and currents of thought in which the sociologists were attempting to establish a new field of study are indicative of the tensions in American liberalism. The writings of Simmel, for instance, speaking of individuality and secrecy, could be opposed to those of the American sociologists concerned with social disorganization and the reconstruction of community, and the views of James, who had a passionate belief in the freedom of the will, could be set against those of Durkheim with his relentless insistence on the primacy of the social. Neo-Kantian liberalism, sociological conservatism, pragmatism and American progressivism are all present as strands in their appraisal of urban living. Bramson uses the inconsistencies and tensions evident in the work of Park, and subsequently Wirth, as demonstration of his thesis that sociology, with inherently conservative concepts and interests, must sit uneasily in a liberal-individualist tradition:[15]

> sociological concepts with their load of anti-liberal freight derived from European social and political philosophies, undergo a subtle transformation at the hands of American sociologists when imported into the liberal American context. Or, . . . these concepts are uncritically applied to the American realities without regard to the deficiencies of the resulting analysis.

However, the argument is suspect when the writings of the Chicago School are examined more closely. Park, for instance, takes the writings of Simmel, the classic liberal, and redirects his analytical sociology to more specifically moral, and conservative, questions consistent with the concern of his contemporaries for the future of community in a rapidly changing society.[16]

Park took over the programme for sociology initiated by the first generation of sociologists, Giddings at Columbia, Small and Thomas at Chicago—the city is still the 'natural laboratory of social science' —but in his hands the emphasis is subtly, but certainly, shifted from social disorganization and personal demoralization to the opportunities for personal achievement, innovation and creativity. City living enabled the individual to realize his own potential untrammelled by primary group affiliations and customary restraints. Park's phrase is 'the mobilization of the individual man'.[17] This new, urban, freedom was gained in the very separation of the city's population into different communities. Not only could the individual pass from one group to another changing his conduct as he did so, thereby gaining a personal freedom of action, but the sheer diversity of social experience, of expected standards of behaviour, meant that it was impossible to impose one standard of behaviour, and out-legislate deviance. The segmentalization of city life, in circumstances of social diversity, had an outcome of personal freedom.

Park attempted to bring together two opposed views of urbanization. In his own mind sure of the progressive and liberative consequences of urbanization in history,[18] he was equally open to the view of his contemporaries that urbanization meant the loss of community and with it the collapse of democracy and an irrevocable degeneration of social existence. And so he presents an image of a pluralistic, segregated city in which the individual is to be free to move and act, although the existing divisions in the urban community are to be intensified and the distance between social groups increased, as one answer to the liberal problematic. Individuality will be enhanced, and yet community remains a living force in everyday life. In fact he turns the contemporary fear and rejection of cities on its head: the city, all appearances to the contrary, is the only basis for a free community. The gap between Park, the European sociologists, and his own student, Wirth, is complete. Tönnies and Weber could neither see the city as a community nor community in the city, Simmel had only been able to point to contrived individualism in face of an overwhelming environment, and Wirth was to reject Park's prediction of the emergence of local communities in the city. Park alone among the sociologists holds out a future for community in the city.

Park's theory of community development resolved the antinomy of community and freedom, if certain ecological postulates were accepted. His statement had the added attraction that the problem of urbanization was presented as one that could be resolved on a local basis through the manipulation of the physical environment. Just as the middle classes of late Victorian London had posed the problem of their metropolis as one of slums, not poverty, and therefore one to be solved by the provision of housing, the equivalent social classes of urban America phrased the issue of urbanization as one of community, not inequality or exploitation. A good society could therefore be planned for through the rearrangement of the city's neighbourhoods without challenging the existing distribution of power and wealth. Booth's survey of London, identifying the urban problem as that of poverty, was therefore very much more radical than Park's theory of urbanization. Although Park was prepared to go against a widespread assessment of urbanization, he did not challenge the terms in which the issue of urbanization was posed for sociology. His statement of the city as an ecological community, a constellation of natural areas, is an answer to the problematic posed by a successful group of professionals and intellectuals. If now that statement is in doubt, it should be necessary to reassess the terms in which the problem was initially set. Was the demoralization and degradation evident in Chicago symptomatic of urbanization as such; was the problem of the city that of community?

The derivation of human ecology

Human ecology now seems an anachronism in urban studies; there would be little support from any quarter for explanations of urbanization in naturalistic categories which take economic action outside of comprehension or control. The most constructive interpretation would be to regard the ecological concepts as analogies which have the merit of drawing attention to the powerful effect of changes in the urban property market on community life. Further, Park's model for community development can be regarded as a counter to the main trend of sociological thinking on urbanization popularized by Wirth, and as an imaginative justification for neighbourhood planning. More commonly it is argued that it 'interprets as a disjunction between Nature and Culture what is the effect of a particular social matrix, determined by the dominant relations of production'.[19] But in Park's terms human ecology enabled him to answer to certain problems posed for sociology by the critical, informed opinion of his own social world.

The major issue posed for sociology by the professional groups with which Park and his colleagues had the closest affiliation has already been outlined—the maintenance of individuality and autonomy of action in the new society and the reconstruction of community. In many ways Park's treatment of community in the city parallels Dewey's discussions on education for democracy, directed at a not dissimilar public. A democratic society, declares Dewey, 'must have a type of education which gives individuals a personal interest in social relationships and control, and the habits of mind which secure social changes without introducing disorder'.[20] The formulation of human ecology provided an answer to this statement of the problems of American society; it was therefore in James's pragmatic terms 'profitable', 'good for so much', 'passing muster in practice'. In that human ecology allowed for a statement of city life in which community was a natural outcome, and yet freedom of action was not imperilled, it offered a provisional solution to the issues posed for sociology by a certain stratum in American society. Park's theorizing, however misleading it appears now, represents an attempt to 'reduce the tension between a social event or process which he takes to be real and some value which this has violated'.[21] Urbanization did not mean an end to community or democracy. Human ecology's success in resolving these issues must be considered an important factor behind its acceptance, elaboration and transmission by Park and his associates.

Additionally, the sociologists were presented with the more personal problem of their status as representatives of a new profession. There were strong pressures from the academic community to carve out an easily described subject matter, and to justify their claims to scientific standing. The consequent interest in statistical investigation of social behaviour was intensified by demands for information and guidance from the many welfare associations in the city. Community development organizations, playground groups, settlement houses, all looked to the sociologists. Although Park conducted a relentless battle against the use of statistics in sociology,[22] he did nevertheless feel it incumbent to make way for these demands for precision. Again ecology provided an answer, in that it would enhance the reputation of sociology as science, and permit full use of the growing statistical armoury.

The gains from human ecology as a theory must be considered, for, given the thinking then current in sociology, related subjects such as geography and psychology, and in philosophy, its derivation at this time presents a problem. It reversed the position previously taken by the department at Chicago under Small, in which interests and economic action had been considered as subordinate to sentiments and the social order, went counter to the whole tenor of work in the closely associated department of psychology,[23] and went further even than geography in its emphasis on the autonomous forces in the man-made environment. There was also a major modification to the pragmatic philosophy then regnant at Chicago. The philosophy initiated by Dewey and Mead was a philosophy of possibilities, innovation, control. Man was seen as an active agent, manipulating the environment and creative of his own future, the universe as unfinished, and the situations confronting the individual as indeterminate. Park takes the basic equation of organism and environment, and the naturalistic view of man, but turns round pragmatic philosophy to consideration of a universe which is outside any individual's control, and which sets the terms both for his survival and his social development. Whereas Dewey, for instance, had sharply criticized the 'natural' conception of competition, Park puts the struggle for existence back into central place. Pragmatism forms part of a mainstream of progressive thinking and social criticism, while Park shows a tacit acceptance of established institutions. The anomaly is not the biological terms of reference, for James and Dewey had demonstrated the creative possibilities of biology for philosophy and social science, but the reversion to the language of nineteenth-century Social Darwinism, as seen in the writings of Spencer and Sumner.

Social Darwinism

A starting impulse for sociology in America had been the impact of Spencer's application of the biological scheme of evolution to human society in *The Study of Society*.[24] The law of natural selection and the survival of the fittest, ideas of evolution and progression in an organism, are carried over from biology to social studies. In Sumner's hands the same doctrines had yet more explicit application to the cut-throat world of American industrialization. He was prepared to argue for a direct analogy between animal struggle and human competition, natural selection and the success of the few, thus presenting a biological apology for *laissez faire* and an unequal society. Society was immutable, a superorganism changing at geological tempo. Hofstadter sums up his career as 'he came to preach the predestination of the social order and the salvation of the economically elect through the survival of the fittest'.[25] But by the twentieth century, the impact of Darwinism on sociology was receding, although the legacy was still evident. In Small's hands, for instance, it is transmitted in the assumption of conflict involving the collision of interests, as the primary, normal, pattern of association. This conflict resolves itself into co-operation through accommodation and socialization. Society is therefore an emerging process against a background of conditions supplied by nature.

The influence of Small on the direction taken by sociology at Chicago has been greatly underestimated. Not only is his influence on the 1921 textbook easily discernible, but he anticipated Park's move into human ecology in his own *General Sociology* (1905). A series of quotations from this demonstrates his position: 'human beings are virtually one with the plants, and the animals in depending upon nature for their own organic conditions and qualities and conduct. . . . Life is an affair of adjusting ourselves to matter-of-fact inexorable nature'; 'Social life must square with the requirements of physical surroundings';[26] and,[27]

> any competent theory of human associations . . . must square with knowledge about those physical and vital relationships upon which the later social phenomena rest. In a word, *some of the social forces are not social at all* . . . they are themselves phenomena neither of consciousness nor of association.

Park's novelty lay in the extension of 'the real world' which for Small had meant 'the grudging soil, the cruel climate and the narrow space of the region'[28] to include the city. This now became the 'natural habitat of civilized man'. Small, however, unlike Park, had been very careful to draw attention to the limitations of environmental explanations of human behaviour. The physical environment was something that had to be considered in sociology, but only as one among other factors.

Park was of the same generation as Small, and like him heir to the great intellectual controversy of the nineteenth century. Although dissatisfied with the speculative style of the first period of sociology, he never shook off the intellectual freight of that century. Throughout his thirty productive years, 1914–44, he continued to draw on its literature—names like Bryce, Teggart, Bagehot, Emerson, Gibbon, Spencer and Adam Smith—the library of an educated gentleman of the old school. Darwinist categories of thought were something that had to be reconciled with the sociology gathered from France and Germany and hybridized in Chicago. In the textbook of 1921 there is an extraordinary amalgam of the two traditions: Simmel is juxtaposed with Darwin, neo-Kantian formalism is encrusted with Small's brand of Social Darwinism. But for the conception of the city as an ecological community Park goes back beyond Small, to Spencer. The city becomes a superorganism in which competition provides the rationale for the ordering of social groups; the resolution of the process of urbanization as an equilibrium, a constellation of natural areas, can be regarded as an application of Spencer's principles of evolution: 'matter passes from an indefinite incoherent homogeneity to a definite coherent heterogeneity'. The only concessions made to criticisms of Darwinism and the mood of critical reform was in the substitution of 'competitive co-operation' for the tooth-and-claw world of Spencer and Sumner, and the recognition of social order that might intervene in the process of natural selection.

European sociology

An impressive characteristic of the Chicago School in this period was its openness to a wide range of ideas. Under Small's long régime no attempt was made to stamp the department with a distinctive brand-image—the writings of American, French, German and English social scientists were as worthy of consideration as the next. The same tolerance and open-mindedness is shown in their sympathetic treatment of deviance and eccentricity among the urban population, and their preparedness to drop *a priori* categorization in favour of the individual's own definition of the situation. In this they show themselves at one with the dominant pragmatic temper. Thomas's famous dictum, 'If men define situations as real, they are real in their consequences', is a clear expression of pragmatic criteria of knowledge. James had declared pragmatism to be 'friendly, democratic, undogmatic', it 'unstiffens all our theories'; all previously held views are liable to be cast into the melting pot of experience, for pragmatism had 'no prejudices whatsoever, no obstructive dogmatism, no rigid canons of what shall count as proof. She is completely genial. She will entertain any hypothesis. She will consider any evidence'.[29] Park, once a student of James's, and temperamentally at home in this easy-going philosophy, shows to the full its consequences for sociological discussion in his lack of rigour in argument, indeterminacy in statement, and his eclecticism.

Park's borrowings from the European sociologists were highly selective and he used their ideas with little regard for the intellectual context from which they had been derived. In this fashion he is able to bring together the writings of Durkheim and Simmel, using the former's 'sociologism' and concern with the moral regulation of society to transform Simmel's science of the forms of association into the science of collective behaviour. So although Park can be regarded as transmitting Simmel's thought into American academic sociology, he succeeds in drawing up a sociology which was 'concrete, dynamic and oriented to social consensus, whereas Simmel's was abstract, structural and oriented to sociological dualism'.[30] Park evidently combed European sociology for answers to questions posed in the aftermath of the controversy over Darwinist interpretations of man in society: how was human collaboration possible, what were the mechanisms by which the individual was induced to remain in society, how was social order maintained in face of the undeniable self-interest of the struggle for existence?

Although Weber's early methodological writings were of interest to Park, the main bulk of his writings were published too late to be of influence. In any event, the division of labour in the university at Chicago, in which there was a strong political studies department and where sociology's focus had been restricted to 'the study of men considered as affecting and affected by association',[31] would have prevented any absorption of Weber's writings on power and authority into sociology by Park. At the same time, the ideas of Marx, with which Small had been engaged, were allowed to drop; the stubborn question of the relations between social classes was no longer of interest.[32] The issues with which Weber or Marx had been concerned were remote to Park and his contemporaries. The ideas of Simmel and Durkheim, both concerned with the designation of sociology as a science with subject matter distinct from that of psychology, history or economics, were more readily assimilable into the direction of inquiry already instigated at Chicago. The field of sociological investigation became restricted to the small-scale world of men in groups, a world of socialization, shared culture, and 'collective representations', linked with that of interaction, and mutual association. Simmel's statement of society as constituting 'forms of reciprocal influencing'—'only when an influence is exerted whether intentionally, or through a third party, from one upon another, has society come into existence in place of a mere spatial juxtaposition or temporal contemporaneousness of succession of individuals'[33]—would have had a ready hearing in an intellectual circle accustomed to the idea that 'the destinies of human beings are always bound up with the fate of the groups of which they are members'.[34]

Such an attitude was not incongruent with pragmatic philosophy, which although strongly individualist in derivation moved steadily towards a conception of man as a social being whose very personality was rooted in his community. Mead, for instance, worked from consideration of the relation of individual to environment, to the assignation of meaning to the individual act in co-operative group action; his is 'a socialized individualism.'[35] Dewey similarly consistently referred to the role of sociality in securing participation and collective action by which individuality is to be realized. 'In social feeling', he wrote in 1887, 'we merge our private life in the wider life of the community, and in doing so, immediately transcend self and realize our being in its widest way.'[36] The interest of both sociologists and philosophers lay in the study of consensus, and for both the most fruitful line of investigation seemed to be the examination of the social processes engendered by the interaction of individuals.

Pragmatism

Sociology, at Chicago, in fact had its origins in a convergence of the working vocabulary of European sociology and pragmatism. On the face of it this seems an improbable cross-fertilization: the emergence of classical sociology in the period 1880–1920 in the hands of Tönnies, Durkheim, Simmel, Weber, is either interpreted as a conservative reaction against the changes in European society, or seen as a rephrasing of some dominant ideologies in nineteenth-century thought—Marxism, rationalism, idealism and positivism.[37] Pragmatism, while critical of industrial-urbanization, was a philosophy of acceptance and moderate reform, and Darwinism, not Marxism, the inspiration for inquiry. But in practical terms this gulf seemed to matter very little. To take one example: Simmel's delineation of sociology from other social sciences by a distinctive method, the analysis of the forms of 'sociation', would have found a ready hearing in Chicago from men accustomed to thinking of science as the practice of science rather than as a set of established theories. Similarly, the rather exceptional empiricism set out by James, 'radical empiricism', came very close to the *verstehen* method for sociology set out by Weber. The general drift of sociology gave force to the pragmatists' concern for the study of individual experience as mediated by social interaction.

Pragmatic philosophy, particularly as set out in the statements of James, seems to have been a major inspiration behind the methodology advanced by Park for sociology. The basic postulate of

Pragmatism (1907) and *Essays in Radical Empiricism* (1912)[38] was the primacy of experience; reality was 'an experience continuum'. Knowledge was drawn from what James variously terms 'the world of concrete experience', the 'immediate flux of life', 'pure experience', and the great philosophical debates as to substance, consciousness, truth, had to adopt this as their testing ground. Knowledge was to be sought, truth derived, 'among the interactions of men's experience in detail'.[39] Philosophy had to be brought out of the classroom into the street, and knowledge was to be democratized. The radical element to this methodology came in James's insistence that 'the relations between things . . . are just as much matters of direct particular experience . . . than the things themselves'.[40] This gave express sanction to the Chicago School's redefinition of sociology as the study of the processes of interaction between men in groups.

In an earlier statement James had made a distinction between two forms of knowledge, 'acquaintance with' and 'knowledge about'.[41] The former, undirected, and largely unconscious, he later takes as the basis for the theory of knowledge set out in *Radical Empiricism*. Knowledge was a process derived in 'working living experience', always in transition, yet always continuous: 'Knowledge of sensible realities thus comes to life inside the tissue of experience. It is *made*, and made by relations that unroll themselves in time.'[42] James rejects the dualism of subject and object hitherto implicitly accepted in philosophy, and argues that knowing is a particular relationship between two portions of what he terms 'pure experience'. Consciousness is a relationship and not a 'special stuff or way of being'; in some circumstances a given portion of experience is a knower, and in another something that is known. Knowledge is no matter of observation but of a fusion between the object studied and the experiencing subject. This means that knowledge is located in the person of the knower; each man has his own universe, based on his experience. James paints the picture of the world as a 'quasi-chaos' in which[43]

> innumerable thinkers, pursuing their several lines of physically true cognition, trace paths that intersect one another only at discontinuous perceptual points and the rest of the time are quite incongruent; and around all the nuclei of shared 'reality' . . . floats the vast cloud of experiences that are wholly subjective.

Park's own career as a journalist must have strengthened his conviction of the need for personal and first-hand encounters with the day-to-day world. James's arguments gave weight to an already held sense that the direction for sociological investigation must be towards empirical study of this world of experience, a world of human actions in which every gesture is 'anticipated, checked, inhibited, or modified by the gestures and intentions of his fellows'.[44] Park, like Thomas, was not a methodologist, and never developed this 'phenomenological empiricism'.[45] He considered simply that sociology must investigate 'the individual experiences of men and women'[46] and to this end he recommended 'detailed and local studies of man in his habitat and under the conditions in which he actually lives',[47] the use of documents, particularly the life history, and intensive fieldwork. There was no substitute for tramping the streets, for direct acquaintance with the many worlds of the city, if possible by participation in the life of those under investigation. Only in this way could one begin to acquire an understanding of man's dual existence, his public and his private life; the task of sociology was akin to that of philosophy in acquiring knowledge that was relevant to man's action and experience, so enhancing awareness of his own self.

Park's attitude to sociological investigation sets him in sharp opposition both to the tough-minded temper of his colleagues at Chicago, intent on gathering yet more information on facets of urban life that would enable control over its disorganization and evident pathologies, and to the dominant mood of urban investigation initiated in England. Booth's mammoth survey of poverty in London had triggered off similar surveys in America, but Park considered that its merit lay not in Booth's statistics, but 'his realistic description of the actual life of the occupational classes . . . their passions, pastimes, domestic tragedies, life philosophies'.[48] Hard data, and policy direction, were not his objectives. His methodology differs also from that of his contemporary, Geddes, who set the pace for urban studies in Britain. Geddes was more interested in the place than the people, prepared to consider the city as a real entity subject to rules of direction and control. Under Park, strongly influenced by pragmatic philosophy, knowledge of urban culture was to be sought among 'the interactions of men's experience in detail'; urban sociology became the study of this world of personal experience. The legacy of Park's direction of research at Chicago is seen in the urban community studies that have come to be sociology's most popular contribution to urban studies.

A main current to urban sociology was therefore initiated in the application of the vocabularies and rules of inquiry formulated in the early years of the twentieth century. Elements from European sociology and American pragmatism were drawn together by Park into a set of background assumptions as to the domain of sociological investigation and the nature of society. The resultant restriction of investigation to 'the patterned collaboration and accommodation among persons',[49] and the social to the consensual, must have presented Park with a theoretical problem when seeking to deal with the city in sociological terms. He had had a career as a reporter, a man trained to recount what he found on the ground: what he encountered in Chicago was a city that extended far beyond any one man's experience or consciousness, which in the terms of his sociology had no basis for the establishment of the consensus on which society depended. Its jostling cosmopolitan population with its diverse interests and experience appeared more akin to the fluid inchoate mass described much later by social anthropologists in the African towns, than the bounded social groups and established hierarchies sociology led him to expect of an ongoing society. There was a lack of congruence between his own sense of what was real in Chicago and the domain of social theory. The derivation of human ecology filled this theoretical void, so helping to restore confidence in the rationality of what had seemed unreasonable and unnatural, thereby bringing the city back into the realm of the permitted, the normal, and by implication the controllable.

In pragmatic terms, ecological theory answered to 'the needs of the situation'. Urbanization had become a reality of everyday life, an element of the experiences of an increasing number of individuals, so the city as the context of experience had to be brought into comprehension. In Dewey's terms an indeterminate situation was to be transformed into a determinate one through inquiry, so furthering the mutual adjustment between the individual and his environment. For a number of reasons, the exclusion of the issues of power and scarcity from sociology, the restriction of the social to the immediate and the consensual, the adoption of personal experience as the point of reference, sociology could not answer to the posed problem of the city's continuance as a social entity. Human ecology, on the other hand, did not take sociology into economics, allowed for competition without denying man's social nature, and provided an explanation for the city's continued existence as a community. In the terms in which the 'urban problem' was posed by Park and his colleagues, human ecology was a highly successful social theory.

To summarize: human ecology as a theory answered to a specific issue posed by 'a protest philosophy of the minority intellectual',[50] i.e. pragmatism. Its derivation must be related to the concerns of liberal democracy at a period of rapid social change. The world had swung round this group too fast for them to assimilate the new environment to the older social theories. The vocabulary taken from sociology could not encompass this new world; there was a dissonance between experience and this intellectual tradition, which human ecology was to bridge. Human ecology reconciled sociology and Darwinism, and provided for the scientific investigation of the city in terms which encouraged reform, without questioning the basic tenets of capitalistic society developing so assuredly in the USA.

But for its continued influence in urban studies, human ecology would now be of strictly academic interest as a digression from the mainstream of sociology. But its key assumptions have been incorporated into the conventions of urban sociology. Its influence is still felt, even in the 'new wave', in three respects: (i) the maintained belief in the integrity of the city as an interconnected system; (ii) the focus on the effect of spatial arrangements on social life; and (iii) the restriction of 'urban' sociological investigation to issues of community living, as a world apart from that of production.[51] The terms of reference for urban sociology have not been shifted far from those derived by Park and his associates for sociology. In that environmental resources, and community, remain key issues, it would seem that the terms of debate have not shifted far from those posed in the vocabulary of liberalism, in Chicago, 'the city of destiny'.

3 A continuing tradition?

The intellectual enterprise at Chicago in this 'Golden Age' affords an opportunity for assessment of the implications of a philosophy for the conduct of sociology. The linkages between the sociologists and the pragmatists, James, Dewey and Mead, were direct and personal, and in many respects their sociology can be regarded as working out its postulates. They accepted its individualistic frame of reference: in contrast to Marxist analyses in terms of objective existence and the impersonal force of history, their sociology took as its concern personal experience. Although they modified the organism-and-environment equation in human ecology, they similarly took the bio-physical world as their starting point, so locating 'all problems between man and nature, instead of between men and men'.[1] Pragmatic contempt for all forms of metaphysics is shown in the sociologists' impatience with abstract theory, and the philosophers' restoration of the world of everyday experience as the basis for knowledge gave force to the methodology advanced by Park. Both groups shared a concern for what Dewey termed 'the immediate community of experience' and saw this as the basis for the good society. At every point sociology and pragmatism converged and interpenetrated.

This closeness of interests, vocabulary and methodology leaves them open to the same range of criticisms. Chief among the reservations about pragmatism is the adoption of naturalistic categories for explanation of thought and action. The biological terms in which it is couched not only serve to minimize the cleavages and power divisions within society by locating all problems between man and nature, but also aid the tacit assumption of cultural sanctions for activity of a certain kind '. . . value-decisions as value-decisions are assimilated into the biological and hidden by formality'.[2] The broad outlines of the existing social order were not called into question, and action was limited to the adaptation of immediate circumstances. A main driving force behind pragmatism was a belief in the power of man's intelligence to control his destiny, remedies for maladjustment between man and environment were in the individual's hands.

These criticisms are perhaps more pertinent now for planning than for urban sociology. There is a strong case for seeing town planning as heir to human ecology either as taught by Park, or by his contemporary, Geddes. In systems planning there is the same sense of the city as a complex organism in relation to a given environment. In McLoughlin's model 'the human eco-system is "driven forward" by modifying actions taken by individuals and groups in society'[3] in much the same way as Park's urban organism derived its characteristic pattern of land uses from competition. The methodology of town planning is that of human ecology as outlined and taught by Park and Burgess: population characteristics, land use, land values, population mobility, the network of communications, and the differentiation into social areas are the object of study. Planning looks at men's activities in space rather than their involvement with one another—'the universe of discourse becomes limited to externalities'.[4] And although the explicit biological categories are dropped there is the same search for universal scientific statements, this time in the language of cybernetics and systems theory.

In many respects the planning profession can be seen as latter-day pragmatists. They too come from the socially mobile middle classes, are liberal in outlook and have a belief in the need for step-by-step adaptation and reform to permit a more democratic society. There is something of the same ethical content as in Dewey's philosophy: the moral and the scientific have frequently become merged in theories of planning. Although planning has its theories, it, like pragmatism, is essentially practical in temper, dealing with concrete difficulties, more concerned with action than with reflection, and impatient with abstract theory. Above all, planning and pragmatism share the same belief in the power of intelligence and reasoned debate, for not only is it considered that individuals have power to change their environment, but public consensus in the need for changes can be achieved through communication of reasoned argument as to 'the practical possibilities'.[5] Criticisms of the Chicago School and pragmatism are therefore of real interest in the debates over the role of planning.

A second line of criticism at pragmatism is addressed to its theories of knowledge. Kolakowski refers to their 'pedantic nominalism':[6] only when experience obliges the individual to admit a thing into his consciousness is it allowed to exist. The pragmatists, and with them the sociologists, made a choice as to what was most real—personal experience and the world of concrete objects, the individual's immediate milieu. Events, existences, beliefs, outside this domain were not admitted. Knowledge is immediate and personal. Russell refers to this as 'subjectivistic madness' and comments: 'unless we are able to infer things not experienced from things experienced, we shall have difficulty in finding grounds for belief in the existence of anything but ourselves'.[7]

The resultant difficulties for Park in seeking to deal with the city have already been described. Given the fragmentation, isolation and sheer diversity of urban experience, adoption of these canons of knowledge meant the restriction of sociological inquiry to the city's social worlds. On these terms a sociology of the modern metropolis simply was not possible. But whereas Weber, adopting not dissimilar criteria of action and meaning to the individual for sociology, had been prepared to abandon the city as an entity worthy of sociological investigation, Park was not prepared to do so. Human ecology was the result.

Central to Dewey's philosophy was the concept of inquiry. This he defined as 'the controlled or direct transformation of an indeterminate situation into one that is so determinate in its constituent distinctions and relations as to convert the elements of the original situation into a unified whole'.[8] However confused and fragmentary a situation appeared to be, underlying it there was an ordered reality, which intelligence was to establish. The whole process of inquiry entailed this adjustment of a situation into something congruent with reason. A similar attitude is expressed in these terms: 'That there is more order in the world than appears at first sight is not discovered till the order is looked for.'[9]

Important in the Chicago School's orientation is the same belief in 'seeing things whole'. Even though the 1921 textbook had as its focus processes of social interaction, there is none of Simmel's preoccupation with facets of social transactions, and when they turn to the city their total, or organic, approach becomes clear. Although they were fascinated by the cosmopolitanism of the city their quest was to understand it in its entirety, as a 'cultural area' (Park) and 'social entity' (Wirth). The fragmented, kaleidoscopic quality of urban experience was no indication as to the organic character of the city as a product of nature. The driving force behind the sociological enterprise at Chicago was the discernment of the ordering principles underlying a state of seeming disorder: the city had to be more than an aggregate, 'a congerie of individual men and social conveniences'; it was an ordered entity, a determinate environment.

The attraction of Park's conception of the city as an ecological community to subsequent generations lies in its statement of the city as an ordered entity in terms which, while opening up the question of its social coherence, do not presume this. Order is there to be established, perhaps through systems analysis, or ekistics, an order intrinsic to the urban situation. Consequently increasing research effort goes into deriving models for the interpretation of urban behaviour. The challenge from sociologists in a long succession from Weber to Castells, asserting, first, the dependence of the urban process on the nature of the social formation for which cities act as productive and controlling centres, and, second, the breakdown of the city as a community, has been largely disregarded. Stated baldly, cities are considered as real entities, and as structured wholes, entities in space which command their own future. Urban students have been very loathe to concede that perhaps cities are indeterminate environments in which there is no intrinsic coherence and whose seeming regularity of activities is either deliberately imposed, through, for instance, the administration of collective services, or derives from beyond the urban community. The pragmatists' assumption of interconnection and unity to the environment confronting man, and their quest for control of that environment through science, lives on in the urban studies research industry.

Since Park's head-on confrontation with the problem, urban sociologists have remained ambivalent in their attitude to cities as objects of scientific study. Well aware of their colleagues' scepticism towards an urban frame of reference, they have yet been committed by their very specialization into demonstration of the validity of spatial analysis in sociology. No longer content with random and piecemeal empiricism, renewed effort has gone into the reformulation of the ecological arguments for consideration of the city as an entity, either in terms of its markets in housing,[10] its structure of access to community resources,[11] or as 'a unit of collective consumption'.[12] An urban frame of reference is justified on either, or both, of two counts: first, that there are situations of interdependence stemming from the distribution of population and activities in space, and, second, that resources, whether in land, housing or community services, are allocated locally. While therefore productive activity is localized, so tying the labour force to a certain land area, and resources are allocated by local rather than national agencies, a certain unity of interest and experience is imposed on the population.

Tempting though this line of argument may be, for certainly there are situations in which it is possible to describe markets in property focused on a single centre, with relatively determinate boundaries, caution is necessary. Even in the provincial community unity is arbitrary and subject to change with every administrative reorganization, and in the metropolitan region identity is lost in a blurred fragmentation of interest and responsibility. It may be necessary to accept that urbanization entails loss of an ordered environment and that the logic of space is inconclusive.

It is clear that the terms and guidelines for urban study set out by the Chicago sociologists live on in both planning and urban sociology, and that human ecology is a continuing tradition: the language of the original statements is important, therefore, as an object lesson, a check to over enthusiastic adoption of an exclusively urban frame of reference. Can no more be said for their other major innovation in sociological studies, their insistence on intensive investigation into the 'interactions of men's experience in detail'? The application of pragmatic canons of knowledge to sociological research initiated a tradition of community studies which rejected the atomization inherent in standard survey methodology. Park, backed by the more sophisticated teaching of Mead, took as axiomatic the need to understand the everyday lives of men and women, as experienced in their social settings. The small-scale, bounded horizons of the common man were the fit subject for a democratic sociology. That so much was omitted is now well known—the contours of economic and political life in America or in Chicago were disregarded, and the social worlds described became, literally, in the best of the studies to be carried out in this tradition, the world of the street. In *The Hobo*, *The Gang*, the later *Street Corner Society*, and Suttles's *Social Order of the Slum*,[13] observation is restricted to the public worlds of streets, playgrounds and bars. For these, and other reasons, such studies have been heavily criticized for their triviality, their dilettante, unscientific character, and their voyeurism—in short the dissipation of energies in the collection of information with no scientific or political objective.

In pioneering these urban community studies, the group at Chicago were furthering a concern inherent to sociology, differing from advocates of survey methodology in their rejection of positivism in its cruder forms, and from the Harvard School in their refusal to make open use of social theory. In that these studies are personal presentations of the issues of private milieux, the product of a unique interaction between an 'experiencing subject' and a 'vast cloud of experiences that are wholly subjective', they are more than other forms of social investigation open to charges of concealed value premises and ideological distortion. Advocates of the methodology have been able to offer only weak apologia for the one-sidedness of studies.[14] But the pragmatists' insistence on the primacy of personal experience as the basis of knowledge and subject of science was resonant of a main strand in Western philosophy, realized in sociology, more particularly American academic sociology. Social theory, even, becomes 'a search for the meaning of the personally real, that which is already assumed to be known through personal experience'.[15] This methodology is that of a humanistic liberal study at its least doctrinaire but, all too frequently, most fragmented.

In fact, not all such studies have been of such limited import. There is a long precedent for the political uses of community studies in a succession of investigations to stimulate changes in social welfare policy. Case histories, anecdotal material, graphic portrayals of the circumstances of day-to-day life, have been used to spur on reform in housing, planning, welfare, poverty and race relations programmes. Very few such studies have been carried out in Britain without such political objectives.[16] Their influence on public policy is at very least subject to dispute. From a Marxist standpoint they would be considered as 'a component part of bourgeois social thought'[17] and subject to the deformations of the social order in whose terms they are couched and to whose ruling interests they are directed. The recent radical critics of mainstream sociology could have no hesitation in designating all aspects of the Chicago School's work as bourgeois obfuscation of the issues of urbanization in capitalist society. In all respects, theory (human ecology), leading assumptions (symbolic interactionism), domain of study (urbanism), methodology (community and small group studies), this group of academics is seen as providing an ideological smokescreen for the issues confronting the societies of advanced capitalism then, and now.

And yet there are certain tempting parallels between pragmatism and aspects of Marxism, particularly as drawn from the early writings of Marx. For both Marx and Dewey, 'experience' derives from the totality of the individual's interaction with his environment; it signifies the 'rich living, sensuous, concrete activity of self-objectification';[18] 'action' in Dewey's vocabulary has echoes of 'praxis' in Marxist terminology, although the former is limited to the contiguous environment and is void of political import. It would seem that the methodology derived by the sociologists for the apprehension of the world of action and self-realization could have application to a sociology concerned with the critical analysis of a given mode of production. Localized studies could be of strategic importance in the appraisal of the 'banal, primitive, simple, everyday life and wishes of the *broadest* mass of the people in all the specificity of their situation in society'.[19] Intensive investigation of the world of daily chores, trivial routine, local gossip, and immediate concerns, could take account of 'how individuals in the welter of their daily experience, often become falsely conscious of their social positions'.[20] This is the world we live in, a world far removed from the grand structural issues which influence daily conduct so remotely.

From different positions, therefore, a case can be made out for the continuation of the research tradition initiated in Chicago a half-century ago. Not only does it enable us to transcend the separate worlds in which individuals in cities are trapped and so achieve some understanding of the diversity and sheer inequality of urban experience, but it can be of direct political import. Apprehension of the world of experience, a world remote from the employer, the politician or the national media, is perhaps the strongest justification for the continuation of investigations based on locality rather than occupational or interest group.

The risk is that research would fall apart into the random accumulation of well-documented evidence as to the diversity of existence in local communities, urban or rural, without prior investigation of the political economy of that place. It would scarcely be possible to investigate the situation of women in Lancashire, for example, without an appreciation of the traditions of employment in that area; similarly community activity in towns in the North East must be placed in the context of the particular political traditions of that region. All of this requires reference outside of the immediate world of experience, beyond the limits of the ecological community, to an appraisal of the relation of that particular place to the national society over generations. An empiricist research programme based on the canons of knowledge advanced by James and Dewey will neither advance understanding as to the process of urbanization, nor assist in the comprehension of the activities from which 'experience' derives. The pragmatists, and with them the Chicago School, and subsequent generations of urban sociologists, were not concerned to establish why experience was so, and not otherwise; their causal chain stopped with the immediate environment confronting the individual. If this tradition of research is to continue at all it must be incorporated into a methodology more searching than either pragmatism or classical sociology permitted.

This perhaps is an unacceptable argument—that a tradition of investigation initiated on certain philosophical premises can be adapted to a philosophy, Marxism, wholly opposed in its methodology to a-historical, uncritical and individualistic research. The plea for 'an exact science based on observation and experiment'[21] relates to the data of history, the objective contradictions in capitalist society, and its political struggles, in which the sociological microscope into the atomized consciousness of the members of that society is not required. On this argument, local studies, if conducted at all, would have to be designed in different terms. There are few models of such research: one might be *Class Consciousness in the Industrial*

Revolution,[22] others to which reference could be made are *Middletown* or *Coal is Our Life*.[23] The first of these relates to three towns only, in a limited epoch in British development; in *Middletown*, the Lynds attempted to capture the essential characteristics of a local world, its work, politics, family life, religion, leisure, in the transition from craft to mass production in the American economy; the last-named was a demonstration of the brutalization of total experience in the dependence of a local community on the coal mine.

This is not the sociology of the Chicago School. The world of family and neighbourhood is not divorced from the sphere of production; political life is an intrinsic aspect of existence; the local social grouping is specifically placed in the context of transformations in the mode of production on a societal scale; a sense of history is a prerequisite to their construction. Undoubtedly a major appeal of the Chicago School in its 'Golden Age' was the excitement of fieldwork—practical engagement with a novel living world whose accurate documentation was sufficient task, but it was this uncritical and limited empiricism that formed the fatal straitjacket to their endeavours.

Epilogue
The city or the town?

> the existence of the town implies at the same time the necessity of administration, police, taxes, etc., in short of the municipality and thus of politics in general.[1]

Urban issues, urban problems, never catch the main headlines: the bankruptcy of New York or Tokyo is of interest, but no more; the *coup de grâce* for London's motorway box merits a few column inches only in the serious press. Towns are too familiar, the grey muddle of the older areas, the redbrick geometry of the new estates, are as interesting as our breakfast toast, and as such disregarded. The general public's blindness to the urban environment is something academics, professional students of the cities and the regions they dominate, are unable to grasp. And yet specialization of study can render the professional equally blind—with intensity of observation, a spotlight on a few themes—and very soon we are dealing with a different world from the one known to the layman. It is not only the language, though this alone can protect the academic from undue cross-examination, but the consensus that is established in an area of professional specialization as to the significant issues for debate.

Conventionally urban sociology has dealt with a narrow range of issues. In Chicago interest was with the city as urban community first, and then with issues of neighbourhood turnover, social disorganization and immigrant assimilation. And until distress and unrest in the central American cities forced attention on housing and urban renewal and community development issues in the late 1960s, interest remained with the subcultures of the city—black, white, street-corner, neighbourhood association, slum, suburb. The resultant image of the American city as a constellation of social worlds was readily accepted by the 'public' of the sociologists, the professional middle classes to which their students graduated. It was a quest for the ecological bases of social integration. A similar consensus as to what was to be taught, discussed and hopefully researched, did not begin to emerge in British urban sociology until the late 1960s. Here the statements of Rex and Pahl were the precipitants in the maze of uncertainty. In providing theoretical bridgeheads from the mainstream of Weberian sociology, they certified the subsequent concentration on welfare issues—housing, planning, poverty, community development and race.

While most members of the general public would readily acknowledge the importance of these issues, they are not the ones that most readily spring to mind as urban problems. For the working class, unemployment and industrial relocation policies have as great a salience as housing and slum clearance; education and public transport services are more important than community development programmes. For the middle classes, education, rates, local government reorganization, environmental conservation and traffic congestion are the recognized issues. Traffic in towns is a problem every car-owner or businessman would readily identify with; few among them would understand the concentration by urban sociologists on the restricted range of issues indicated here. Why do we not talk about the town centres, the industrial estates, football crowds, street gangs? How is it we debate the morality of town planning, the parameters controlling built form, without ever examining the buildings themselves? We discuss the future of the city without stopping to examine the towns we live in.

A first explanation could be the peculiar status of sociology within the social sciences, and the uneasy demarcation lines between branches of sociology. Education is excluded from the debates in urban sociology, for instance, because of the parallel specialization in educational sociology, as well as its tangential impact on urban space. More importantly, sociology is still the residual science, taking for study areas of existence not pre-empted by other social sciences. Urban politics thereby fall into the lap of the political scientist, land questions go to the economist, and issues of employment to the planners. The effective division of labour between social scientists restricts the urban sociologist to social affairs. But this does not explain the virtual exclusion of so much urban social life—its professional associations, its discothèques and youth clubs, its pubs and its football crowds—from discussion. There is no sense of excitement with urban life, no puzzlement over its variegation, or its monotony. There is instead a sense of intellectual passivity and near-boredom.

Perhaps urban life presents methodological problems for the empirical investigation of social behaviour. Neither a discothèque nor a crowded pub are settings conducive to systematic observation or sympathetic discussion; their hurly-burly provides privacy in the midst of the throng into which the social investigator would be wise not to intrude. Similarly, the football crowd presents problems in collective behaviour irreducible to survey and questionnaire. And it could be argued that, strictly speaking, traffic in towns is not social behaviour at all.[2] On the roads social intercourse is mediated by headlight and horn, each car load a social bubble forced to acknowledge the others around only in so far as trajectories in public space converge. The car has carried onto the roads what the nineteenth-century observers remarked on the pavements—the intersection of many paths, the coexistence of many individuals in the thick of the crowd, mutual recognition in the use of space but in no other respect. To the man in the street, towns are signified by their public activity, a world amenable to sociological analysis only in terms of 'phenomenological situationalism'.

However, there has been so little empirical study of aspects of British towns by sociologists that the methodological arguments are unconvincing. The point is simply that the sociologist and 'general public' are aware of different urban worlds. Those who make their living writing or teaching about towns debate themes which are not the most apparent 'urban' issues to those who live in them. Who, then, are the urban sociologists talking to? Themselves, other sociologists, other social scientists, the students they teach, all members of the academic world of which they are a part. To this extent the debates are artificially contrived, not phoney perhaps, but without the push of conviction or the need to survive. There is none of the urgency or the immediacy of the planners' office or a crowded housing centre. Is it, then, just another academic treadmill, is there no discourse with the public?

Directly there is very little—a few popular monographs hit the bestseller lists, some joint conferences are organized between planners and sociologists, architects take note of their consultant sociologists, and so on. But there is an expectancy by sociologists that they will be listened to, that what they write and earnestly debate will have bearing for decisions taken about our urban future. In their own minds they have established an *alter ego* with whom to debate urban questions—the government official or local government officer who acts as arbiter in decisions on urban investment—and tacitly restricted the range of urban sociology to issues of mutual interest. There is an emphasis on buildings, on the form of the cities, on the allocation of resources in land, all matters which directly concern government and its agents, the local authorities. That so little attention is given to the commercial uses of urban space is indicative of the restricted public addressed by urban sociologists:

attention is directed at that which the State can encompass. Realistically, therefore, the profit-seeking investors and their stake in the towns have been excluded from discussion.

Urban sociology in Britain, as in America, grew out of the interest and concern of the middle classes in the living conditions and social welfare of the population of the towns. Sociology as systematic investigation took its direction from the desire of the middle classes to ameliorate the squalid conditions by which they were surrounded, and, in turn, gave impetus to reform movements. Diffuse sentiment was harnessed to determine goals, gratuitous goodwill was directed to political ends; the new science, sociology, was to provide weapons with which to muster support for reform. Just as Chadwick's inquiries in the 1840s had helped to push forward sanitary legislation, so the inquiries into poverty and housing, carried out by public-spirited citizens in many towns, gave good grounds for welfare legislation before and after the First World War. The findings of this social research were the unwritten footnotes to these enactments, in every way comparable to the citations in the New Deal legislation to expert commission and research. Once the capabilities of private philanthropy were exhausted, as they soon were, arguments had to be directed to the State, its agencies and its officials.

There were certain definite episodes in which urban study, urban social research, has attracted interest. One such period in Britain was that before the First World War; in America the Golden Age of Chicago sociology ran into the American Depression; in all the so-called developed societies there was a resurgence of interest through the late 1960s. In each case an interest in urban sociology coincided with a period of uncertainty in the political culture, and economic instability, subsequent to a phase of relatively rapid development. And in each episode the themes which shaped discussion and research in urban sociology echoed the debate among the middle classes. Underlying the interest in urban studies in Britain before 1914 was a concern for citizenship; anxious debates about poverty and housing overlaid uncertainty as to the future of the liberal democratic order in an era of mass (male) franchise. In America, similar concerns were expressed in the arguments about community in the cities. The sociologists had pinpointed the disorganization of the populations of the large cities and their detachment from the consensus held to by middle America: could community development through education and planning give a solution? And to take one example from the self-searching of the post-war world—French urban sociology—does not its theoretical brilliance and Marxist terminology not express, in perhaps unfamiliar guise, the widespread unease at the growing power of the state in modern society?

Although overtly hostile to the capitalist state and opposed to its continuance, urban sociologists in France, as in America and Britain, are part and parcel of the apparatus of state planning. The rhetoric may be Marxist, but the message is reform; social science acts 'to persuade resistant or undecided segments of the society that such problems do exist and are of dangerous proportions . . . the state requires social researches that can express those social problems with which the state is ready to deal'.[3] The difference in language, maybe in abstractness of level of analysis, is indicative of the differing status of the 'intellectual proletariat' in the different national social structures. In Britain, where this group have always had the sense of access to the 'bourgeois political arena', social theory has typically been linked to detailed and concrete proposals for practical social reform.

The public to whom the urban sociologists have directed their debates and research is, if this argument is accepted, the 'professional proletariat'[4] from whom are recruited the personnel of government offices (and increasingly the legislature itself). The restricted compass of field of study has therefore reflected the consensus in this stratum as to the proper area for state intervention—namely issues of housing and planning. It was the environment of the towns, not their economy, which was deemed to be manageable. The assumptions made by the Fabians as to the nature of power in a modern society, the accreditation given to 'superior brains', the weight of rationality and professional expertise, live on in much British social research. Urban sociology remains élitist in orientation, bureaucratic in tone, and, interestingly, centralist in leanings. Arguments are consistently directed at central government. Despite the localism of subject matter, the belief in the rationality of administration and scientific policy-making implies the non-local control of local affairs.

The relationship between local and central government is one that attracts little public interest. It also goes unremarked in most academic discussions of the role of the State in modern British society. And yet it is estimated that local authorities dispose of approximately 20 per cent of national income.[5] They act as a 'supplementary apparatus of agencies by which the State might operate some of the services for which it has assumed responsibility'.[6] Contrary to popular mythology and local historiography, local authorities are of recent origin and national provenance. Their charter for action was set out by central government, they are legally bound (in the *ultra vires* rules) by the statutes of parliament, and increasingly financed by central funds and not local rates. Although many a county borough (now county district) can trace its ancestry to medieval charter and corporation, the boroughs were divested of these ancient rights in 1835 and formally reconstituted as agents of central government charged with the maintenance of law and order, as well as local services. And so it has continued. Central government has the power to grant or withhold authority or finance, the final right to dismantle the administrative apparatus devolved over a century and a half, as seems most fitting in the national interest.

One of the weaknesses in the position adopted by British urban sociologists in recent years is the restriction of debate not just to welfare issues but to those which are locally adjudicated. The 'managers' to whom reference is made are the individuals encountered in local offices, not the government department or regional committee. This is to ignore the switch of responsibility for planning and administration of services from local authority to statutory body in post-war years. Hospitals and water supply are two such examples; trunk roads, port administration and national airports have been taken into the hands of central government. On any criterion these are life-supporting services, essential to production and circulation as well as the 'reproduction of the labour force', managed on neither a local nor an urban basis. And yet, at the same time, many of the activities of local government, the mundane and trivial duties of, say, rat-catching, or dustbin-collecting, are unmentioned. The designation of field of study as that of the collective means of consumption sounds all-embracing, but in practice it is resolved as the consideration of some of the more contentious activities with which local authorities and their officers are still charged, i.e. housing and planning. The sociologists probe where policy directives are ambiguous and local discretion possible.

The division of responsibility between national and local government has been subject to continual alteration. Throughout the nineteenth century local authorities were not the prime agents of local government. The administration of the poor law was in the hands of commissioners answerable directly to London; responsibility for compulsory education was vested in school boards elected for that purpose; initially public health controls were in the hands of sanitary boards. Innovatory policies, in some cases threatening local interests, were handled by specialized agencies answerable directly to central government. The great years of the local authorities were to come after 1888 when the counties (and county boroughs) became the principal units of administration. The ever widening range of social legislation was to be handled by bodies answerable in name to local electorates and not to national government. All this at a period of London's social, political and cultural supremacy.

The panoply of local government—the ceremonial council chamber, the robes and chain of office, the monumental town hall—is Victorian in its solemnity and inconsequentiality. The old boroughs had been divested of sometimes considerable wealth and political responsibilities, purged, reformed; new townships, such as Oldham, were handed the trappings of authority and divested of the power of control over, for example, the police force. The newly constituted local authorities were almost entirely dependent upon central government. The councillors, aldermen, mayors and sheriffs were unpaid agents of government, caught up in the obligations of public service by their ambitions or standing in the local community. The blunt businessmen of the northern towns, the new breed of professional men, had much to gain from local prominence: arguably, local affairs benefited from the application of acumen and skill. In this way the new bourgeoisie were incorporated into the state, the gulf between these new classes and traditional élites bridged by their co-optation into the machinery of government. The extension of the powers of government in the nineteenth century was made easier by the retention of the shell of existing institutions. New policies were rendered more acceptable for not only might they originate in the interchanges on parish pump affairs, but the very parochiality of debate could blunt the edge of novel policies.

Then, as now, local authorities as institutions were the mediators of government policy. Their intricacies of internal organization and reputed particularity clouded the injustice inherent in the allocation of services when resources were inadequate. Increasingly, however, in the welter of directives and circulars from central government, the execution of social welfare and public services must fall into the hands of the functionaries, professional servants of the public interest. Now it is more important to recruit the best qualified officers and devise the most efficient system of administration than to interlock with local élites. Reorganization of local government has been one answer, together with the passing of responsibilities to statutory bodies. In these circumstances, the current lack of interest by sociologists in local community power structures is understandable; the officers of the local authorities, so-called managers, have more power over investment decisions than local notables. But there has been little scrutiny of the interplay between politician, professional and public in the local community. Indeed, in the examination of professional ideologies it is implicitly assumed that the criteria employed by the officers are given nationally rather than by local considerations. The relationship between national and local, between higher and lower levels of government, similarly is taken as given, and the structural position of the local authority in the chain of command not explored.

There is no sociology of local authorities, and yet an argument presented for an 'urban' sociology has been the variability in the administration of services by these institutions.[7] This, it was suggested, justified local studies. The egalitarian principles of the Welfare State had foundered on the intractability of local politics and unevenness in distribution of resources. So, although in sociological terms there might be no theoretical justification for urban studies (recognizing the proclaimed partnership between central government and local authority as an administrative fiction), empirically the boundaries outlined units that merited study in themselves. This assumed an immutability and essential continuity to local government to which, claims of civic patriotism notwithstanding, recent changes give the lie. The city is a product of the State, assembled and dismantled at the will of central government, and the academic enterprise, urban sociology, on this argument, is built up on conceptual quicksand.

Ostensibly local government reorganization in 1973 was carried out to improve the efficiency of the authorities, thereby justifying their greater autonomy. In this way their standing in the eyes of the ratepayers could be raised and their democratic functions enhanced. However, the sheer size of the new counties militates against involvement of constituents and may reduce the influence of elected representatives. In that the number of authorities has been reduced and the authority of professional officers increased, their surveillance by central government is easier. It is argued that 'the effect of reform has been to lock them more effectively into the mechanisms of central administration'.[8] Local idiosyncrasies are less pronounced, embarrassing deviations from government policy less likely to occur. The very feature of local life emphasized by social scientists—its variability in social structure and political culture—is for administrative purposes nullified. Is it then ironic that at the very period when there is interest by social scientists in the intricacies of local communities, their independence is further erased?

Urban sociology exists in an intellectual world in which 'social scientists . . . find it most difficult to formulate [a significant problem] in terms of any unit smaller than the nation state'.[9] Much of urban sociology, therefore, has had to do with national policies, and local study has been considered as instrumental to criticisms of these. Dramatic accounts of local injustice are used to highlight their absurdities, as well as the maladroitness of local authority implementation. Sociology becomes academic populism:[10] bureaucracy is the target, official bumbledom aggravated by the social distance between central executive and local community, and the communication problems inherent in an organizational over-reach. In these criticisms social scientists repeat a generalized disquiet with vested administrative interests, one voiced by the many pressure groups independent of the hierarchy of government. They can be said to confer academic authority on the arguments for reorganization.

Does, therefore, the 'Welfare State require at very least a limited sort of critical sociology',[11] which not only exposes the existence of social problems, but also the inadequacy of previous arrangements for dealing with it, and in so doing undermines the local élites formerly in charge of these arrangements? So writes Gouldner of the relationship between higher and lower levels of government in the United States. On this basis urban sociology had been instrumental in its own academic downfall. However, it can be claimed that the administrative overburden placed on the central executive, and the political inconvenience of widespread organized protest against state policies, were the precipitants to government moves on reorganization and devolution.

It is hard to see urban sociology as intrinsic to 'the special means by which the ruling class maintains its hegemony'.[12] The arguments have not only been hostile to official policy, but notably ineffective. For instance, neighbourhood units continue to be built despite sociological criticisms, the residents of the zones in transition have not been preferred over those from traditional cottage housing, and housing policies create yet new situations of shortage. All this is a measure of the powerlessness of the sociologist, without support from those whose interests he claims to represent, hammering at aspects of state policy which from the centre seem trivial, those whom he or she might seek to influence in local government being without power themselves. In what sense can urban sociology relate to 'the political necessities of the dominant fundamental group'?[13] It neither relates to the social necessities of production, nor offers an ideological justification of authority. Essentially it is inoffensive, a 'radicalism without teeth . . . a critique without dangerous consequences',[14] and as such can be dismissed politically.

Or can it? Can it be dismissed so cavalierly? Urban sociology as applied social science does not offer 'weapons of the intellect': it proffers information on how things actually work at ground level, picks up discontent, suggests remedies—remedies in the hands of the State. It does not pretend to social engineering, there is too much scepticism about the possibilities of moulding social relations through the manipulation of the environment, but adopts a managerial stance. Irrationality and absurdity offend, equality in the distribution of social welfare is sought, and efficiency in the conduct of social policy. The assumption is that there are remedies, and that present ills may be administered away. It therefore builds up an atmosphere of reform. However iniquitous present policies, they may be changed, so that, for instance, housing can be seen to respect the needs of the inhabitants, slum clearance can be effected gracefully, and so on. Reform is possible within the existing social order, even if the reform effected is not that sought.

Now that continued spending on social services, in fact all public expenditure, has to be set against claims on government funds from industry, the reformist position adopted by so much urban research appears tenuous. In a situation of international recession, local housing and planning issues seem of no account, and there is little interest in land as a source of national productivity or absorber of national wealth. Sociologists are only too aware that the general public is more concerned with unemployment than with the fortune won or lost in property booms. The ground has shifted from under their feet, and the sense of achievement, the momentum in academic terms felt as courses in higher education were set up throughout the 1960s, has faded away. The coherence sought for an academic subject seems impossible to attain now that the Welfare State consensus is under open attack. The deterrent to a retreat to field research in the current impasse of urban sociology rests in a sense of the futility of policy research, and a shortage of government funds. Nor is a movement to pure theory probable given the concrete and applied terms with which urban sociologists have tended to operate.

It is, after all, the environment of cities that has attracted and repelled, and it is to this that the theoretical statements of urban sociology must be returned. It is in the plans and pronouncements on urban form by writers such as Howard, Geddes, Mumford or Le Corbusier that urban social theory reaches its widest audience. Demonstrably the ideas of Park for a civilized cosmopolitan city built up on the division of labour were expressed in the organic forms advocated by Mumford and his mentor, Geddes. Biological analogies find expression in parkways as well as neighbourhood units. In the same way the theories of Tönnies, Simmel or Wirth—the city as the mass society with no common cause or interest other than physical proximity—are realized in the concrete towers of Le Corbusier's Radiant City. In these last conceptions of the city man is all alone, technology rules supreme, all social life depends on bureaucracy. But that bureaucracy with its experts and power to command resources can sweep away the clutter and injustice of bourgeois society to convey a sane rational environment for all. Similarly, Howard's modest proposals for a garden city in the Home Counties sought to show that purposeful organization, i.e., planning, of investment in land, houses and community resources, could give a more humane and democratic society. Like so many of his contemporaries in Britain and elsewhere, Howard sought to establish the preconditions for national democracy. The garden city proposals were of immediate practical application, and also political reference, in an era of mass suffrage. But whereas Weber, the sociologist, in his discussion of the city, saw participation through association as the necessary condition, Howard presented an environmental solution to the liberal impasse.

If there is a common theme to the scattered and diverse writings considered as urban sociology, it is that of democracy. Individualism/community, anomie/association, disorganization/communication, bureaucracy/participation, conflict/consensus, these are the debating themes to which theories have been addressed. The practical concern with the redress of environmental injustice stems from a belief as to the rights of the citizen in the democratic state. Those rights, once conferred by property ownership, now derive from residence; therefore access to the benefits of membership in the Welfare State are set by environmental conditions. The inference is that the modern state will function as democracy once these inequalities on the ground are removed. The towns, as towns, aggregates of population and activity, concentrations of capital and enmassed labour, buildings on the ground, are not the object of study. They are a fact of existence, their buildings, traffic and daily life common to all. It is from this datum line the sociologists seek in their theories and their prescripts for planning to recapture, build afresh, re-create, an image of a democratic community. We are held in thrall to the idea of the city.

As Weber or Wirth recognized, the idea of the city is central to Western culture. The Greek city state of Aristotle or Plato, and the medieval guild corporations, enshrine an image of urban life far removed from what men have experienced in the modern big city, and all the more appealing for that. In the surroundings of the Victorian city, an image of an existence in which private interest could be harnessed to public service was gleaned from classical texts and historical scholarship. In a situation which permitted of no sense of control over one's personal environment, the city state of classical Greece or Rome, or medieval Europe, fired the imagination. These were the images in whose terms the new men of modernizing Europe shaped the world they lived in. The metropolis, in its size, organization, impersonality of economic, social and political life, was the antithesis of this idea of the city as the free community-association, and everyday experience flatly contradicted democratic ideals. But scholarship, the arts, architecture, strengthened the image of bourgeois democracy, counterface to the mundane round. Urban social theory answered to this theme and spoke to this audience, the administrators and professional servants of empire and nation-state. The political vitality of the city, its ideals of citizenship and organization as a community were at issue, not its productive interests nor mode of capital accumulation.

As Williams argues, the ideas of country and city are often used as 'forms of isolation and identification of more general processes . . . the ideas . . . express . . . human interests and purposes for which there is no immediately available vocabulary'.[15] To this particular social stratum, then and subsequently, the environment of the towns, and the administrative machinery associated with their establishment as agencies of government, had a more powerful 'reality' than the productive activities or class relationships from which urban growth derived. The municipality demanded their service and the conduct of urban politics became their charge. To the working class a different vocabulary was available—that of class and state. The main power over its existence was a productive system that allowed no respite from degrading toil, not the urban environment resented by the middle classes. Work, wages, national representation, were the themes which informed working-class culture. In these terms the great towns were welcome in that they provided greater opportunity for livelihood and organization than underdeveloped rural economies. Similarly, the interests of capital could not be contained in urban imagery. The constructs (myths?) of 'city' and 'country' indicate the life experience of a middle stratum represented by writers as diverse as Clare or Dickens, Morris or Howard, or, for that matter, William James or Park.

The discontinuities in the professional development of urban sociology are marked, but criticisms of its restricted field of discussion and limited analytical penetration must still be criticisms of liberalism, a liberalism in which the state is accepted as a natural ally representing the interests of the national community. The emphasis on empirical research and practical proposals for a reformed environment has muffled the idealism of philosophy and method. The bluntness of research design of, say, Booth was inadmissible—'phenomenological empiricism' took its place; and the directness of purpose of the practical planner would have betrayed the spirit which took urban sociologists out of the uncertainties aroused by mass urbanization. Both in substantive orientation and language, urban sociology has remained quite distinct from, for instance, human geography with, from the days of Mackinder and the Royal Geographical Society, more straightforwardly conservative concerns which strengthened a natural materialism. Despite the subsequent cross-recruitment and hybridization of method between survey research, urban planning, geography and sociology, the domain assumptions of urban sociology were not challenged until recently. Politics, not economic activity; community, not class; housing and domestic buildings and not industrial development; the city, and not towns and regions—these constituted the bill of fare for students and teachers alike. The construction of this book is evidence of the vitality of this convention of inquiry.

It is now generally accepted among urban sociologists that urbanization cannot be considered as a thing in itself. For Pahl, for example, the main reason for 'the confusion, aridity and lack of development in the field . . . is that sociologists have as their main objective the sociological understanding of the city *per se*. Paradoxically, the fundamental error of urban sociology was to look at the city for an understanding of the city'.[16] And certainly the denuded urban sociology we have relates to the short circuit of observation and explanation referred to here. In effect it is a brand of functionalism, sharing in the latter's formalism of language and lack of specificity in orientation. Unacknowledged, however, is the ahistorical idealism underpinning the endless discussions about the city and where it stands in modern society, which did give the subject a certain coherence. The confusion of recent years is indicative of the uncertainty among this 'professional proletariat' as to the market for their ideas: local authority or national legislature; the 'state' and its functionaries or resistance groups (urban social movements); the lay public or academic colleagues. In that the State is addressed there has been a shift from issues of allocation to those of production, so widening out the domain of urban sociology. Towns and regions have been rediscovered as the unevenness in opportunities for work and development in the society have become more pronounced.

Political economy is now the catchword. At its simplest it is taken to mean locating urbanization (or urbanism) in 'the central political and economic dynamic of the social system'.[17] Towns and regions—the disposition of a society on the ground—are then incorporated into an anatomy of advanced capitalism. In this way, to take one example, the buildings of the town are related to conditions of capital circulation and the nature of state intervention in property investment. Marxist categories of analysis are implied but by no means obligatory. There is the implication that there are *urban* institutions, phenomena, call them what one will, which simply need to be related to a broader context to make sense. But political economy, in the hands of Smith or Ricardo, or historical materialism, were all-embracing methods of analysis in which societies were to be studied at a certain level of abstraction, in their entirety. On this precedent community life and buildings cannot be extracted from the productive relations and political structure to whose contradictions, local as well as global, they bear witness. If Marxist social science is to supplant liberalism in the study of urban (civil) society, a change of vocabulary is not enough. There should be different themes, concerns alien to the tradition of urban sociology. At very least, towns must be regarded as signifying 'the existence and development of capital independent of landed property . . . property having its basis only in labour and exchange'.[18]

In many respects it is easier to write about urban sociology than about towns and regions. For everyday professional purposes, the city, the political community, the socio-ecological system, can be treated at a middle range level of abstraction which permits the discussant to move between generalities and particular cases at random. Definition on the ground is not necessary; conceptually 'the idea of "place" is only a rough practical approximation: there is nothing logically necessary about it, and it cannot be made precise'.[19] By contrast, in political economy towns and regions are granted an objectivity not readily substantiated empirically. Method of study and involvement with specific empirical questions would push the social scientist into the classic issues of science—analytical clarity, precision of information, theoretical comprehensiveness, and capacity to inform action. Urban sociology provides none of these; Marxism cannot, as yet, bridge the gap between political conviction and knowledge. There would seem to be no alternative but painstaking, inconclusive, empirical investigation into substantive issues of urbanization and regional development.

In the present conjuncture of international recession and diminution in the authority of the state, there is an impatience with politics and with ideas. The material circumstances governing spending on housing, or choice of location, or planning policies, are too blatant to be lost in discussions of myth, imagery or professional creed. Accordingly, suburban expansion is analysed in terms of land economics and State subsidy, high rise housing in the light of the economics of construction, and regional development interpreted in terms of a renewed exploitation of national resources. Study of the images of suburban living which might spur households into a move there, the professional creeds which justified point blocks and streets in the air, and the regional and nationalist movements which ventilated peripheral discontent are regarded as an unwarranted luxury, or ideological deformation. But the ideas of democracy and community have shaped our towns, and given form to the environment in which people spend their lives. An unknown number of households now live in neighbourhood units; how many more are forced to commute from out-country estates through greenbelts and past low density 'suburbs in the city'? Certainly there are now more than 1 million residents in the new towns, a group equivalent to that housed in high rise developments whose construction was spurred on by visions of urbane living. And many of these latter were displaced from their homes under the compelling argument of slum clearance. Reformers' slogans have solidified into conventional wisdom and sociologists' case studies used as warrant for the execution of state policy. The critical tradition of urban sociology is an important one.

Perhaps this is a gloomy survey of urban sociology. The literature on urbanization which has proved of enduring interest over the years, that of Mayhew, Booth, Park, the Lynds, Willmott and Young, Dennis, Henriques and Slaughter, for instance, was produced by writers who were ignorant of town life, personally intrigued by urbanization—the way cities grew, changed their activities, generated new classes, new riches, and extremes of poverty, different forms of political organization—and also were concerned with an issue, an argument, a campaign. (Even a writer as intently academic as Park.) If not polemical, their writings show a concern with a specific issue, a definite aspect of urban or local life which gave direction to their research. This tradition is coming to life again with the interest of non-specialists—journalists, English critics, Marxists, historians, economists, or even plain sociologists—in urban and regional questions. Hopefully their position as outsiders to urban sociology drawn to investigation of 'things in space' out of sheer puzzlement will abate the tendency to 'focus our consciousness upon, to analyse, and to combine our ideas'.[20]

Notes

Preface

1 H. G. Wells, 'The so-called science of society', *Sociological Papers*, 1906, pp. 360–3.
2 M. Castells, 'Is there an urban sociology?' (1968), translated by C. G. Pickvance (ed.) in *Urban Sociology*, Tavistock, London, 1976, pp. 33–59.
3 Centre for Environmental Studies, *Proceedings of the Conference on Urban Change and Conflict*, London, January 1975 (Capes CP 14), pp. 9–14.
4 K. Marx and F. Engels, 'The German Ideology' (1845–6), included in *Karl Marx, Selected Writings*, ed. T. Bottomore and M. Rubel, Penguin Books, Harmondsworth, 1963, p. 87.
5 N. Harris 'Urban England', *Economy and Society*, vol. 3, no. 3, p. 349.
6 R. E. Park, 'The city: suggestions for the investigation of human behaviour in the urban environment'(1915), included in *The City*, ed. R. Park and E. Burgess, University of Chicago Press, 1926, p. 1.
7 M. Castells, *La Question urbaine*, Maspero, Paris, 1972, pp. 442–3:

> Il y des unités urbaines dans la mesure on il y a des unités de ce processus de reproduction. . . . Une telle spécificité de l'urbain est historique: elle découle de la domination de l'instance économique dans la structure sociale—l'espace de la production étant l'espace régional et celui de la reproduction étant appelé espace urbain.

8 L. Wirth, *On Cities and Social Life*, University of Chicago Press,1964, quoted in the introduction by A. Reiss, p. xv.
9 W. Morris, 'News from nowhere' (1890), *William Morris: Selected Writings and Design*, Penguin Books, Harmondsworth, 1962, p. 245.

Introduction

* Previously published in the *British Journal of Sociology*, vol. XXVI, no. 3, September 1975.

1 M. Weber, *The City* (1921), Free Press, Chicago, 1958.

2 M. Castells, *La Question urbaine*, Maspero, Paris, 1972.

3 R. Glass, 'Urban sociology in Great Britain: a trend report', *Current Sociology*, vol. 4, 1955, p. 5.

4 B. E. Coates and E. M. Rawstron, *Regional Variations in Britain*, Batsford, London, 1971.

5 G. Moorhouse, *The Other England*, Penguin Books, Harmondsworth, 1964, pp. 13–28.

6 Recent textbooks include H. W. Richardson, *Urban Economics*, Penguin Books, Harmondsworth, 1971; and B. Goodall, *The Economics of Urban Areas*, Pergamon, Oxford, 1972.

7 Note H. V. Wiseman, *Local Government in England 1958–69*, Routledge & Kegan Paul, London, 1970; N. Boaden, *Urban Policy Making*, Cambridge University Press, 1971; J. C. Banks, *Federal Britain?*, Harrap, London, 1971.

8 C. W. Mills, *The Sociological Imagination*, Oxford University Press, New York, 1959, p. 5.

9 Reich's full statement reads:

> An all embracing economic and state politics, operating on a wide historical perspective . . . must find the connection with the banal, primitive, simple everyday life and wishes of the *broadest* mass of the people in all the specificity of their situation in society. Only thus is it possible to unify the objective sociological process and the subjective consciousness of men, and to abolish the contradiction, the schism between them.
>
> W. Reich, *What is Class Consciousness?*, a pamphlet reproduced by Socialist Reproduction, London, 1971.

10 R. Williams, *The Country and the City*, Chatto & Windus, London, 1973.

11 Glass first drew attention to the marked 'anti-urbanism' of these groups in 1955, *op. cit.*, but more full exploration of the context in which urban problems were defined as housing and planning issues is more recent. Reference should be made to G. S. Jones, *Outcast London*, Oxford University Press, London, 1971; and V. Kiernan, 'Victorian London', *New Left Review*, 1972, no. 76, pp. 73–90.

12 Explicit reference is made here to L. Wirth, 'Urbanism as a way of life', *American Journal of Sociology*, 1938, pp. 1–24.

13 Castells, *op.cit.*, p. 103.

14 A. Small's statement is taken from an appraisal of the state of American sociology in 1915. 'Fifty years of sociology in the United States', reprinted in *American Journal of Sociology*, 1947, index to vols. I–III. He, and the Chicago School, had been greatly influenced by Simmel, here writing in *Soziologie*, reprinted in K. Wolff, *The Sociology of Georg Simmel*, Free Press, Chicago, 1950, p. 16.

15 D. Martindale, 'The formation and destruction of communities', in *Explorations in Social Change*, ed. G. Zollschan and W. Hirsch, Routledge & Kegan Paul, London, 1964.

16 R. Glass. These quotations are taken from pp. 487 and 497 respectively of 'Urban sociology', included in *Society: Problems and Methods of Study*, ed. A. T. Welford, Routledge & Kegan Paul, London, 1962.

17 J. Rex and R. Moore, *Race, Community and Conflict*, Oxford University Press, London, 1967, p. 273.

18 R. Pahl, *Whose City?*, Longmans, London, 1970, p. 218.

19 Pahl's more recent statements—'The sociology of urban and regional development as a problem in political economy', paper prepared for the Eighth World Congress of Sociology, Toronto, 1974; and 'Urban managerialism reconsidered', unpublished paper circulated August 1974—indicate a substantial revision in the position previously adopted and a radically different approach to the problem of 'the modern big city'.

20 W. James, *Pragmatism*, Longmans Green, London, 1907.

21 C. Halsey comments on D. Donnison, *Whose is the Good City?* in *New Society*, December 1973, p. 720.

22 A. Vidich, J. Bensman and M. Stein, *Reflections on Community Studies*, John Wiley, New York, 1964, p. xi.

23 N. Harris, *Beliefs in Society*, Penguin Books, Harmondsworth, 1968, p. 35.

24 Frank's writings include: *Capitalism and Underdevelopment in Latin America: Historical Studies of Chile and Brazil*, 1967; *Latin America: Underdevelopment or Revolution*, 1969; *Lumpenbourgeoisie: Lumpendevelopment—Dependence, Class and Politics in Latin America*, 1972, all Monthly Review Press, New York; and James D. Cockroft, Andre Gunder Frank and Dale L. Johnson, *Dependence and Underdevelopment: Latin America's Political Economy*, Doubleday, Anchor Books, New York, 1972.

25 M. Stein, *The Eclipse of Community*, Princeton University Press, 1960.

26 Frank, *Latin America: Underdevelopment or Revolution*, p. xxi.

27 Tönnies described the metropolis as 'typical of *gesellschaft* in general'. In it

> money and capital are unlimited and mighty. It is able to produce and supply goods and science for the entire earth as well as laws and public opinion for all nations. It represents the world market and world traffic; in it world industries are concentrated. Its newspapers are world papers, its people come from all corners of the earth.

(*Community and Society* (1887), Harper & Row, New York, 1963, p. 228.)

28 D. Booth, 'Andre Gunder Frank: an introduction and appreciation', in I. Oxaal *et al.*, *Beyond the Sociology of Development: Economy and Society in Africa and Latin America*, Routledge & Kegan Paul, London, 1975.

29 K. B. Clark in *Dark Ghetto*, Harper & Row, New York and London, 1965, p. 11; and S. Carmichael in *Black Power*, included in *Dialectics of Liberation*, ed. D. Laing, Penguin Books, Harmondsworth, 1968. Both use the description of colony for the ghetto. M. Oppenheimer, in

Urban Guerrilla, Penguin Books, Harmondsworth, 1969, on proposals for black control of the cities: 'most of the economic disadvantages of the plan would be to the black community' (p. 107). W. Tabb, in *The Political Economy of the Black Ghetto*, Norton, New York, 1970, while sympathetic to its designation as a colony as an 'organizational construct', nevertheless concludes: 'The black communities can gain only a limited independence. . . . They do not live in a territory which could become independent' (p. 31).

30 J. Foster (ed.), 'Nineteenth century towns—a class dimension' included in *The Study of Urban History*, ed. H. J. Dyos, Edward Arnold, London, 1968; and N. Dennis, F. Henriques and C. Slaughter, *Coal is Our Life*, Eyre & Spottiswoode, London, 1956.

31 R. Williams, 'Literature and the city', *The Listener*, vol. 78, no. 2017, 1967.

32 G. S. Jones, *op. cit.*, p. 349.

33 L. Althusser, *For Marx* (1966), Allen Lane, The Penguin Press, London, 1969, p. 113.

34 W. Christaller, *Central Places in Southern Germany* (1933), republished by Prentice-Hall, Englewood Cliffs, N.J., 1966; and A. Lösch, *The Economics of Location* (1940), Yale University Press, New Haven, 1954.

35 N. Dennis, *People and Planning*, Faber & Faber, London, 1970; and *Public Participation and Planners' Blight*, Faber & Faber, London, 1972.

36 M. Harloe, in 'Inner urban areas—a review article', published as Centre of Environmental Studies, CES WN 356, London, 1973, argues as forcibly that their approach means 'an almost total abandonment of the attempt to understand and analyse the institutions which face and even oppress these people' (p. 7), and of 'turning their backs on any attempt to analyse the structural reasons for the relative powerlessness of the majority in modern society' (p. 8).

37 The most incisive criticisms of Frank's writings are those by E. Laclau, 'Feudalism and capitalism in Latin America', *New Left Review*, no. 67, 1971, pp. 19–38; F. H. Cardoso, 'Dependency and development in Latin America', *New Left Review*, no. 74, 1972, pp. 83–95; and B. Warren, 'Imperialism and capitalist industrialization', *New Left Review*, no. 81, 1973, pp. 1–44. Cardoso argues that there occurs 'a kind of *dependent capitalist development*' (p. 89), and Warren furnishes the empirical evidence for *in*dependent industrialization in Latin America. P. J. O'Brien in 'A critique of Latin America theories of dependency', included in Oxaal *et al.*, *op. cit.*, comments on the eclecticism of a theory which 'can straddle petty bourgeois nationalism and socialist revolution', and asks specifically, 'In a world of increasing interdependence is it important to know what makes a dependent country dependent?'

38 Glass, *op. cit.*, p. 7.

39 Some interesting discussions of 'urbanism' in Britain are to be found in R. Hoggart, *The Uses of Literacy* (1957), Penguin Books, Harmondsworth, 1958; A. Briggs, *Victorian Cities* (1963), Penguin Books, Harmondsworth, 1968; B. Jackson, *Working Class Community*, Routledge & Kegan Paul, London, 1968; J. Seabrook, *City Close Up*,

Allen Lane, The Penguin Press, London, 1971; Williams, *The Country and the City*; Jones, *op. cit.*; and Kiernan, *op. cit.*

40 Moorhouse, *op. cit.*, p. 22.

41 L. Wirth, *On Cities and Social Life*, University of Chicago Press, 1964, p. 2.

Chapter 1 The British experiment

1 Uneven development in cities and regions

* Based on a paper previously published in the *Proceedings of Conference*, 1975, Centre for Environmental Studies, York.

1 R. Williams, *The Country and the City*, Chatto & Windus, London, 1973, p. 291.

2 F. Tönnies, *Community and Society* (1887), Harper & Row, New York, 1963, p. 234.

3 K. Marx and F. Engels, 'The German Ideology' (1845), in *Precapitalist Economic Formations*, ed. E. J. Hobsbawm, Lawrence & Wishart, London, 1966, p. 127.

4 F. Engels, *The Housing Question* (1872), Progress Publishers, Moscow, 1970, p. 30.

5 M. Castells, *Luttes urbaines et pouvoir politique*, Maspero, Paris, 1973, p. 15.

6 A. G. Frank's writings include *Capitalism and Underdevelopment in Latin America*, 1967; *Latin America: Underdevelopment or Revolution*, 1969; *Lumpenbourgeoisie: Lumpendevelopment—Dependence, Class and Politics in Latin America*, 1972, all Monthly Review Press, New York.

7 K. Korsch, *Karl Marx*, Russell & Russell, New York, 1963, p. 75.

8 I. Oxaal, 'The dependency economist as grassroots politician in the Caribbean' in I. Oxaal *et al.*, *Beyond the Sociology of Development*, Routledge & Kegan Paul, London, 1975, pp. 28–49.

9 University of Durham, North East Area Studies working paper no. 6, *et seq.*

10 R. Pelham, 'Fourteenth century England', in H. C. Darby, *An Historical Geography of England before 1800*, Cambridge University Press, 1961, pp. 230–5.

11 Quoted by E. G. East, 'England in the eighteenth century', in Darby, *op. cit.*, p. 494.

12 A writer quoted by Stow in 1598; the full context given by E. G. Taylor, 'Leland's England', in Darby, *op. cit.*, p. 335.

13 S. Bindoff, *Tudor England*, Penguin Books, Harmondsworth, 1950, p. 40.

14 E. A. Wrigley, 'London's importance in changing English economy and society, 1650–1670', *Past and Present*, no. 37, 1967.

15 Barrington Moore, *Social Origins of Dictatorship and Democracy* (1966), Allen Lane, The Penguin Press, London, 1967, p. 512.

16 C. Hill comments

This victory of South and East over North and West in the civil

> war led, paradoxically, to a reversal of the growing economic importance of London, against which the outports had been struggling since 1604 . . . the proportion of English shipping tonnage owned by Londoners was 43% in 1702, 30% in 1788.

(*Reformation to Industrial Revolution* (1967), Penguin Books, Harmondsworth, 1967, pp. 137–8.)

17 J. Foster, 'Nineteenth century towns—a class dimension', included in *The Study of Urban History*, ed. H. J. Dyos, Edward Arnold, London, 1968.

18 E. J. Hobsbawm, *The Age of Revolution* (1962), Sphere Books, London, 1973, p. 52.

19 E. Gaskell, *North and South* (1854–5), Penguin Books, Harmondsworth, 1970, p. 510.

20 V. Kiernan, 'Victorian London: unending purgatory', *New Left Review*, no. 76, 1972, p. 78.

21 London's financial supremacy was based on the amalgamation of the private banks into the joint stock banks. Pollard comments 'it is noteworthy that Lancashire was the most bitterly opposed to the amalgamation of its banks with London based firms'; S. Pollard, *The Development of the British Economy 1870–1967*, Edward Arnold, London, 1969, p. 15. For the decline of the provincial press, see A. Briggs, *Victorian Cities* (1963), Penguin Books, Harmondsworth, 1968, pp. 356–60.

22 F. Engels, *The Condition of the Working Class in England in 1844*, Allen & Unwin, London, 1892, p. 24.

23 A. Briggs, 'Public health: the health of the nation', *New Society*, February 1968, pp. 267–70.

24 Engels, *The Condition of the Working Classes*, Preface of 1892, p. xix.

25 G. S. Jones, *Outcast London*, Oxford University Press, London, 1971.

26 *Ibid.*, p. 27.

27 N. Harris, 'Urban England', *Economy and Society*, 1974, p. 349.

28 Engels, *op. cit.*, p. 24.

29 Engels, *The Housing Question.*

30 R. E. Park, 'The city: suggestions for the investigation of human behaviour in an urban environment' (1915), reprinted as chapter 1, R. Park and E. W. Burgess, *The City*, University of Chicago Press, 1926, pp. 10–11.

31 A. Briggs, *Victorian Cities*, p. 84.

32 D. Pinkney, *Napoleon III and the Rebuilding of Paris*, Princeton University Press, 1958, pp. 151–73.

33 J. P. Bury, *Napoleon III and the Second Empire*, English Universities Press, London, 1964, p. 58.

34 Hobsbawm, *op. cit.*, p. 218.

35 Pinkney, *op. cit.*, notes that fifty–four out of eighty–six departments lost population between 1851 and 1856 alone, and there were widespread complaints from the provinces as to the effect on local economies. In 1859, the Préfect of the Creuse, where there was a tradition of stone-working, reported that half the able men of the department migrated annually to Paris.

36 S. E. Rasmussen, *London: the Unique City*, Cape, London, 1948.
37 S. E. Rasmussen, *Towns and Buildings* (1949), MIT Press, Cambridge, Mass., 1969, p. 114.
38 W. Ashworth, *An Economic History of England*, Methuen, London, 1960, chapter 3.
39 H. J. Gayler, 'Land speculation and urban development, contrasts in south east Essex 1880–1940', *Urban Studies*, vol. 7, 1970, p. 21.
40 B. Thomas, *Migration and Urban Development*, Methuen, London, 1972, p. 114.
41 L. Ginsberg, 'Shaping the future of Greater London', *Chartered Surveyor*, September 1972, p. 134.
42 R. Unwin, 'Nothing gained by overcrowding' (1911), reprinted in W. L. Creese, *Raymond Unwin: A Human Pattern for Planning*, MIT Press, Cambridge, Mass., 1967, p. 124.
43 E. Howard, *Garden Cities of Tomorrow* (1902), reprinted Faber & Faber, London, 1946, pp. 45–9.
44 Pollard, *op. cit.*, pp. 92–161.
45 N. Branson and M. Heinemann, in *Britain in the Nineteen Thirties* (1971), Panther Books, St Albans, 1973, p. 86, note that factories were relieved of 75 per cent of their rate obligations from 1929.
46 *Royal Commission on the Distribution of the Industrial Population* (Barlow Report), HMSO, London, 1941.
47 Branson and Heinemann, *op. cit.*, p. 92.
48 *Ibid.*, p. 57. See also E. Wilkinson, *The Town that was Murdered*, Gollancz, London, 1939.
49 For a summary of post-war social geography see B. E. Coates and E. M. Rawstron, *Regional Variations in Britain*, Batsford, London, 1971.
50 *Long Term Population Distribution in England and Wales*, HMSO, London, 1971, p. 13.
51 Department of Economic Affairs, *A Case for Regional Employment Premium*, HMSO, London, 1967, p. 11.
52 A good example of this new pattern of industrial development is Lincolnshire, where long-distance commuting from the villages and small towns to the new plant along the Trent or Humber is common, and a disaster such as that of Flixborough affects a whole region. This is the 'new countryside'; J. Clammer, 'Flixborough: a rural-industrial disaster', *New Society*, June 1974.
53 The question of the precise impact of the housing boom of the 1930s remains unresolved. H. W. Richardson and D. H. Aldcroft, *Building in the British Economy Between the Wars*, Allen & Unwin, London, 1968, conclude that the importance of building in the recovery from the slump had been over-estimated (pp. 280–90), although it accounted for a quarter of the increase in employment 1932–5, whereas Pollard, *op. cit.*, p. 239, considers it as 'a critical mechanism for stimulating British recovery'.
54 Richardson and Aldcroft, *op. cit.*, p. 272.
55 Pollard, *op. cit.*, p. 96.
56 *Ibid.*, pp. 289–96.

57 M. Bowley, *Housing and the State*, Allen & Unwin, London, 1945. See also G. D. H. Cole, *Building and Planning*, Cassell, London, 1945, chapter X.
58 Richardson and Aldcroft, *op. cit.*, p. 71.
59 E. J. Cleary, *The Building Society Movement*, Elek Books, London, 1965.
60 Richardson and Aldcroft, *op. cit.*, pp. 206–10.
61 Pollard, *op. cit.*, p. 99.
62 N. Taylor, *The Village in the City*, Temple Smith, London, 1973.
63 Branson and Heinemann, *op. cit.*, p. 18.
64 The co-operation of planning authorities and private developers has been well publicized in, for example, Counter-Information Services, *The Recurrent Crisis in London*, London, 1972. (The independent impact of property speculation can be overstated: in the five years prior to proposals for the redevelopment of north Southwark there had been 8,600 redundancies in ninety firms. *Ibid.*, p. 35.)
65 D. E. C. Eversley, 'Urban problems in Britain today', *GLC Intelligence Unit Quarterly Bulletin*, no. 19, June 1972, pp. 43–51. Note also D. Eversley, *The Planner in Society*, Faber & Faber, London, 1973, pp. 102–7.
66 R. H. Best and A. G. Champion, 'Regional conversions of agricultural land to urban use in England and Wales 1945–1967', *Transactions of the Institute of British Geographers*, no. 49, 1970, pp. 15–32.
67 C. Buchanan and Partners, *The Prospects for Housing*, Nationwide Building Society, London, 1971, pp. 43–4.
68 In France there has not only been an increase of 10 million in population in 1945–70, but the proportion of population employed in farming fell from 35 per cent to 15 per cent. Towns such as Grenoble or Toulouse increased their population by 20 per cent in 1962–8. H. D. Clout, *The Geography of Post-War France*, Pergamon Press, Oxford, 1972, pp. 1–20.
69 E. Laclau, 'Feudalism and capitalism in Latin America', *New Left Review*, no. 67, 1971, p. 25.

Chapter 2 The modern big city

1 Regional cities: a suburban diaspora

1 F. Tönnies, *Community and Society* (1887), Harper & Row, New York, 1963, p. 226.
2 W. Morris, 'News from nowhere' (1890), *William Morris: Selected Writings and Designs*, Penguin Books, Harmondsworth, 1962, pp. 183–304.
3 J. Gottmann, *Megalopolis*, 20th Century Fund Inc., MIT Press, Cambridge, Mass., 1964, pp. 7–23.
4 The writings of M. Webber include 'The urban place and the non-place urban realm' in *Explorations into Urban Structure*, ed. M. Webber, University of Pennsylvania Press, Philadelphia, 1964; 'Order in diversity: community without propinquity' in *Cities and Space*, ed.

L. Wingo, Johns Hopkins University Press, Baltimore, 1963, pp. 23–54; 'The post city age' (1968) in *The Internal Structure of the City*, ed. L. S. Bourne, Oxford University Press, New York, 1971; 'Beyond the Industrial age' and 'Permissive planning', Centre for Environmental Studies, London, 1969.

5 M. Webber, *op. cit.*, 1964, p. 28.

6 The central aim of the *Milton Keynes Development Plan* (1970), vol. I, was stated to be to allow for a 'wide variety of patterns of life and the greatest possible choice for the future inhabitants'; neighbourhood planning was to allow for overlapping catchment areas.

7 Centre for Environmental Studies, *Developing Patterns of Urbanisation*, London, 1969. Refer also to *The Future of the City Region*, vols 1 and 2, London, 1968.

8 For a graphic portrayal of the organization and way of life in the metropolitan city-region see P. Hall, *London 2000*, Faber & Faber, London, 1969.

9 L. Mumford, *The City in History* (1961), Penguin Books, Harmondsworth, 1966, p. 553.

10 It is estimated that 40 per cent of the population increase of England and Wales during 1951–61 had been absorbed in the Outer Metropolitan Ring to London; P. Hall, *The World Cities*, Weidenfeld & Nicolson, London, 1966, p. 31.

11 The rate of loss of population in Manchester and Liverpool trebled in the decade 1961–71, as compared with 1951–61; inner London boroughs such as Islington suffered dramatic declines, 3 per cent during 1951–61, 24 per cent during 1961–71.

12 M. Ash, 'The quiet urban revolution', *Town and Country Planning*, January 1972, p. 8.

13 Mumford, *op. cit.*, p. 561.

14 There are many accounts of the resultant antagonisms, as residents in the middle-class suburb attempt to ward off the status decategorization of their neighbourhood. For America see G. Osofsky, *Harlem: the Making of a Ghetto*, Harper & Row, New York, 1968, part 2, pp. 71–105; and W. I. Firey, *Land Use in Central Boston*, Harvard University Press, Cambridge, Mass. For Britain see R. Durant, *Watling*, King, London, 1939; and D. White, 'Metroland', *New Society*, July 1971, pp. 5–7.

15 A good summary of the American literature is found in W. M. Dobriner, *Class in Suburbia*, Prentice-Hall, Englewood Cliffs, N.J., 1963.

16 M. Young and P. Willmott, *Family and Kinship in East London* (1957), Penguin Books, Harmondsworth, part II, p. 129.

17 Dobriner, *op. cit.*, p. 65.

18 J. Rex and R. Moore, *Race, Community and Conflict*, Oxford University Press, London, 1967, p. 9.

19 P. Hall, 'The spatial structure of metropolitan England and Wales' in M. Chisholm and G. Manners, *Spatial Policy Problems of the British Economy*, Cambridge University Press, 1971, p. 124.

20 S. Fava, 'Suburbanism as a way of life', *American Sociological Review*,

vol. 21, 1956, pp. 34–43, reports that 60 per cent of residents moving from inner city to suburban locations in Milwaukee and Columbia, Missouri, had originated in rural districts.

21 Refer to R. Hoggart, *The Uses of Literacy* (1957), Penguin Books, Harmondsworth, 1958, pp. 326–30.

22 R. Williams, *The Country and the City*, Chatto & Windus, London, 1973, p. 40.

23 D. Riesman, 'The suburban dislocation', *Metropolis: Values in Conflict*, ed. C. E. Elias, Wadsworth, Belmont, Calif., 1964, pp. 71–8.

24 By 1796, a village eight miles west of Hull, then a town little over 25,000, was described as 'full of handsome buildings belonging to several wealthy merchants of Hull, who at their ease here . . . enjoy . . . in a pure atmosphere free from the noise and the hurry of the town', J. Tickell, *History of the Town and County of Kingston upon Hull*, Hull, 1796, p. 884.

25 R. Hurd, *The Principles of City Land Values* (1903), Arno Press, New York, 1970, pp. 1–21.

26 See D. J. Olsen, *Town Planning in London: the Eighteenth and Nineteenth Centuries*, Yale University Press, New Haven, Conn., 1964; and S. E. Rasmussen, *Towns and Buildings* (1949), MIT Press, Cambridge, Mass., 1969, pp. 103–16.

27 K. Marx, *Capital*, vol. II, p. 234, in the edition Foreign Languages Publishing House, Moscow, 1957, quoted by J. P. Lewis, *Building Cycles and Britain's Growth*, Macmillan, London, 1965, p. 87.

28 H. J. Dyos, *Victorian Suburb*, Leicester University Press, 1961, pp. 124–5.

29 D. R. Denman, 'Land in the market', *Hobart Paper*, no. 30, Institute of Economic Affairs, London, 1964, gives the example of a $\frac{1}{4}$-acre plot sold for £15,000 to the owner of an adjacent 20-acre site previously worth £20,000, access to which is controlled by the small plot; the combined development value would be £200,000. He notes only, 'He has paid £1,728 per annum for an estate worth approximately £10,000 per annum' (p. 11).

30 For accounts of the involvement of building societies with estate financing, see M. Bowley, *Housing and the State*, Allen & Unwin, London, 1945, p. 175; E. J. Cleary, *British Building Societies*, Elek Books, London, 1965, pp. 184–201; and N. Branson and M. Heinemann, *Britain in the Nineteen Thirties*, Panther Books, St Albans, 1973, pp. 204–10.

31 F. C. Pennance and H. Gray, *Choice in Housing*, Institute of Economic Affairs, London, 1968, p. 58.

32 M. Clawson, *Suburban Land Conversion in the United States*, Johns Hopkins University Press, Baltimore, 1971.

33 R. L. Harrington, 'Housing—supply and demand', *National Westminster Bank Quarterly Review*, May 1972, pp. 43–54.

34 P. Willmott, *The Evolution of a Community*, Routledge & Kegan Paul, London, 1961, p. 38.

35 Berger, confronted with a persisting image of mass suburbia which no empirical evidence substantiated, argued that it was a myth, which

both served certain interests in the marketing of consumer goods and also revealed an ambivalence over America's cultural pluralism. B. M. Berger, 'Suburbs, subcultures and the suburban future' in S. B. Warner, *Planning for a Nation of Cities*, MIT Press, Cambridge, Mass., 1966, pp. 143–62.

36 M. Stein, *The Eclipse of Community*, Harper & Row, New York, 1960, p. 284.

37 D. Lockwood, 'Sources of variation in working class images of society', *Sociological Review*, vol. 14, 1966, p. 258.

38 R. Frankenberg, *Communities in Britain*, Penguin Books, Harmondsworth, 1966, p. 200.

39 Young and Willmott, *op. cit.*, p. 164.

40 J. Mogey, *Family and Neighbourhood*, Oxford University Press, London, 1956, p. 156.

41 Gans found, for instance, that there had been a reduction in self-assessed loneliness; husbands and wives spent more time together, and half (56 per cent) unexpectedly found themselves engaged in local activities. H. J. Gans, *The Levittowners*, Allen Lane, The Penguin Press, London, 1967, pp. 256–7.

42 Williams, *op. cit.*, p. 59.

43 British studies all indicate an absence of the sociability that characterized the American suburban estates. Willmott had to conclude 'the people of Dagenham, when they are not at work, have in the main elected to stay at home', *op. cit.*, p. 89. J. Goldthorpe and D. Lockwood found that the family group was in a 'state of near isolation', *The Affluent Worker in the Class Structure*, Cambridge University Press, 1969, vol. III, p. 107.

44 A succinct summary of the European and American literature is presented by Goldthorpe and Lockwood, *op. cit.*, pp. 1–29.

45 Both Gans and Dobriner provide evidence of the tensions between classes in the estates they studied, particularly over education. In a period of ten years 1951–61 the latter found that Levittown, New Jersey, had changed from a predominantly middle-class community to one in which blue-collar workers predominated. Dobriner, *op. cit.*, pp. 85–126.

46 N. J. O'Neill, 'Class and social consciousness', Ph.D. thesis, University of Hull, 1973, in a comparative study of dockers in established districts and post-war council estates in Hull, found that there was evidence of substantial differences in family life, political orientation and attitudes on social issues, and argued that dislocation from the traditional community had precipitated these changes.

47 Throughout the last quarter of the nineteenth century working-class emigration to the suburbs was encouraged in parliamentary insistence on the provision of working men's trains on new railway lines into the cities. H. J. Pollins, 'Transport lines and social divisions' in *London: Aspects of Change*, ed. Centre of Urban Studies, MacGibbon & Kee, London, 1964, pp. 29–61. Booth was to give authority to the policy, as in a paper, 'Improved means of locomotion as a cure for the housing difficulties of London', cited in F. W. Lawrence, 'The

housing problem' in *The Heart of the Empire*, ed. C. Masterman, Fisher, London, 1901.

48 F. Engels, *The Housing Question*, Progress Publishers, Moscow, 1970, p. 89.

49 Mumford, *op. cit.*, p. 553.

50 C. Bauer, *Modern Housing*, Houghton Mifflin, Boston and New York, 1934, p. 215.

51 The key statements of these writers are included in R. Sennett, *Classic Essays in the Culture of Cities*, Meredith, New York, 1969.

52 Their principal statements are found in J. Rex, *Race, Colonialism and the City*, Oxford University Press, London, 1973; R. Pahl, *Whose City?*, Longmans, London, 1970; and M. Castells, *La Question urbaine*, Maspero, Paris, 1972.

53 N. Long, 'The local community as an ecology of games', *American Journal of Sociology*, vol. 64, 1958.

54 E. A. Gutkind, *The Twilight of Cities*, Free Press, New York, 1962, p. 112.

2 The urban problem: the inner cities

1 L. Benevolo, *The Origins of Modern Town Planning* (1963), Routledge & Kegan Paul, London, 1967, p. 31.

2 In the Columbia University Statement (1894) it was declared that 'by general sociology is meant the scientific study of society as a whole . . . leading up to a more particular study of the phenomena of modern populations and their concentration in large cities'. The statement continued, 'It is in the city that the problems of poverty, of mendicancy, of intemperance, of insanitary surroundings, and of debasing social influences are met in their most acute form. Hence the city is the natural laboratory of social science.' Quoted by H. Odum, *American Sociology*, Longmans Green, New York, 1951.

3 C. Abrams, *The City is the Frontier*, Harper & Row, New York, 1965, p. 19.

4 H. Zorbaugh, *The Goldcoast and the Slum*, University of Chicago Press, 1926, p. 9.

5 A. Morrison, *A Child of the Jago* (1896), Penguin Books, Harmondsworth, 1946.

6 *Housing Act*, 1957, part II, section 4, HMSO, London, 1957.

7 For examples of such discussions see A. L. Schorr, *Slums and Social Insecurity*, Nelson, London and Edinburgh, 1964; J. Seeley, 'The slum: its nature, use and users', *Journal of American Institute of Planners*, vol. 25, 1959, pp. 7–14; and H. J. Gans, *The Urban Villagers*, Free Press, New York, 1962, p. 308.

8 For extracts from Mayhew and Booth, see H. Mayhew, *Voices of the Poor*, ed. A. Humphries, Frank Cass, London, 1971; C. Booth, *On the City*, ed. H. W. Pfautz, University of Chicago Press, 1967; A. Mearns, *The Bitter Cry of Outcast London* (1883), is reprinted by Frank Cass, London, 1970.

9 C. Bauer, *Modern Housing*, Houghton Mifflin, New York, 1934, p. 244.

10 Source of quotation mislaid.
11 G. S. Jones, *Outcast London*, Oxford University Press, London, 1971, estimates that between 1830 and 1900, street clearance schemes displaced *c*. 100,000 people (p. 169), and railways between 1859 and 1867, alone, a further 37,000 (p. 162).
12 A. S. Wohl, 'The housing of the working classes in London, 1815–1914' in *The History of Working Class Housing*, ed. S. D. Chapman, David & Charles, Newton Abbot, 1971, p. 24, estimates that in 1801 there had been an average of 7.03 persons per house, by 1851 this had risen to 7.72, and by 1896, to 8.02 persons per house.
13 A. Briggs, *The Age of Improvement*, Longmans Green, London, 1959; see also A. Briggs, 'Public health: the sanitary idea', *New Society*, February 1968, vol. 11, pp. 229–31.
14 H. J. Dyos, 'The slum observed', 'The slum attacked', *New Society*, February 1968, vol. 11, pp. 151–5 and 192–5.
15 Jones, *op. cit.*, p. 151.
16 Mearns, *op. cit.*, p. 18.
17 Jones, *op. cit.*, p. 230.
18 Mayhew, *op. cit.*, p. 5.
19 F. W. Lawrence, 'The housing problem' in *The Heart of the Empire*, ed. C. Masterman, Fisher, London, 1901, pp. 80–1.
20 T. Marr, *Housing Conditions in Manchester and Salford*, Manchester, 1901, argued (p. 55)

> though a high death rate and frequent cases of illness are perhaps to be looked for in some parts of Ancoats owing to the high proportion earning low wages and therefore poorly nourished . . . yet some share of the blame for the high mortality must be thrown on the close packing of the dwellings on the land and also on the poor character of many of these dwellings.

21 N. Dennis, 'Mass housing: a reformer's myth', *Planning Outlook*, vol. 5, 1969, p. 8.
22 N. Dennis, *People and Planning*, Faber & Faber, London, 1970, p. 366.
23 M. Young and P. Willmott, *Family and Kinship in East London* (1957), Penguin Books, Harmondsworth, 1962, p. 198.
24 Gans, *op. cit.*, p. 4.
25 One of the few sociologists to attempt to penetrate the 'urban jungle' has been O. Lewis in *Five Families*, Basic Books, New York, 1959; *The Children of Sanchez*, Random House, New York, 1962; and *La Vida*, Secker & Warburg, London, 1967.
26 G. D. Suttles, *The Social Order of the Slum*, University of Chicago Press, 1969, p. 6.
27 Young and Willmott, *op. cit.*; N. Dennis, *Public Participation and Planners' Blight*, Faber & Faber, 1972.
28 N. Dennis *et al.*, *Coal is Our Life*, Eyre & Spottiswoode, London, 1956.
29 J. B. Mays, *Growing up in the City*, Liverpool University Press, 1954.
30 W. F. Whyte, *Street Corner Society*, University of Chicago Press, 1943; Gans, *op. cit.*, in their respective methodological appendices both detail the difficulties in 'entering' the neighbourhood social

groupings. The former found his analysis restricted to the street corner gangs, the latter to the 'respectable' Italian working class, on his estimation only half the neighbourhood population.

31 Seeley, *op. cit.*, presents an elaborate classification of a slum population, using two scales, 'permanent/temporary' and 'necessitarian/opportunist', and indicates considerable divergence of interest in the area under study between different groups, as for instance between 'entrepreneurs' and 'trapped necessitarians'.

32 R. Mellor, 'Planning for housing: market processes and constraints', *Planning Outlook*, vol. 13, 1973.

33 J. Mogey, *Family and Neighbourhood*, Oxford University Press, London, 1956.

34 Young and Willmott, *op cit.*, p. 177.

35 K. Coates and R. Silburn, *Poverty: the Forgotten Englishmen*, Penguin Books, Harmondsworth, 1970, p. 99. Their presentation of life in this district can be interpreted as redressing the image of working-class communities drawn from earlier studies both in America and Britain, in particular that of Young and Willmott, *op. cit.* The latter do not draw attention to the housing conditions of their sample—in 1946, 89 per cent of households in the borough of Bethnal Green had been estimated to be without a bath, and 75 per cent without a hot water supply, nor the conditions of housing shortage which enforced matrilocality. (Half the couples in their sample had started off married life with their in-laws.) Elsewhere demand for better housing and aspirations for suburban living have been found to be widespread. R. Wilkinson and D. M. Merry, 'A statistical analysis of attitudes to moving', *Urban Studies*, vol. 2, 1965, reported that 81 per cent of a sample of households from seven neighbourhoods in Leeds were in favour of slum clearance.

36 J. B. Cullingworth, 'New towns for old: the problems of urban renewal', *Fabian Research Series*, no. 229, London, 1962.

37 J. Rex and R. Moore, in their account of 'Sparkbrook 2', *Race, Community and Conflict*, Oxford University Press, London, 1967, pp. 43–83, recount the sense of loss of neighbourliness and communtiy identity felt by older residents; K. Coates and R. Silburn, *op. cit.*, describe a similar situation in St Ann's, Nottingham.

38 J. Seabrook, *City Close Up* (1971), Penguin Books, Harmondsworth, p. 45.

39 S. R. Arnstein, 'A ladder of citizen participation in the USA', *Journal of the Town Planning Institute*, vol. 58, 1971, p. 177.

40 W. Reich, *What is Class Consciousness?* (1933), Socialist Reproduction Press, London, 1971, p. 23.

41 R. McKie, 'Housing and the Whitehall bulldozer', *Hobart Papers*, no. 52, Institute of Economic Affairs, London, 1971, p. 58.

42 For the development of these arguments see Dennis, *People and Planning* and *Public Participation and Planners' Blight*; McKie, *op. cit.*; and J. G. Davies, *The Evangelistic Bureaucrat*, Tavistock, London, 1972.

43 F. C. Pennance and H. Gray, *Choice in Housing*, Institute of Economic Affairs, London, 1968, p. 6.

44 T. MacMurray, 'Comments on the concepts and techniques of local planning', *Planning Outlook*, vol. 9, 1970, pp. 19–32.

45 For two exponents of this view see R. F. Muth, 'Slums and poverty' in *The Economic Problems of Housing*, ed. A. Nevitt, Institute of Economic Affairs, Macmillan, London, 1967, pp. 12–26; and W. Bunge, *Fitzgerald: Geography of a Revolution*, Schenkman, Cambridge, Mass., 1971.

46 For accounts of the latter see M. Castells, *La Question urbaine*, Maspero, Paris, 1972, pp. 355–96. By the end of 1967 it was estimated that in post-war slum clearance, 870,000 dwellings had been demolished in Britain, 404,000 in the USA, and 40,000 in France. National Economic Development Office, *New Homes in the Cities*, London, 1971.

47 'Groupe de sociologie urbaine de Nanterre', 'Paris 1970: reconquête urbaine et renovation—déportation', *Sociologie du Travail*, vol. 12, 1970, p. 488.

48 Gans, *op. cit.*, p. 291.

49 Rex and Moore, *op. cit.*, p. 9ff.

50 E. Gauldie, *Cruel Habitations*, Allen & Unwin, London, 1974, p. 17. Dennis's argument that bringing the problems of the underprivileged into the sphere of legislation 'depended in large measure upon the propaganda and the decisive energies of an aggressive minority', 'Mass housing: a reformer's myth', p. 8, is unconvincing.

51 Like the term 'slum' that of the 'twilight area' has been used indiscriminately to refer to very different areas. It has been used in *The Deeplish Study*, HMSO, London, 1966, to refer to the cottage housing of Rochdale, and also to the areas of 'housing stress' identified by the Milner Holland Report: *Housing in Greater London*, HMSO, London, 1965. See R. Mellor, 'Structure and process in the twilight areas', *Town Planning Review*, vol. 44, no. 1, 1973, p. 55. The present discussion relates to the metropolitan twilight areas.

52 T. Blair, 'Social systems analysis', *Official Architecture and Planning*, May 1969, p. 580.

53 Rex and Moore, *op. cit.*, pp. 272–85; see also J. Rex, 'The social segregation of the immigrant in British cities', *Political Quarterly*, vol. 39, 1968.

54 Plowden Report, *Children and their Primary Schools*, HMSO, London, 1967, chapter 5, pp. 50–68.

55 J. Palmer, 'Introduction' to R. Goodman, *After the Planners*, Penguin Books, Harmondsworth, 1972, p. 25.

56 *Ibid.*, p. 39.

57 A full definition of this term is presented by L. Althusser, *For Marx*, Allen Lane, The Penguin Press, London, 1965, p. 67:

> What actually distinguishes the concept of the *problematic* from the subjectivistic concepts of an idealist interpretation of the development of ideologies is that it brings out within the thought *the objective internal reference system of its particular* themes, the system of *questions* commanding the *answers* given by the ideology.

> If the meaning of the ideology's answers is to be understood at this internal level it must first be asked *the question of its own questions*. But this problematic is *itself an answer*, no longer to its own internal questions—problems—but to *the objective problems posed* for ideology by its time. A comparison of the problems posed by the ideologue [his problematic] with the *real problems* posed for the ideologue by his time, makes possible a demonstration of the truly ideological element of the ideology, that is, what characterises ideology as such, its *deformation*. So it is not *the interiority of the problematic* which constitutes its essence but its relation to real problems: *the problematic of an ideology* cannot be demonstrated without relating and submitting it to the real problems to which its deformed enunciation gives a false answer.

58 Milner Holland Report, *op. cit.*, p. 228. The brief given to the committee in 1963 was limited to the 'survey of the housing situation in Greater London with particular reference to the use, maintenance and management of rented accommodation whether privately or publicly owned'. So although the report acknowledges that the housing crisis was the outcome of 'a long-term national pattern of economic development' (p. 22) the committee were unable to shift the statement of the problem from 'housing' to questions as to London's economy. The problem posed, the questions asked, and the answers commanded, in the report, then became the basis for policy discussions as to the future of the twilight areas.

59 R. Pahl, Runnymede Lecture, 1973, reported in *New Society*, no. 554, 1973, p. 354.

60 D. Eversley, 'The core of the housing problem', *Built Environment*, June 1973, p. 323, noted that the GLC area had lost 500,000 actual or potential households through migration, and that 750,000 new dwellings had been added to the housing stock since 1945, and declared,

> If only the critics . . . looked at the realities of housing stock management and the effects of purchasing power on housing conditions, they might stop their irrelevant proclamations of new gimmickry, to deal with what is increasingly a distribution of effective powers to command a home in society; not a shortage of homes as such, nor a deterioration of stock.

61 F. Engels, *The Housing Question* (1872), Progress Publishers, Moscow, 1970, p. 74.

62 The peculiarity of London's situation, now, as in the nineteenth century, is reflected in the debates about preferential income policies for its service workers. The undermanning of London Transport by 10–15 per cent, the police force by some 20 per cent, postal services by 15 per cent, and the extreme turnover in teaching staff, among other essential workers, raised the question of 'London's case for special treatment' in *Financial Times*, 6 November 1973: 'the only answer left open when all the foreclosed options have been considered must be to effect a transfer of income from those in the cities who demand these services to those who provide them.'

Chapter 3 The housing question

1 The house as home

1 K. Millett, *Sexual Politics* (1969), Sphere Books, London, 1971, p. 158.
2 F. Tönnies, *Community and Society* (1887), Harper & Row, New York, 1963.
3 Engels directed the first part of *The Housing Question* (1872) against the Proudhonist, Mülberger, whom he attacked for declaiming: 'in the big cities 90 per cent and more of the population have no place they can call their own. The real nodal point of moral and family existence, hearth and home, is being swept away by the social whirlpool.' Progress Publishers, Moscow, 1970, p. 22.
4 J. Seeley, R. H. Sim and E. W. Loosley, *Crestwood Heights* (1956), John Wiley, New York, 1963, p. 54.
5 F. L. Wright, *An Autobiography*, Faber & Faber, London, 1945.
6 Seeley *et al.*, *op. cit.*, p. 58.
7 *Ibid.*, p. 46.
8 E. Goffman, *The Presentation of Self in Everyday Life*, University of Edinburgh Social Science Research Centre, 1958, pp. 66–86.
9 For instance, Mogey commented of an inner district of Oxford: 'St Ebbe's has a tradition of never entering other people's homes: good neighbouring is a matter for the street or public places like the local shops.' J. Mogey, *Family and Neighbourhood*, Oxford University Press, London, 1956, p. 92. Rainwater, in a study of lower income housing projects, argues that the house was a haven which 'acquires a sacred character from its complex inter-twining with the self and from the symbolic character it has as a representation of the family'. L. Rainwater, 'Fear and the house as haven in the lower class' (1960) in *Urbanism, Urbanization and Change*, ed. P. Meadows and E. Mizruchi, Addison-Wesley, Reading, Mass., 1969, p. 291.
10 A wide range of architectural writers and critics have drawn attention to the possibilities of new design principles in housing. Among these are S. Chermayeff and C. Alexander, *Community and Privacy* (1963), Penguin Books, Harmondsworth, 1966; and O. Neumann, *Defensible Space*, Architectual Press, London, 1973. For an early investigation into the divisions within a relatively isolated housing estate see E. D. Mitchell and T. Lupton, *Neighbourhood and Community*, University Press, Liverpool, 1954.
11 Engels, *op. cit.*, p. 34.
12 *Ibid.*, p. 29.
13 Le Corbusier, *The Radiant City* (1933), Faber & Faber, London, 1964, p. 27.
14 D. V. Donnison, *The Political Economy of Housing*, Penguin Books, Harmondsworth, 1967, p. 64.
15 F. Allaun, *No Place Like Home: Britain's Housing Tragedy*, André Deutsch, London, 1972, p. 191, argues, 'when the Government builds a motorway or battleship it does not pay a penny in interest. It builds it out of income . . .'.

16 R. Goodman, *After the Planners*, Penguin Books, Harmondsworth, 1972, p. 247.

2 Architects and housing

1 E. Gauldie, *Cruel Habitations*, Allen & Unwin, London, 1974, p. 45. Ruskin, in describing the design of a pitched roof for a woodland cottage, commented, 'Of course we are not thinking of interior convenience: the architect must establish his mode of beauty first and then approach it as nearly as he can.' J. Ruskin, *The Poetry of Architecture*, George Allen, London, 1893, p. 85.

2 RIBA Research and Statistics Office, *Observer Survey on the Architectural Profession in Britain 1964*, quoted by T. Blair, 'Architects' dilemma', *Official Architecture and Planning*, August 1970.

3 B. Kaye, *The Development of the Architectural Profession in Britain*, Allen & Unwin, London, 1960.

4 M. MacEwan, *Crisis in Architecture*, RIBA Publications, London, 1974, pp. 6–10.

5 Pawley comments: 'Sketches on the backs of envelopes may be all right for the Parthenon but for a new university . . . they simply will not do. The official client demands value for money.' 'Architecture and the philosopher's stone', *New Society*, April 1971, p. 920.

6 An outstanding expression of this viewpoint is that by J. Summerson in the collected essays, *Heavenly Mansions*, Cresset Press, London, 1949.

7 Le Corbusier sums up the radical impact of rationalism on architecture in the conclusion to *The City of Tomorrow* (1929): 'Things are not revolutionized by making revolutions. The real Revolution lies in the solution of existing problems. . . .' The Architectural Press, London, 1947.

8 Le Corbusier, *op. cit.*, pp. 51–7.

9 C. Bauer, *Modern Housing*, Houghton Mifflin, Boston and New York, 1934, writing of the experience of the 1930s: 'What the architect was planning for as fruitful leisure in his world within a world descended finally in the shape of miserable unemployment' (p. 215).

10 A. Kopp, *Town and Revolution*, Thames & Hudson, London, 1970, p. 12.

11 L. Martin, 'Address on receipt of gold medal award', *Journal of Royal Institute of British Architects*, vol. 80, no. 8, 1973.

12 Le Corbusier, *The City of Tomorrow*; and *The Radiant City*, Faber & Faber, London, 1964.

13 A. Cox, 'Le Corbusier—his impact on four generations: the thirties', *Journal of Royal Institute of British Architects*, vol. 72, no. 10, 1965, p. 498.

14 MacEwan, *op. cit.*, p. 15.

15 *Ibid.*, p. 18.

16 C. Jencks, *Le Corbusier and the Tragic View of Architecture*, Allen Lane, The Penguin Press, London, 1973, p. 144.

17 'Housing and the environment', *Architectural Review*, vol. 162, no. 849,

1967, provides ample illustration of the distortion and brutalization of Le Corbusian designs in public housing, by architects without his 'sensibility' and resources.

18 Le Corbusier, *op. cit.*, p. 51.
19 Cox, *op. cit.*, p. 498.
20 Kopp, *op. cit.*, p. 12.
21 Jencks, *op. cit.*, p. 25.
22 Le Corbusier, *op. cit.*, makes it explicit that 2 million out of 3 million inhabitants of his proposed city were to live in garden city suburbs.
23 All these statements are taken from *The City of Tomorrow*, pp. 74–81.
24 Le Corbusier, *The Radiant City*, p. 36.
25 *Ibid.*, p. 96.
26 J. Jacobs, *Death and Life of Great American Cities* (1961), Penguin Books, Harmondsworth, 1964, p. 32.
27 Jencks, *op. cit.*, pp. 112–15.
28 Jacobs, *op. cit.*, p. 680.
29 L. Mumford, *The Urban Prospect: Yesterday's City of Tomorrow*, Secker & Warburg, London, 1968, pp. 121–2.
30 S. Crockford and F. Fromer, 'When is a house not a home?' in *The Body Politic*, ed. M. Wandor, Stage 1, London, 1972, p. 123.
31 Mumford, *op. cit.*, p. 118.
32 Bauer, *op. cit.*, pp. 276–8.
33 A. Nevitt, *Housing Taxation and Subsidies*, Nelson, London, 1966, pp. 92–112.
34 MacEwan, *op. cit.*, p. 20.
35 The programme of high rise construction was brought to a rapid halt after 1968. Although the Ministry of Housing had introduced 'cost yardsticks' in 1963, it was not until 1968 that Exchequer grant was made conditional on exact adherence to these standards which were framed in such a way as to render impossible further expensive projects.
36 In some circumstances, 'modern housing' schemes can be satisfactory to their residents; Parkhill, Sheffield, seems to have been one such case. In 1962, only 20 per cent of the tenants were dissatisfied with the estate (*Official Architecture and Planning*, February 1966, p. 234), and a large number had asked to move there. Banham comments, 'Parkhill is first and foremost an example of the City of Sheffield rehousing people of Sheffield', *New Society*, October 1973, p. 156, in recounting the care taken over the development and local pride in it.
37 Bauer, *op. cit.*, p. 221.
38 P. and P. Goodman, *Communitas* (1947), Random House, New York, 1963, p. 3.

3 Housing finance

1 D. H. Pinkney, *Napoleon III and the Rebuilding of Paris*, Princeton University Press, N.J., 1958, comments of nineteenth-century Paris, 'Critics complained of crowded shanty towns, masses of hovels fashioned of planks and tar-paper, growing up on the extremities of the city in the sixties' (p. 166), a picture remarkably like that of the

Paris *bidonvilles* now, or the Third World shanty towns. For a discussion of the latter see J. Turner, 'Architecture that works', included in *Human Identity in an Urban Environment*, ed. G. Bell and J. Tyrwhitt, Penguin Books, Harmondsworth, 1972; and S. Brett, 'Low income urban settlements in Latin America' in *Sociology and Development*, ed. E. Kadt, Tavistock, London, 1974. For a detailed case study in an Athens suburb see the report in *Architects' Association Quarterly*, April 1970.

2 A. A. Nevitt, *Housing, Taxation and Subsidies*, Nelson, London, 1966, pp. 13–23, presents a general account. For more detailed discussion of development in particular localities see H. J. Dyos, *Victorian Suburb*, Leicester University Press, 1961; and C. W. Chalklin, *The Provincial Towns of Georgian England*, Arnold, London, 1974.

3 For a general account see J. P. Lewis, *Building Cycles and Britain's Growth*, Macmillan, London, 1965; for a detailed presentation of the boom from 1895 to 1904, see J. Blackman and J. Sigsworth, 'The home boom of the 1890s', included in 'Studies in the British economy 1870–1914' ed. J. Saville, *Yorkshire Bulletin of Economic and Social Research*, 1965, pp. 75–97.

4 C. Bauer, *Modern Housing*, Houghton Mifflin, New York, 1934, appendix to part III, pp. 260–302.

5 G. D. H. Cole, *Building and Planning*, Cassell, London, 1945, p. 88.

6 These figures are taken from G. D. H. Cole and R. Postgate, *The Common People 1716–1945*, Methuen, London, 1969, pp. 344, 446–7, 496–7; and from M. Kidron, 'Imperialism: highest stage but one' (1962), reprinted in *International Socialism*, no. 61, 1973. The latter estimates that foreign investment in 1914 accounted for 50 per cent of savings, of which 90 per cent was in private portfolios.

7 For analysis of the economic impasse confronting private house builders see M. Bowley, *Housing and the State*, Allen & Unwin, London, 1945, pp. 1–10; and Cole, *op. cit.*, pp. 88–109.

8 D. V. Donnison, *The Government of Housing*, Penguin Books, Harmondsworth, 1967, p. 94, relates the experience of Switzerland, Belgium and Britain, in this respect. S. Pollard, *The Development of the British Economy 1870–1967*, Arnold, London, 1969, comments on 'the extraordinary care taken to shelter the money market' (p. 72) in the immediate aftermath of the First World War.

9 F. Allaun, *No Place Like Home: Britain's Housing Tragedy*, André Deutsch, London, 1972, p. 191.

10 A. Briggs, *Victorian Cities* (1963), Penguin Books, Harmondsworth, 1968, pp. 228–30.

11 The administrative machinery set up after 1945 had to be used to *restrict* housing production after the economic débâcle of mid–1947. P. Foot, *Aneurin Bevan 1945–1960*, Davis-Poynter, London, 1973, pp. 60–101.

12 Cole, *op. cit.*, p. 226.

13 The costs per dwelling (including land) for housing estates in construction during 1969 for Camden Borough Council, London, ranged from £5,805 to £9,734; the economic rental after Exchequer subsidy

would have ranged from £6 9s. 3d. to £9 0s. 2d.; after rent rebates, a household of man, wife and two dependent children would have required an income of £37 10s. to £47 14s. *Housing in Camden*, Camden Borough Council, 1969, vol. 1, tables V and VI. The costs of the housing revenue account were then apportioned in the ratio: 42 per cent, tenants, 40 per cent, ratepayers, and 18 per cent, Exchequer (p. 14).

14 By 1967, twice as many dwellings had been demolished in Britain by the local authorities as in America, under urban renewal schemes; twenty times as many as in France; National Economic Development Office, *New Homes in the Cities*, London, 1971, p. 2.

15 'Who makes money out of council housing?', *Community Action*, no. 5, p. 28. Housing income in the year up to March 1973 was £975.5 million, interest charges amounted to £556.1 million, and another £93 million was spent on paying off loan capital. Rent income was £625 million. Department of Environment, *Housing and Construction Statistics*, HMSO, London, 1974.

16 The history of the building society movement is covered by E. J. Cleary, *British Building Societies*, Elek Books, London, 1965.

17 Bowley, *op. cit.*, p. 175.

18 Bristol and West Building Society, 'Factual Background', July 1974, Table 5, details the movement of savings between insurance companies, pension funds, the major banks, savings banks, local authorities and unit trusts, between 1965 and 1974. The same predicament is experienced in the USA where 'high Treasury bill yields are attracting more small investors away from the savings banks', *The Times*, 28 August 1974, forcing government intervention in direction of finance to housing.

19 Statistics drawn from H. Ashworth, 'Perennial problems of interest rates', *Investors Chronicle*, 5 June 1959; Bristol and West Building Society, *op. cit.*; and *Building Societies Year Book*.

20 Foot, *op. cit.* p. 77.

21 There is considerable disagreement as to the effect of planning controls over land availability. The arguments as to insufficiency of land for building were presented in Shankland Cox and Associates, *Land Availability for Residential Development*, Housing Research Federation, London, 1972, and countered in the report of the Standing Conference on London and South East Regional Planning, *The Housing Land Situation*, London, 1972 (LRP 1984). In the latter it was argued, first, that ample land was allocated for development, and the total of outstanding planning permissions for new dwellings was substantial; second, that 'the rise in land prices has resulted from rather than been a cause of rising house prices' (p. 5). The best conclusion on the events of 1972–3 would seem to be that 'Delay factors upset the programme's flow of land through the process, a flow which is in any event characterised by a very imperfect relationship between land considered available in planning terms, and land actually available for development'; C. Rawlinson, 'Housing land availability in the South East', *The Planner*, vol. 61, 1975, p. 51.

22 Nationwide Building Society, *Occasional Bulletin*, 110, 109, July 1972; the differential between regions subsequently has narrowed, see bulletin 121, July 1974.
23 W. R. Lethaby, *Form in Civilization: Collected Papers of Art and Labour*, Oxford University Press, London, 1922, p. 35.

4 Housing in urban sociology

1 Discussions on the weaknesses of British housing policy in the light of experience of immigration over the last twenty years are found in the Milner Holland Report: *Housing in Greater London*, HMSO, London, 1965; J. Rex and R. Moore, *Race, Community and Conflict*, Oxford University Press, London, 1967; E. Burney, *Housing on Trial*, Oxford University Press, London, 1967; E. Rose, *Colour and Citizenship*, Oxford University Press, London, 1969, chapter 12, pp. 120–48; R. Haddon, 'A minority in a welfare state: the location of West Indians in the London housing market', *New Atlantis*, 1970, vol. 1, pp. 80–133.
2 American urban studies have taken the same direction in the last decade, as for instance in discussions of the ghetto. A recent study of the finance of housing and its bearing on ethnic segregation is that by D. Harvey, *The Political Economy of Urbanization in Advanced Capitalist Societies: the Case of the United States*, Centre for Metropolitan Planning and Research, Johns Hopkins University, Baltimore, 1974.
3 The 'sociology of contraint' as a distinctive contribution to urban studies and sociology was elaborated by R. Pahl, *Whose City?*, Longmans, London, 1970, chapter 13, 'Urban social theory and research', pp. 215–18. See also R. Pahl, 'Poverty and the urban system', included in *Spatial Policy Problems of the British Economy*, ed. M. Chisholm and G. Manners, Cambridge University Press, 1972, pp. 126–45.
4 J. Rex, 'The sociology of the urban zone of transition', included in *Readings in Urban Sociology*, ed. R. Pahl, Pergamon Press, Oxford, 1968, p. 211.
5 The concept was introduced by Rex first in Rex and Moore, *op. cit.*, p. 9, and subsequently explored in a series of papers: 'The sociology of the urban zone of transition'; 'The social segregation of the immigrant in British cities', *Political Quarterly*, vol. 39, 1968; 'The concept of housing class and the sociology of race relations', *Race*, vol. 12, 1971; and papers included in *Race, Colonialism and the City*, Routledge & Kegan Paul, London, 1973, chapters 10–11, pp. 111–26.
6 Rex and Moore, *op. cit.*, 1967, p. 273.
7 Rex, 'The sociology of the urban zone of transition', p. 215.
8 Weber seems to have considered that classes could be positively or negatively privileged in relation to domestic property as much as in the disposal of their skills in the market place, so that ownership of housing and its use as income was one explanation of the 'typical chance for a supply of goods, external living conditions, and personal

life experiences'; M. Weber, 'Class, status, party' in *From Max Weber: Essays in Sociology*, ed. H. Gerth and C. W. Mills, Routledge & Kegan Paul, London, 1948, p. 181. 'Class situation' becomes a position in relation to property in a given empirical context, and not an on-going relationship in the mode of production.

9 In the paper 'Urban and other elements in race relations theory', Rex, *Race, Colonialism and the City*, he would seem to discard the housing class concept in admitting that racialism has to be placed in a wider context than that of an urban crisis, 'in particular, the advantages to the entrepreneurs or planners of a metropolitan society in having alongside the normal unionized and protected free labour force, another more disposable force . . .' (p. 154).

10 For criticisms, note J. Lambert and C. Filkin, 'Race relations research: some issues of approach and application', *Race*, vol. 12, 1971, pp. 329–35; and J. G. Davies and J. Taylor, 'Race, community and no conflict', *New Society*, July 1970, pp. 67–9.

11 T. Blair, 'Social systems analysis', *Official Architecture and Planning*, May 1969, p. 583.

12 Haddon, *op. cit.*, p. 133, concluded that the principal value of the housing class model was in the insights it had provided into the local neighbourhood situation in Sparkbrook, p. 133.

13 S. Castles and G. Kosack, 'Twelve hypotheses on labour migration and class struggle', unpublished paper circulated May 1971; see also their paper, 'The function of labour immigration in Western European capitalism', *New Left Review*, no. 73, 1972, pp. 1–21.

14 F. Engels, *The Housing Question* (1872), Progress Publishers, Moscow, 1970, p. 65.

15 G. S. Jones, *Outcast London*, Oxford University Press, London, 1971, p. 243.

16 J. Foster, *Class Consciousness and the Industrial Revolution*, Weidenfeld & Nicolson, London, 1973.

17 M. Castells, *Luttes urbaines et pouvoir politique*, Maspero, Paris, 1973, p. 12.

Chapter 4 The practice of planning

1 The town-planning movement

1 D. L. Foley, 'Idea and influence: the Town and Country Planning Association', *Journal of the American Institute of Planners*, 1962, pp. 10–17.

2 For an account of German 'municipal capitalism' before 1914, see C. Bauer, *Modern Housing*, Houghton Mifflin, New York, 1934, pp. 97–101. Extensive municipal land ownership and efficient management meant that city planning 'was already in full swing in Germany in 1900' (p. 99).

3 R. Wade and H. Mayer, *Chicago: Growth of a Metropolis*, Chicago University Press, 1969.

4 Creese suggests that Howard's stay in Chicago between 1872 and 1876 was an influence on his later promotion of garden cities. Not only

were there already several privately promoted suburban townships along the railways into the city, but Chicago itself, before the fire, was known as the 'Garden City'. W. L. Creese, *The Search for Environment*, Yale University Press, New Haven and London, 1966, pp. 150–7.

5 W. Ashworth, *The Genesis of Modern British Town Planning*, Routledge & Kegan Paul, 1954, chapters 4 and 5.

6 P. Alden and E. Hayward, *Housing*, Headley Bros, London, 1907, p. 3.

7 Ashworth, *op. cit.*, p. 181.

8 *Ibid.*, p. 182.

9 W. Morris, listed by W. L. Creese, *The Legacy of Raymond Unwin*, MIT Press, Cambridge, Mass., 1967, p. 40, as a quotation kept by Unwin among his personal papers.

10 For a critical presentation of the debate see G. S. Jones, *Outcast London*, Oxford University Press, London, 1971, pp. 126–51.

11 Quoted by B. Semmel, *Imperialism and Social Reform*, Allen & Unwin, London, 1960, p. 62.

12 E. Howard, 'Report of discussion at a meeting of the Sociological Society', *Sociological Papers*, 1904, pp. 119–20.

13 J. Saville, 'Henry George and the British Labour movement', *Science and Society*, vol. 24, 1960, p. 323.

14 *Ibid.*, p. 325.

15 *Ibid.*, p. 330.

16 H. George, *Progress and Poverty: the Remedy*, Kegan Paul, London, 1881, p. 149.

17 S. Pollard, *The Development of the British Economy*, Arnold, London, 1969, p. 41.

18 H. Samuel, quoted by Semmel, *op. cit.*, pp. 159–60.

19 Quoted by D. Foot, 'New support for the old idea of land nationalization', *The Times*, 14 September 1974.

20 L. Benevolo, *The Origins of Town Planning* (1963), Routledge & Kegan Paul, London, 1967, terms Buckingham's plan a 'form of blueprint for mass production' (pp. 129–30).

21 E. Howard, *Garden Cities of Tomorrow*, Faber & Faber, 1946, pp. 118–27. See also W. A. Eden, 'Ebenezer Howard and the Garden City movement', *Town Planning Review*, 1947, p. 123; Saville, *op. cit.*, p. 323; and D. MacFadyen, *Sir Ebenezer Howard and the Town Planning Movement*, Manchester University Press, 1933, pp. 20–1. The latter draws attention to the impact of E. Bellamy, *Looking Backward*, on both Howard and Morris.

22 Eden, *op. cit.*, p. 132.

23 L. Mumford, *The Urban Prospect*, Secker & Warburg, New York, 1968, terms Howard's plan 'closer in spirit to Joseph Paxton than to William Morris' (p. 144).

24 Howard, *Garden Cities of Tomorrow*, p. 45.

25 *Ibid.*, p. 127.

26 Report of the Cities Committee, 'Towards the third alternative', *Sociological Review*, vol. 11, 1919, pp. 62–5.

27 R. Unwin, quoted by Creese, *The Legacy of Raymond Unwin*, p. 154.

28 S. Gurney, 'Civic reconstruction and the garden city movement', *Sociological Review*, vol. 3, 1910, p. 39.

29 P. Self, 'Introduction' to *The New Towns: the British Experience*, ed. H. Evan, Town and Country Planning Association, London, 1972.

30 G. D. H. Cole and R. Postgate, *The Common People 1716–1946*, Methuen, London, 1949, p. 471.

31 P. Abercrombie, 'One of the functions of planning is to recover the community spirit, and provide for its proper working with a centre and a boundary' (p. 55), and 'the centre provides for a continuity of this tide of civic idealism' (p. 68). *Plan for Kingston upon Hull*, A. Brown & Sons, London and Hull, 1945.

32 J. Palmer, 'Introduction' to R. Goodman, *After the Planners*, Penguin Books, Harmondsworth, traces the development of town planning in the Welfare State as a co-ordinated response to industrial and technological change, and relates it to 'government pump-priming to restart the profitability cycle'.

2 The town-planning profession

1 D. L. Foley, 'British town planning: one ideology or three?', *British Journal of Sociology*, vol. 11, 1960, pp. 211–31.

2 There were fewer than 200 societies affiliated to the Civic Trust in 1960, over 1,000 by 1974; membership gains were equally impressive—in the two years 1967–9 alone, it rose from 130,000 to 180,000. *Civic Trust Newsletter*, London, 1970.

3 T. MacMurray, 'Comments on the concepts and techniques of local planning', *Planning Outlook*, vol. 9, 1970, p. 19.

4 J. G. Davies, *The Evangelistic Bureaucrat*, Tavistock, London, 1972, p. 222.

5 C. Buchanan, *The State of Britain*, Faber & Faber, London, 1972, p. 60.

6 *Ibid.*

7 In 1945 there were 1,600 members of the Town Planning Institute, by 1962 there were 2,350, by 1973 there were 9,500 (all figures include student members). In 1962, among qualified local authority planning staff, 14.5 per cent were previously qualified as architects, 11.5 per cent as engineers, and 15 per cent as surveyors; 54 per cent had qualified as planners; in 1963–4 the biggest single group of students on post-graduate planning courses were architects, and by 1972 recruits with a first degree in geography predominated. Source: Town Planning Institute membership survey, April 1962, reprinted in A. Goss, *The Architect and Town Planning*, Royal Institute of British Architects, London, 1965; S. Adamson, 'Happy birthday?', *The Planner*, vol. 59, no. 8, p. 350.

8 N. Harris, *Beliefs in Society* (1968), Penguin Books, Harmondsworth, 1971, p. 27.

9 K. Mannheim, *Ideology and Utopia*, Routledge & Kegan Paul, London, 1936, p. 3.

10 Foley, *op. cit.*, p. 212.

11 P. Boardman, *Patrick Geddes: Maker of the Future*, University of North Carolina Press, Chapel Hill, 1944, p. 479. Another review of Geddes's work is L. Mumford, 'Geddes and Branford', included in H. Barnes, *Introduction to the History of Sociology*, University of Chicago Press, 1948, pp. 677–97.

12 P. Geddes, 'Civics: as applied sociology', *Sociological Papers*, 1904; 'Civics: as concrete and applied sociology', *ibid.*, 1905; 'A suggested plan for a civic museum', *ibid.*, 1906. In the first of these papers Geddes criticizes Booth's work for leaving London 'a less or more foggy labyrinth' (p. 106), and defines civics 'as that branch of Sociology which deals with cities—their origin and distribution; their development and structure: their functioning, internal and external, material and psychological; their evolution, individual and associated' (p. 111).

13 P. Geddes, *Cities in Evolution* (1915), ed. J. Tyrwhitt, Williams & Norgate, London, 1949, p. 53.

14 Boardman, *op. cit.*, p. 241.

15 Geddes, *Cities in Evolution*, p. 39.

16 *Ibid.*, p. 47.

17 Report of the Cities Committee, 'Towards the third alternative', part II, *Sociological Review*, vol. 11, 1919, p. 67.

18 J. Tyrwhitt, 'Introduction' to Geddes, *Cities in Evolution*, p. xxvi.

19 W. Ashworth, *The Genesis of Modern British Town Planning*, Routledge & Kegan Paul, London, 1954, p. 193.

20 R. Unwin, 'Regional planning with reference to Greater London' (1930), included in W. L. Creese, *The Legacy of Raymond Unwin: A Human Pattern for Planning*, MIT Press, Cambridge, Mass., 1967, p. 165.

21 R. Unwin, 'Town planning in practice' (1909) in Creese, *op. cit.*, p. 71.

22 R. Unwin, 'Higher building in relation to town planning' (1924) in Creese, *op. cit.*, p. 150.

23 R. Unwin, 'Regional planning with reference to Greater London' (1930) in Creese, *op. cit.*, p. 161.

24 W. L. Creese, 'The ideals and assumptions of Sir Raymond Unwin' in Creese, *The Legacy of Raymond Unwin*, p. 16.

25 R. Unwin, 'Housing and town-planning lectures at Columbia University' (1936–7) in Creese, *The Legacy of Raymond Unwin*, p. 170.

26 Creese, *The Legacy of Raymond Unwin*, p. 27.

27 L. Reissman, *The Urban Process*, Free Press, New York, 1964, p. 45.

28 *Ibid.*, p. 40.

29 Unwin, 'Housing and town-planning lectures at Columbia University', p. 194.

30 Foley, *op. cit.*, pp. 216–18.

31 L. Keeble, *Principles and Practice of Town and Country Planning* (1952), The Estates Gazette, London, 1959, p. 9. The emphasis of the book is on 'legally-practicable planning'.

32 Under the 1947 Act, 'the price of land was reduced to existing use value. Owners could no longer receive the benefit of potential or

speculative value for future development'. C. M. Haar, 'Planning Law', ed. Department of Civic Design, University of Liverpool, *Land Use in an Urban Environment*, Liverpool University Press, 1961, p. 106. Under the 1953 Act, all payments out of the Betterment Fund of £300 million were stopped, and the development charge dropped, and under the 1959 Act local authorities once more had to acquire land for planning purposes at market value. Glass termed this an 'Anti-Planning Bill', in that it finally denationalized development rights and values (contrary to the previously bi-partisan agreement in Parliament), so opening the floodgates to a free speculative market in land. R. Glass, 'The mood of London' in *London: Urban Patterns, Problems and Policies*, ed. D. V. Donnison, Heinemann, 1973, p. 408.

33 Ministry of Housing and Local Government, *The Future of Development Plans*, a report by the Planning Advisory Group (PAG), HMSO, London, 1965.

34 Royal Commission on Local Government in England and Wales; Chairman, Lord Redcliffe-Maud; HMSO, London, 1969.

35 Skeffington Committee Report, *People and Planning*, HMSO, London, 1969.

36 Ministry of Housing and Local Government, *South Hampshire Study* (Colin Buchanan & Partners), HMSO, London, 1966, pp. 9–10.

37 J. B. McLoughlin, *Urban and Regional Planning: a Systems Approach*, Faber & Faber, London, 1969, p. 86.

38 The methodology proposed is very similar to that which Medawar terms 'hypothetico-deductive' in that the plan (hypothesis) is derived, and then subject to verification procedures in an alternation of intuitive leap and critical testing. P. B. Medawar, *Induction and Intuition in Scientific Thought*, Methuen, London, 1969, pp. 42–59.

39 J. Ferris, *Participation in Urban Planning: the Barnsbury Case*, Occasional Papers in Social Administration, no. 48, G. Bell & Sons, London, 1972, for the Social Administration Research Trust, p. 15.

40 Buchanan, *op. cit.*, p. 84.

41 E. Reade, 'Contradictions in planning', *Official Architecture and Planning*, October 1969, p. 1185.

42 S. Ward, 'The Town and Country Planning Act, 1932', *The Planner*, vol. 60, no. 5, 1974.

43 Buchanan, *op. cit.*, p. 85. For a full review of the state of local planning see *The Planner*, vol. 60, no. 4, 1974.

44 The debate about participation in planning has been lengthy. A full issue of the *Journal of the Town Planning Institute*, vol. 57, no. 4, 1971, was devoted to 'Participation in planning'; a series of case studies were presented in *ibid.*, vol. 59, no. 4, 1973. For two useful review articles see Ferris, *op. cit.*, Introduction, pp. 13–19, and L. B. Ginsburg, 'Can participation be positive?', *Town and Country Planning Summer School*, 1973, Report of Proceedings, pp. 18–21.

45 W. Bor, 'The Town and Country Planning Act, 1968', *The Planner*, vol. 60, no. 5, 1974, p. 702. See also his 'Towards new planning', *Official Architecture and Planning*, March 1972, pp. 146–50. In the same issue, D. Senior, 'Lessons of the GLDP inquiry', terms the public

examination of the Greater London Council's proposals 'a self-defeating charade' (p. 157).

46 P. Self, 'The planner's future', *Town and Country Planning*, March 1972, p. 160.

47 J. B. McLoughlin and J. Thornley, 'Some problems in structure planning: a literature review', Centre for Environmental Studies, London, 1972 (CES IP 27). In 1972, it was found that 90 per cent of the staff engaged in the preparation of structure plans, were drawn from planning departments: J. B. McLoughlin, 'Structure planning in Britain', report of a survey conducted for CES, and *Royal Town Planning Institute Journal*, vol. 59, no. 3, 1973.

48 J. Palmer, 'Introduction' to R. Goodman, *After the Planners*, Penguin Books, Harmondsworth, 1972, p. 38.

49 D. E. C. Eversley, 'Planning in an age of stagnation', *Built Environment*, January 1975, p. 15.

50 'Town-planners and their future, a discussion paper', the Royal Town Planning Institute, 1971, p. 12. (See also the later document of the same title, September 1973.)

51 W. R. Lethaby, *Form in Civilization: Collected Papers on Art and Labour*, Oxford University Press, London, 1922. 'Town tidying' (1916), pp. 17–21.

52 *American Behavioural Scientist*, vol. 14, 1971, 'Profession in contemporary society'. See also T. Parsons, 'The profession and social structure', included in *Essays in Sociological Theory*, Free Press, Chicago, 1954, pp. 34–69, in which he argues that in addition to a key attribute of 'disinterestedness' or altruism ('service'), the professions are denoted by their rationality, functional specificity and universalism.

53 Breakdown in the immediate post-war consensus as to the public interest is described by Glass, 'The mood of London'. She suggests that resentment against the arbitrary element to planning is indicative both of the opening up of alternatives, and its diminished responsibility: 'as it has been deprived of a plausible definition of the public interest and of tangible means of serving it, it has lost its sense of direction and reputation' (p. 409).

54 F. J. C., Amos, 'The development of the planning process', *Journal of the Royal Town Planning Institute*, vol. 58, no. 6, 1972, p. 305.

55 M. Webber, 'The prospects for policies planning', included in *The Urban Condition*, ed. L. Duhl, Basic Books, New York, 1963, p. 320.

56 McLoughlin, *Urban and Regional Planning*, p. 7. For a subsequent elaboration of his views on cybernetics and planning, see *Control and Urban Planning*, Faber, London, 1973.

57 G. Chadwick, 'A systems view of planning', *Journal of the Royal Town Planning Institute*, vol. 52, p. 186. See also his *A Systems View of Planning*, Pergamon Press, Oxford, 1972. Much of the discussion about planning theory and methodology in the last decade must be related to the move of town-planning education into the universities and polytechnics. Planning education, at undergraduate and post-graduate level, was a growth industry in the 1960s: in the academic year 1965–6 there were approximately 1,000 full- and part-time

students; by 1968–9, there were an estimated 1,800. 'Report on Planning Education', *Journal of Royal Town Planning Institute*, vol. 59, no. 1, 1973, p. 21.

58 M. Hebbert, 'Ideologies in planning' (1973), quoted by P. Healey, 'The problem of ideology', *The Planner*, vol. 60, no. 3, 1974, p. 604.

59 J. Rex, 'Some theses on sociology and planning' (1971), *Race, Colonialism and the City*, Routledge & Kegan Paul, London, 1973, p. 71

60 Davies, *op. cit.*, p. 105.

61 B. Needham, 'Concrete problems: not abstract goals', *Journal of the Royal Town Planning Institute*, vol. 57, p. 317.

62 D. V. Donnison, review article, 'Systems and policies', *Town and Country Planning*, vol. 40, no. 4, 1972, p. 242.

63 P. Goodman, 'The world of Paul Goodman', *Journal of the Royal Institute of British Architects*, February 1973, p. 70. The reaction against professionalism is reflected in the debate about the future of the institute: should it remain a professional body whose primary concern is with qualifications and employment, or should it become a 'learned society', 'an acknowledged centre for fostering and advancing the understanding of cities, regions, and their governance, . . . with dominant interests in learning and in values'? J. B. McLoughlin, 'The future of the planning profession', included in *The Future of Planning*, ed. P. Cowan, Heinemann, London, 1973. On this argument it would be very similar to the Outlook Tower of Geddes in Edinburgh.

64 T. Blair, 'Twilight area renewal', *Official Architecture and Planning*, November 1969, p. 1367.

65 *Community Action*, vol. 1, no. 1, 1972.

66 A. L. Altshuler, *Community Control*, Pegasus (Bobbs-Merrill), Indianapolis, 1970, pp. 15–16.

67 H. J. Gans, 'Planning for people, not buildings', *Environment and Planning*, vol. 1, 1969, pp. 33–46.

68 R. Goodman, *After the Planners*, Penguin Books, Harmondsworth, 1972, p. 212.

69 T. Blair, 'Advocacy Planning', *Official Architecture and Planning*, February 1971, p. 132.

70 Altshuler, *op. cit.*, p. 205.

71 D. McGonaghy, 'The limitations of advocacy', *Journal of the Royal Institute of British Architects*, February 1972.

72 Goodman, *op. cit.*, p. 247.

73 *Ibid.*, p. 221. Subsequently, Goodman would seem to have shifted his position. He now dissents from the position adopted by C. A. Reich, *The Greening of America* (1970), Penguin Books, Harmondsworth, 1971, namely that change will come about through consciousness, and emphasizes the growing 'inadequacies of the economic structures'. *Building Design*, 22 February 1974.

74 The full statement reads,

> How therefore can we bring about desirable changes? How can we think about and formulate a new philosophy to guide this development of the total environment? It is of course already

happening. The powerful anti-pollution lobbies; the articulate writers in the quality newspapers and journals on environmental matters; vociferous local residents and action groups; television documentaries; 'Doomwatch'; the threat of voting pressures on elected members. All these and others are fragmented, but locally powerful and persuasive. . . . Our duty as professional planners is to reorientate ourselves, and to offer leadership to these new forces, and to provide the means for their propagation and development.

(*Journal of the Royal Town Planning Institute*, vol. 58, 1972, p. 292.)

75 D. E. C. Eversley, *The Planner in Society*, Faber & Faber, London, 1973, pp. 342–4.

3 The sociologists' critique

1 S. Greer, 'Policy and the urban future' in W. Bell and J. Mau, *Sociology of the Future*, Russell Sage Foundation, New York, 1971, p. 261.

2 R. Glass, *The Social Background of a Plan: a Study of Middlesbrough*, Routledge & Kegan Paul, London, 1948, pp. 43–4.

3 N. Dennis, 'The popularity of the neighbourhood community idea', *Sociological Review*, vol. 6, 1958, p. 198.

4 J. Rex, 'Economic growth and decline—their consequences for the sociology of planning', Town and Country Planning School, Report of Proceedings, 1968, p. 33.

5 E. Reade, 'Contradictions in planning', *Official Architecture and Planning*, October 1969, p. 1180.

6 N. Dennis, review article, *Urban Studies*, vol. 7, 1970, p. 221.

7 R. Pahl, 'Urban social theory and research', *Environment and Planning*, vol. 1, 1969, pp. 143–53, reprinted in R. Pahl, *Whose City?*, Longmans, London, 1970.

8 See J. Palmer, 'Introduction' to R. Goodman, *After the Planners*, Penguin Books, Harmondsworth, 1972, pp. 9–50; A. H. Halsey, 'Introduction' to N. Dennis, *People and Planning*, Faber & Faber, London, 1970, pp. 21–6; J. Rex, 'Some theses on sociology and planning' (1971) in J. Rex, *Race, Colonialism and the City*, Routledge & Kegan Paul, London, 1973, pp. 64–71.

9 D. Eversley, *The Planner in Society*, Faber & Faber, London, 1973, p. 343.

10 See M. Kidron, *Western Capitalism since the War* (1968), Penguin Books, Harmondsworth, 1970, pp. 104–23; N. Harris, *Beliefs in Society* (1968), Penguin Books, Harmondsworth, 1971, pp. 97–129.

11 J. Carney and R. Hudson, 'Ideology, public policy and underdevelopment in the North-East', North East Area Studies, University of Durham. W.P. no. 9, 1974; see also J. M. Cousins, 'Regional problems or problem regions', Rowntree Research Unit, University of Durham, 1974.

12 The issue of the inner city twilight areas, or 'areas of housing stress', was not of concern to the planning profession until the mid-1960s, specifically after the publication of the Milner Holland Report:

Housing in Greater London, HMSO, London, 1965, which itself was initiated in the aftermath of the Profumo affair in 1963, and the associated publicity given to 'Rachmanism'.

13 Palmer, *op. cit.*, p. 44.

14 Examples of incremental implementation of elements of long outdated master plans abound: one example—a section of an inner ring road proposed for Hull in the Abercrombie Plan for Hull in 1945, incorporated into the Development Plan approved in 1954, and still to be constructed 1975–9.

15 C. Reich, *The Greening of America* (1970), Penguin Books, Harmondsworth, 1972, p. 94.

16 P. Self, 'Show me a structure plan', *Town and Country Planning*, December 1972, pp. 559–60, comments on the South Hampshire Structure Plan 1972.

17 Eversley, *op. cit.*, p. 342.

18 In two recent papers Pahl has presented a refutation of 'managerialism' in urban studies. R. Pahl, 'Urban managerialism reconsidered', prepared for the Simposio do Sociologia Urbana, Barcelona, January 1974; 'The sociology of urban and regional development as a pattern in political economy', prepared for the Eighth World Congress of Sociology, Toronto, August 1974.

19 P. Norman, 'Managerialism—a review of recent work', paper prepared for the Centre of Environmental Studies conference, York, January 1975, p. 19.

20 See D. McCrone and B. Elliott, 'Property relations in the city: the fortunes of landlordism', paper prepared for CES conference, York, January, 1975 for a review of empirical research.

21 P. Self, 'Introduction' to *New Towns: the British Experience*, ed. H. Evans, published for the Town and Country Planning Association, C. Knight & Co., London, 1972, p. 5.

22 Reade, *op. cit.*, p. 1181.

23 Report of the Expert Commission on Compensation and Betterment (Uthwatt Report), HMSO, London, 1942; see also J. Switzer, 'Enterprise and ethics', Town and Country Planning School, *Report of Proceedings*, 1973, pp. 50–6.

24 Eversley, *op. cit.*, p. 303.

25 A. Gramsci, *Selections from the Prison Notebooks*, ed. Q. Hoare and G. Smith, Lawrence & Wishart, London, 1971, p. 271.

26 B. Russell, *The History of Western Philosophy*, Allen & Unwin, London, 1946, chapter 30, 'William James', p. 844.

27 *Ibid.*

28 P. Geddes, *Cities in Evolution* (1915), ed. J. Tyrwhitt, Williams & Norgate, London, 1949, p. xxvi.

Chapter 5 The critique of metropolis

1 Urbanism and Marxism

1 F. Tönnies, *Community and Society* (1887), Harper & Row, New York, 1963, p. 234.

2 Engels had early emphasized the impact of urbanization: 'The great cities are the birthplace of the labour movement' and without 'their forcing influence upon popular intelligence the working class would be less advanced than it is'. F. Engels, *Condition of the Working Class in England in 1844* (1892), Allen & Unwin, 1920, p. 122.
3 K. Marx, 'The British rule in India' (1853), included in *Marx and Engels: Basic Writings on Politics and Philosophy*, ed. L. S. Feuer, Collins, London, 1969, p. 517.
4 K. Marx, *Capital*, vol. III, included in T. Bottomore and M. Rubel, *Karl Marx* (1956), Penguin Books, Harmondsworth, 1963, p. 260.
5 R. E. Park, 'The city: suggestions for the investigation of human behaviour in the urban environment', included in *The City*, ed. R. E. Park and E. W. Burgess, University of Chicago Press, 1926, p. 12.
6 L. Mumford, *The City in History* (1961), Penguin Books, Harmondsworth, 1966, p. 656.
7 *Ibid.*, p. 292.
8 R. A. Nisbet, *The Sociological Tradition* (1966), Heinemann, London, 1967.
9 E. Fromm, *The Fear of Freedom*, Routledge & Kegan Paul, London, 1942; C. W. Mills, *White Collar*, Oxford University Press, London and New York, 1951; H. Marcuse, *One Dimensional Man*, Routledge & Kegan Paul, London, 1964; and F. Fanon, *The Wretched of the Earth* (1961), MacGibbon & Kee, London, 1965.
10 A. G. Frank, *Capitalism and Underdevelopment in Latin America*, Monthly Review Press, New York, 1969. The recent emphasis on cities as the centre of power re-echoes the analysis of Marx and Engels in 'The German ideology' (1845). They wrote, 'the existence of the town implies at the same time the necessity of administration, police, taxes, etc., in short of the municipality and thus of politics in general'. *Precapitalist Formations*, ed. E. J. Hobsbawm, Lawrence & Wishart, London, 1964, p. 127.
11 M. Castells, *La Question urbaine*, Maspero, Paris, 1972, p. 113; in translation, 'the sphere of social domination of a specific class, the bourgeoisie'.
12 A. Gramsci, *Selections from the Prison Notebooks*, ed. Q. Hoare and G. N. Smith, Lawrence & Wishart, London, 1971, p. 103.
13 Tönnies, *op. cit.*, pp. 168–9.
14 M. Weber, *The City* (1920), Free Press, Chicago, 1958, p. 201.
15 Tönnies, *op. cit.*, p. 235.
16 G. Simmel, 'Metropolis and mental life' (1903), included in *The Sociology of Georg Simmel*, ed. K. H. Wolff, Free Press, Chicago, 1950, p. 422.
17 G. Simmel, 'The field of sociology' (1917), in *The Sociology of Georg Simmel*, p. 9.
18 K. Marx, 'Economic and philosophical manuscripts' (1844) in D. McLellan, *The Thought of Karl Marx*, Macmillan, London, 1971, p. 107.
19 K. Marx and F. Engels, 'The German ideology', included in Bottomore and Rubel, *op. cit.*, p. 253.

20 G. Simmel, 'Metropolis and mental life', included in *The Sociology of Georg Simmel*, p. 410.

21 Castells, *op. cit.*, p. 102; in translation, 'that particular ideology which seizes on the modes and forms of social organization as characteristics of a stage in the evolution of society, directly tied to the ecological conditions of human existence, and ultimately with its context of life'.

2 Tönnies: Community and Society (1887)

1 T. Parsons, *The Structure of Social Action* (1968), Free Press, New York, pp. 686–94. Parsons draws attention to the descriptive nature of Tönnies's schema—*gemeinschaft* and *gesellschaft* are 'ideal types of concrete relationship'—and comments on the Marxian influence on his sociology, as evidenced in his inclusion of 'institutional elements' in the classification.

2 R. Redfield, 'The folk society', *American Journal of Sociology*, vol. 52, 1947, pp. 293–308; and *The Little Community* (1953), University of Chicago Press, 1960, chapter 9, pp. 132–48, sets out the fullest statement of the 'imaginary system' of combined opposites. Tönnies's typology based on historical description is transmuted into 'pairs of imaginary contrasts' (p. 146), for comparative community analysis.

3 R. A. Nisbet, *The Sociological Tradition* (1966), Heinemann, London, 1967, p. 19.

4 F. Tönnies, *Community and Society* (1887), Harper & Row, New York, 1963, p. 35.

5 *Ibid.*, pp. 76–7.

6 *Ibid.*, p. 80.

7 *Ibid.*, p. 228.

8 C. P. Loomis, 'Introduction' to F. Tönnies, *Community and Society*, Michigan State University Press, 1957, p. 5.

9 Tönnies, *Community and Society* (1963), p. 165.

10 *Ibid.*, p. 37.

11 The essential difference in quality of analysis of Marx and Tönnies is easily discernible in their discussions of the peasantry. For the latter, peasant life is harmonious and serene, there is a vision of a free people cultivating their land in common, whereas for Marx, writing of France, 'the smallholding of the peasant is now only the pretext that allows the capitalist to draw profits, interest and rent from the soil. . . . Smallholding property, in this enslavement by capital . . . has transformed the mass of the French population into troglodytes'. 'The eighteenth Brumaire' (1852), included in *Marx and Engels: Basic Writings on Politics and Philosophy*, ed. L. S. Feuer, Collins, London, 1969, p. 381.

12 Tönnies, *Community and Society* (1963), p. 233.

13 *Ibid.*, p. 168.

14 *Ibid.*, p. 227.

15 *Ibid.*, p. 230

16 *Ibid.*, p. 179.

17 *Ibid.*, p. 230.
18 *Ibid.*, p. 79.
19 The most thorough commentaries on Tönnies's writings (in addition to Parsons and Loomis, *op. cit.*) are those by R. Heberle, 'The sociological system of Ferdinand Tönnies' in *Introduction to the History of Sociology*, ed. H. Barnes, University of Chicago Press, 1948; D. Martindale, *The Nature and Types of Sociological Theory*, Routledge & Kegan Paul, London, 1961, pp. 81–92; and W. Cahnman and R. Heberle, 'Introduction' to F. Tönnies, *On Sociology: Pure, Applied and Empirical*, University of Chicago Press, 1971.
20 Early in his career, Wirth had prepared a critical review of Tönnies's sociological work: L. Wirth, 'The sociology of Ferdinand Tönnies', *American Journal of Sociology*, vol. 32, 1926. In this he drew largely from the later writings of Tönnies, in which the emphasis had shifted to that of a 'pure' sociology, in which Wirth was unconvinced. He concluded 'that community and society are suggestive and helpful conceptual tools . . . but can only lead to sterile philosophizing if they are to be used as the perennial frames into which the many-sided, complex, and elusive facts of history are to be squeezed' (p. 422). But his interpretation of *gemeinschaft* throughout is ecological—'The community grows out of the organic relationship of man to his environment' (p. 416)—and Tönnies's concept is distorted.

3 Simmel: 'Metropolis and mental life' (1902–3)

1 D. Martindale, *Nature and Types of Sociological Theory*, Routledge & Kegan Paul, London, 1961, pp. 216–25, 236–47.
2 G. Simmel, 'The field of sociology' (1917), included in *The Sociology of Georg Simmel*, ed. K. H. Wolff, Free Press, Chicago, 1950, p. 8.
3 *Ibid.*, p. 5.
4 *Ibid.*, pp. 16–17.
5 *Ibid.*, p. 9.
6 D. Levine, 'Introduction' to G. Simmel, *On Individuality and Social Forms*, University of Chicago Press, 1971, p. xii.
7 G. Simmel, 'Metropolis and mental life' (1903), included in *The Sociology of Georg Simmel*, p. 409.
8 K. H. Wolff, in the 'Introduction' to *The Sociology of Georg Simmel*, draws attention to the lack of clarity in Simmel's formulation of sociology, and argues that it is indicative of his attitude 'that Simmel did *not* want to socialize the spirit: he wished (half-heartedly in his sociology and wholeheartedly elsewhere) to preserve its *autonomy*' (p. xxxvii).
9 A. Giddens, 'Georg Simmel', included in *The Founding Fathers of Social Science*, ed. T. Raison, Penguin Books, Harmondsworth, 1969, p. 138.
10 Weber and Simmel were close colleagues at Berlin, and the indebtedness of the former to Simmel in his interpretation of Western civilization as characterized by rationalization, and his analysis of the forms of social interaction, tends to be underestimated. Levine, *op. cit.*,

p. xiv, quotes Lukács to the effect that Weber's achievement in the sociology of culture 'was possible only on the foundations created by Simmel'.

11 G. Simmel, 'On the significance of numbers for social life' (1902) in *The Sociology of Georg Simmel*, p. 87.

12 L. Wirth, 'A bibliography of the urban community', included in *The City*, ed. R. E. Park and E. W. Burgess, University of Chicago Press, 1926, p. 219.

13 G. Simmel, 'Metropolis and mental life' in *The Sociology of Georg Simmel*, p. 412.

14 *Ibid.*, p. 412.

15 Simmel relates the 'cosmopolitanism' of the city not just to its immediate size, but to its intellectual dominance over its hinterland. He contrasts the small town with the metropolis, in that the 'sphere of life of the former is, in the main, self-contained and autarchic', whereas the latter 'overflows by waves into a far-flung national or international area' (*ibid.*, p. 419). The freedom and individuality of life in the metropolis is explicitly related to its control over subordinate territories.

16 R. Sennett, *Classic Essays in the Culture of Cities*, Meredith, New York, 1969, argues that 'Simmel was willing to explore what good the great, impersonal metropolises of his and our time could serve . . . a visionary of the real world . . . He had the courage to look at routine and impersonality in order to see what could be made of it' (p. 6).

17 G. Simmel, 'Metropolis and mental life' in *The Sociology of Georg Simmel*, p. 422.

18 R. A. Nisbet, *The Sociological Tradition* (1966), Heinemann, London, 1967, pp. 305–12, interprets the paper as a study in alienation, 'the tyranny of objectivism', and concludes that 'Community and alienation are, for Simmel, but the two poles of man's eternal destiny' (p. 312).

19 Quoted in the *American Journal of Sociology*, vol. 63, 1958, p. 641. The letter was written in 1908, by Schaeffer, when Simmel was being considered for a Chair at Heidelberg.

20 P. Sorokin, *Contemporary Sociological Theories*, Harper, New York and London, 1928, pp. 489–91, 495–507.

21 D. Martindale, 'Introduction' to M. Weber, *The City*, Free Press, Chicago, 1958, p. 31.

22 G. Simmel, 'The field of sociology' in *The Sociology of Georg Simmel*, p. 16.

4 Weber: The City (1911–13)

1 L. Mumford, *The City in History* (1961), Penguin Books, Harmondsworth, 1966, p. 606.

2 L. Wirth, 'Urbanism as a way of life', *American Journal of Sociology*, vol. 44, 1938, p. 8.

3 R. Sennett, *Classic Essays in the Culture of Cities*, Meredith, New York, 1969, p. 6.

4 Weber commences his discussion of the definition of the city with the comment, 'This massing of elements interpenetrates the everyday concept of the "city" which is thought of quantitatively as a large locality', only to conclude 'Both in terms of what it would include and what it would exclude size alone can hardly be sufficient to define the city'. M. Weber, *The City*, Free Press, Chicago, 1958, pp. 65–6.
5 *Ibid.*, p. 81. The exact date of the original lecture papers is uncertain: all the evidence points to their elaboration between 1911 and 1913. They were first published in *Wirtschaft und Gesellschaft*, 1920; for an English edition, see G. Roth and C. Wittich, *Economy and Society*, Bedminster Press, New York, 1968, vol. III, chapter 16, pp. 1212–374. Sections in the *General Economic History* (1923), Allen & Unwin, London, 1927, in particular chapter 28, pp. 315–37, parallel the arguments of *The City* more directly and forcefully.
6 T. Parsons, *The Social System*, Free Press, Chicago, 1951, p. 5.
7 R. Aron, *Main Currents in Sociological Thought* (1967), Penguin Books, Harmondsworth, London, 1970, pp. 185–258.
8 Weber, *The City*, p. 102.
9 *Ibid.*, p. 81.
10 *Ibid.*, p. 181. In the *General Economic History*, pp. 315–17, Weber elaborates on this discussion of medieval urbanization as integral to the analysis of the development of capitalism. The medieval cities were unique social associations in that they offered 'citizenship', allegiance to a territorially identifiable social group, and freedom from ascribed groupings. See R. Bendix, *Max Weber* (1959), Methuen, London, 1966, pp. 70–9.
11 Weber, *The City*, p. 197.
12 It is this aspect of Weber's work that D. Martindale, in the 'Introduction' to *The City*, stresses as a valuable corrective to ecological, or socio-psychological, theories of urbanization.
13 Weber, *The City*, chapter 5, 'Ancient and medieval democracy', pp. 197–230.
14 Sennett, *op. cit.*, p. 7.
15 *Ibid.*, p. 7.
16 M. Weber, 'Science as a vocation' (1919), included in *From Max Weber: Essays in Sociology*, ed. H. Gerth and C. W. Mills, Routledge & Kegan Paul, London, 1948, p. 155.
17 M. Weber, *The Methodology of the Social Sciences*, Free Press, Chicago, 1949, pp. 74–5.

5 Wirth: 'Urbanism as a way of life' (1938)

1 The two articles to which reference is usually made are those of L. Wirth, 'Urbanism as a way of life', *American Journal of Sociology*, vol. 44, 1938, pp. 1–24; and R. Redfield, 'The folk society', *American Journal of Sociology*, vol. 52, 1947, pp. 293–308. Wirth's many articles on urbanization, community planning and social welfare are collected in L. Wirth, *On Cities and Social Life*, ed. A. Reiss, University of Chicago Press, 1964, and L. Wirth, *Community Life and Social Policy*,

ed. A. Reiss and E. M. Wirth, University of Chicago Press, 1956.

2 In many respects the essay can be regarded as a popularization of the statements of the European sociologists for an American academic audience. From the outset of his career Wirth had continued the Chicago School's tradition of leavening the parochialism of American sociology. His early essay 'The sociology of Ferdinand Tönnies', *American Journal of Sociology*, vol. 32, 1926; the translation and introduction to K. Mannheim, *Ideology and Utopia*, Harcourt Brace, New York, 1936; and his later efforts in setting up the International Sociological Association of which he was first President in 1950, are all indicative of his dissatisfaction with homespun sociology, shared by many contemporaries, such as Parsons.

3 D. Martindale, 'Introduction' to M. Weber, *The City*, Free Press, Chicago, 1958, p. 28.

4 R. E. Park, and E. W. Burgess, *Introduction to the Science of Society*, University of Chicago Press, 1921, p. 44.

5 L. Wirth, *The Ghetto*, University of Chicago Press, 1926.

6 Mannheim, *op. cit.*, 'Introduction', reprinted in Wirth, *On Cities and Social Life*, p. 133.

7 W. James, *Essays in Radical Empiricism*, Longmans, London, 1912, p. 93.

8 R. A. Nisbet, *Émile Durkheim*, Prentice-Hall, Englewood Cliffs, N.J., 'Introduction', p. 3.

9 A. Small, *General Sociology* (1905), included in Park and Burgess, *op. cit.*, p. 199.

10 *Ibid.*, p. 42.

11 D. Levine, 'Introduction' to G. Simmel, *Individuality and Social Forms*, University of Chicago Press, 1971, p. liii.

12 E. W. Park, 'Human ecology' (1936), included in E. W. Park, *Human Communities*, ed. E. C. Hughes, Free Press, Chicago, 1952, p. 157.

13 L. Wirth, 'Ideological aspects of social disorganization' (1940), included in *On Cities and Social Life*, p. 46.

14 Wirth, 'Urbanism as a way of life', p. 18.

15 *Ibid.*, p. 8.

16 R. E. Park, 'The urban community as a spatial pattern and moral order' (1925), included in *Human Communities*, p. 165.

17 This scepticism is marked in his dissertation, *The Ghetto*; in his summary presentation he concludes 'the ghetto . . . can be completely understood only if it is viewed as a socio-psychological, as well as an ecological, phenomenon; for it is not merely a physical fact but a state of mind'; Wirth, *On Cities and Social Life*, p. 98. In 1945, while paying tribute to the studies in human ecology, he reasserted the view taken by Small forty years previously, 'that physical factors . . . are at best conditioning factors offering the possibilities and setting the limits for social and psychological existence and development'; 'On human ecology', included in *On Cities and Social Life*, p. 187.

18 R. Bendix, 'Social theory and social action in the sociology of Louis Wirth', *American Journal of Sociology*, vol. 59, 1954, p. 523.

19 Wirth, *On Cities and Social Life*, p. 132.

20 *Ibid.*, 'Introduction' by A. Reiss, p. xxvii.
21 H. Becker, 'Introduction' to C. Shaw, *The Jack Roller*, University of Chicago Press, 1966, p. viii.
22 C. W. Mills, *The Sociological Imagination*, Oxford University Press, London, 1961, chapter 3, pp. 50–99.
23 Wirth, 'Urbanism as a way of life', p. 23.
24 Martindale, *op. cit.*, p. 41.
25 L. Wirth, 'The scope and problems of the community' (1933) in Wirth, *On Cities and Social Life*, p. 176.
26 L. Wirth, 'Consensus and mass communication' (1948) in *On Cities and Social Life*, p. 20.
27 Levine, *op. cit.*, p. liii.
28 Wirth, 'Consensus and mass communication', p. 34. His debt to Dewey is here unmistakable.
29 Bendix, *op. cit.*, p. 526.
30 R. Sennett, *Classic Essays in the Culture of Cities*, Meredith, New York, 1969, p. 16.
31 The most thorough criticism of Wirth's thesis for the study of urbanization internationally is that by P. Hauser, 'Observations on the urban-folk and urban-rural dichotomies as forms of Western ethnocentricism' in *The Study of Urbanization*, ed. P. Hauser and D. Schnore, John Wiley, New York, 1965, pp. 503–17.
32 M. Castells, *La Question urbaine*, Maspero, Paris, 1972, p. 113. In transition,

> Now, it only needs certain moments' reflection to realize the absurdity of a theory of social change based on the growing complexity of human association in terms of demographic growth alone. . . . All evolution in size and differentiation of a social group is itself product and expression of a social structure and its laws of transformation.

33 R. Williams, *The Country and the City*, Chatto & Windus, London, 1973, p. 96
34 K. Marx, 'The holy family' (1845), quoted by D. McLellan, *The Thought of Karl Marx*, Macmillan, London, 1971, p. 32.

Chapter 6 The Chicago School: urban experience

1 The ecological city

1 R. E. Park, 'The city: suggestion for the investigation of human behaviour in the urban environment' (1915), included in R. E. Park, *Human Communities*, Free Press, Chicago, 1955, p. 16.
2 E. A. Shils, 'The present situation in American sociology', *Pilot Papers*, June 1947, vol. 2, pp. 12–13.
3 Selections from the studies made in Chicago are included in E. W. Burgess and D. J. Bogue (eds), *Urban Sociology*, University of Chicago Press, 1964; and J. F. Short, *The Social Fabric of the Metropolis*,

University of Chicago Press, 1970. See also, R. E. L. Faris, *Chicago Sociology, 1920–32*, University of Chicago Press, 1970.

4 Park was introduced to the Chicago department by Thomas whom he met at Tuskegee in 1912: E. C. Hughes, 'Robert E. Park', included in *The Founding Fathers of Social Science*, ed. T. Raison, Penguin Books, Harmondsworth, 1969, p. 162. For the first few years his post was unpaid, and not until 1922 was his position in the university regularized. R. Faris, obituary article, *American Sociological Review*, vol. 9, 1944, pp. 322–4. His early career had shown a restless alternation between academic life and active work, first as a journalist and then as assistant to Booker Washington. Hughes writes,

> This dialectic of focus upon the individual case and upon the genus runs through all that Park did. . . . It is in a sense the dialectic of his own life: reform and action against detached observation; writing the news of the unique event as against the discovery of the eternal themes and processes of history.

(R. E. Park, *Race and Culture*, Free Press, Chicago, 1950, p. xiii.)

5 First published in the *American Journal of Sociology*, vol. 20, 1915, pp. 577–612, two years after his arrival in Chicago, this can be regarded as the programmatic statement for sociological research at Chicago.

6 R. E. Park and E. W. Burgess, *Introduction to the Science of Society*, University of Chicago Press, 1921.

7 McKenzie spent only a short time at Chicago, but he and Park remained in continuous communication. Selections from his writings are found in R. McKenzie, *On Human Ecology*, University of Chicago Press, 1968.

8 H. Becker, 'Introduction' to C. Shaw, *The Jack Roller*, University of Chicago Press, 1966, pp. vii–viii.

9 D. Rucker, *The Chicago Pragmatists*, University of Minnesota Press, Minneapolis, 1969, describes the university at the turn of the century as providing 'an atmosphere of energy and breaking with the past' (p. 9). But although the social sciences were breaking new ground in their concern for empirical work in Chicago (as, for example, the work by C. E. Merriam, *New Aspects of Politics*, University of Chicago Press, 1925, perhaps the earliest study by a political scientist of local politics), radical involvement in Chicago was constrained by the university's origin as a Baptist college, and its dependence for finance on local businessmen. Small and Mead's attempt in 1907 to persuade faculty to award Jane Addams an honorary degree failed because of the radical reputation of her settlement, Hull House. J. W. Linn, *Jane Addams*, Appleton Century, New York, 1935.

10 Burgess and Bogue, *op. cit.*, p. 6.

11 A. Briggs, *Victorian Cities* (1963), Penguin Books, Harmondsworth, 1968.

12 Even in 1880 the population of Chicago was only 550,000. Thereafter its growth was phenomenal: 1.1 million in 1900, 1.6 million in 1910, 2.7 million in 1920, 3.3 million in 1930. By the latter date the city was already losing population to the country, whose population was

then 4.675 million. All figures taken from R. Wade and H. Mayer, *Chicago: Growth of a Metropolis*, University of Chicago Press, 1969.

13 Burgess and Bogue, *op. cit.*, p. 275.

14 L. Wirth, *The Ghetto*, University of Chicago Press, 1926; H. Zorbaugh, *The Goldcoast and the Slum*, University of Chicago Press, 1929.

15 R. McKenzie, 'The ecological approach to the study of the human community' (1924), *On Human Ecology*, p. 4.

16 M. A. Alihan, *Social Ecology* (1938), Cooper Square Publishers, New York, 1964, p. 6.

17 Park and Burgess, *op. cit.*, p. 506.

18 McKenzie, *On Human Ecology*, p. 5.

19 A. Small, *Adam Smith and Modern Sociology*, University of Chicago Press, *c.* 1907. R. Bendix, 'Social theory and social action in the sociology of Louis Wirth', *American Journal of Sociology*, vol. 59, 1954, pp. 523–9, suggests that Park, and subsequently Wirth, were exploring the interplay between the themes of competition and communication, originally separated by Adam Smith in *The Wealth of Nations* and *The Theory of Moral Sentiments*. In this respect, therefore, they inherited an issue posed for sociology by the first generation of sociologists, Giddings at Columbia, Small at Chicago.

20 Small, *op. cit.*, p. 24.

21 R. Hurd, *Principles of City Land Values* (1903), Arno Press, New York, 1970.

22 McKenzie brings human ecology very close to human geography: 'the spatial and sustenance relationships in which human beings are organized are ever in process of change in response to the operation of a complex of environmental and cultural forces'. 'The scope of human ecology' (1926), included in McKenzie, *On Human Ecology*, p. 19, differing only in the emphasis on relationships between units in the aggregate. In Burgess's only theoretical statement 'The growth of the city', included in R. E. Park and E. W. Burgess, *The City*, University of Chicago Press, 1926, his indebtedness to the plant ecologists, in particular C. M. Child, and the organismic analogy, is much more marked than in Park's writings. Burgess subsequently emphasized those aspects of human ecology most readily susceptible to statistical analysis, for instance, the concept of 'gradient'.

23 R. E. Park, 'Community and society' (1929), included in Park, *Human Communities*, p. 196.

24 R. E. Park, 'The city as a social laboratory' (1929), included in Park, *Human Communities*, p. 78.

25 M. Stein, *The Eclipse of Community*, Harper & Row, New York, 1960, p. 19.

26 R. E. Park, 'Human ecology' (1936), included in Park, *Human Communities*, p. 148.

27 Park, 'The city: suggestion, for the investigation of human behaviour in the urban environment', p. 17.

28 L. Wirth, *On Cities and Social Life*, University of Chicago Press, 1964, 'Introduction' by A. Reiss, p. xviii.

29 Park, 'Human ecology', p. 158.

30 M. Weber, 'Class, status, party' in *From Max Weber: Essays in Sociology*, ed. H. Gerth and C. W. Mills, Routledge & Kegan Paul, London, 1948, p. 183.
31 Park and Burgess, *Introduction to the Science of Society*, p. 30.
32 H. Barrows, 'Geography as human ecology', *Annals of the Association of American Geographers*, 1923, p. 3. A monograph by J. C. Galpin, *The Social Anatomy of a Rural Community*, University of Wisconsin, 1915, is an early example of geographical study which impressed Park.
33 Park, 'Human ecology', p. 158.
34 J. Dewey, *Democracy and Education* (1916), included in Park and Burgess, *Introduction to the Science of Society*, pp. 185–6.
35 Park, 'Community and society', p. 182.
36 Park, 'Human ecology', p. 158.
37 Alihan, *op. cit.*, pp. 47–9. Park quite disarmingly admitted the validity of her criticisms, and confessed to 'the immaturity of the logical structure on which these studies are based'. *Annals of the American Academy of Political and Social Sciences*, vol. 20, 1939, p. 265.
38 Park, 'Community and society', p. 192.
39 Whereas in The *Ghetto* Wirth was concerned to demonstrate the historical and cultural dimensions to the ghetto and their bearing on the segregated pattern of living in the American cities, he had written earlier, 'land values are the chief determining influence in the segregation of local areas . . .'. L. Wirth, 'A bibliography of the urban community' (1925), included in Park and Burgess, *The City*, p. 203.
40 Park, 'The city: suggestions for the investigation of human behaviour in the urban environment', p. 49.
41 Burgess, 'The growth of the city'. His simple diagram is probably the best known feature of the Chicago School's work. The concept of 'zone' is, however, flatly contradicted by that of 'gradient'; both cut across Park's conceptualization of natural areas.
42 Alihan, *op cit.*, p. 182.
43 G. Osofsky, *Harlem: The Making of a Ghetto*, Harper & Row, New York, 1968.
44 Park, 'The city: suggestions for the investigation of human behaviour in the urban environment', p. 47.
45 Park, 'Community and society', p. 201.
46 R. E. Park, 'The mind of the hobo' (1925), included in *Human Communities*, p. 94.
47 In the 1915 paper Park had introduced a discussion on the division of labour as the basis for new forms of social solidarity in the city, and his line of argument would seem to parallel that of Durkheim. His indebtedness to Durkheim is, at this remove, unascertainable. On his own account, he considered Simmel and Durkheim to be the two greatest sociologists (D. Levine, 'Introduction' to G. Simmel, *On Individuality and Social Forms*, University of Chicago Press, 1971, p. li), but direct citations are rare, and there is uncertainty as to whether this was because Park had so deeply internalized Durkheim's sociology that reference was unnecessary, or, alternatively, because he had

come to his writings after long exposure to the currents of thought in which Durkheim, too, was immersed. In these circumstances, Durkheim is only cited as confirmation of ideas already arrived at by Park—he confers conceptual distinction upon them.

48 R. A. Nisbet, *Émile Durkheim* (1963), Prentice-Hall, Englewood Cliffs, N.J., 1966, p. 36. The gulf between the American sociologists and Durkheim was by no means as great as Nisbet makes out: 'Between American sociological thought and the acceptance of Durkheimian perspectives lay the wilderness of homespun individualism, Pragmatism, and general suspicion of theory . . .' (*ibid.*, p. 4). And although Durkheim repudiated biologism, his designation of social morphology as one of the branches of sociology backed up Park's formulation of human ecology.

49 R. E. Park, 'The urban community as a spatial pattern and moral order' (1925), included in Park, *Human Communities*, p. 177.

50 One example, the well-known study, M. Young and P. Willmott, *Family and Kinship in East London*, Routledge & Kegan Paul, London, 1957, probably gave greater prominence to certain aspects of family and neighbourhood life, in the absence of any overall analysis of London's housing market of East London, or the century-old movement of East Londoners into suburban Essex.

51 Park, 'Community and society', p. 197.

2 Community and freedom

1 'Columbia University Declaration' (1894), quoted by H. Odum, *American Sociology*, Longmans, London, 1951. See also R. L. and G. J. Hinkle, *The Development of Modern Sociology*, Random House, New York, 1954.

2 L. and J. Bernard, *The Origins of American Sociology*, Crowell, New York, 1943, p. 846.

3 'Columbia University Declaration'. The Eugenics movement 'reached dimensions of a fad' by about 1915; R. Hofstadter, *Social Darwinism in American Thought*, Beacon Press, Boston, 1955, pp. 161–7.

4 M. and L. White, *The Intellectual Versus the City*, Harvard University Press, Cambridge (Mass.), 1962, p. 147.

5 *Ibid.*, pp. 147–54. But the settlement rapidly achieved a radical reputation through its association with the trade unions, and its challenge to the corruption of ward politics. The women around Jane Addams in the early years were formidable campaigners for the unionization of women, the introduction of legislation for an eight-hour day for women, the restriction and supervision of child labour, and the institution of juvenile courts. (J. W. Linn, *Jane Addams*, Appleton Century, New York, 1935.) Later the settlement at Hull House did withdraw from the battleground between capital and labour, but they never showed the detachment and disinterest in economic and political life that is so striking a feature of the writings of Park and his colleagues in the university.

6 C. W. Mills, *Sociology and Pragmatism*, Oxford University Press, London, 1966, p. 435.

7 P. Orleans, 'Robert Park and social area analysis', *Urban Affairs Quarterly*, 1966, p. 13. Mills, writing more generally of the sociological profession in America, stresses the farm, or small-town background, of many sociologists, and interprets their stress on community as a reaction to urbanization: 'the notion of disorganization is quite often merely the absence of that *type* of organization associated with the stuff of primary group communities having Christian and Jeffersonian legitimations'. C. W. Mills, 'The professional ideology of social pathologists', *American Journal of Sociology*, vol. 49, 1943, p. 175.

8 L. Bramson, *The Political Context of Sociology*, Princeton University Press, Cambridge (Mass.), 1961, p. 70.

9 P. D. Goist, 'City and community: the urban theory of Robert Park', *American Quarterly*, vol. 23, 1971, pp. 46–50.

10 M. Stein, *The Eclipse of Community*, Harper & Row, New York, 1960, p. 18.

11 R. Sennett, *Classic Essays in the Culture of Cities*, Meredith, New York, 1969, p. 16.

12 R. E. Park, 'Understanding a folk culture' (1934), included in R. E. Park, *On Social Control and Collective Behaviour*, University of Chicago Press, 1967, p. 19.

13 R. E. Park, 'The city as a social laboratory' (1929), included in R. E. Park, *Human Communities*, Free Press, Chicago, 1955, p. 74.

14 R. E. Park, 'Community organization and the romantic temper' (1925), included in *Human Communities*, p. 72.

15 Bramson, *op. cit.*, p. 17.

16 Park's indebtedness to Simmel in delineating a field for sociology is enormous, but in his hands Simmel's sociology is transmuted from a science of the forms of association to the 'science of collective behaviour'. Park shifted Simmel's conception of a social fact from 'the normatively indifferent fact of human association' (D. Levine, 'Introduction' to G. Simmel, *On Individuality and Social Forms*, University of Chicago Press, 1971, p. li) to concerns of social organization, social control, concerted action and community.

17 R. E. Park, 'The city' (1915), included in Park, *Human Communities*, p. 46.

18 In an essay late in his career, Park summarized his views on urbanization in history: cities are melting pots, solvents of existing systems of stratification in the name of economic efficiency. Through natural selection, and an increasingly fine division of labour, productivity, and creativity, are furthered. The breakdown of custom means individuation, releasing energy and ambition for 'new and strange inventions and achievements'. The city is the 'workshop of civilization' 'The city and civilization' (1936), included in Park, *Human Communities*, pp. 128–41. Social Darwinist thinking is allied to Marxist terminology. 'It is said that every civilization carries in itself the seeds of its own destruction. Such seeds are likely to be the technical devices that

introduce a new social order and usher out the old' (p. 121). The theme of the essay runs parallel to the arguments of Tönnies, and is a precursor of the folk-urban continuum subsequently elaborated by Redfield, and as such is a link between nineteenth-century liberalism and twentieth-century theories of development.

19 M. Castells, *La Question urbaine*, Maspero, Paris, 1972, p. 441.

20 J. Dewey, *Democracy and Education*, Macmillan, London, 1916, p. 115.

21 A. Gouldner, *The Coming Crisis of Sociology*, Heinemann, London, 1971, p. 484.

22 R. E. L. Faris, *Chicago Sociology, 1920–1932*, University of Chicago Press, 1970; see also R. E. Park, 'The city as a natural phenomenon' (1939), included in Park, *Human Communities*, pp. 125–7.

23 The head of the philosophy department had been Dewey (1892–1904) and it was only after his departure that a separate psychology department was formed. Subsequently Mead, in his consideration of the realization of self in social situations, pushed psychology yet further from instinctivist or behavioural interpretations of social existence. G. M. Mead, *On Social Psychology*, University of Chicago Press, 1956. All senior students in the sociology department in this period attended Mead's lectures. R. E. L. Faris, 'The social psychology of G. Mead', *American Journal of Sociology*, vol. 43, 1937, p. 509.

24 H. Spencer, *The Study of Society*, published in America 1872–3, had had a tremendous impact on social thought; 'it soon gave Spencer a public influence that far transcended Darwin's', and he became 'the metaphysician of the home-made intellectual'. Hofstadter, *op. cit.*, pp. 31–2.

25 *Ibid.*, p. 66.

26 A. Small, *General Sociology*, University of Chicago Press, 1905, pp. 408–10.

27 *Ib'd.*, p. 420.

28 *Ibid.*, p. 407.

29 W. James, *Pragmatism*, Longmans, London, 1907, p. 79.

30 Levine, *op. cit.*, p. liii.

31 Small, *op. cit.*, p. ix.

32 Small had studied Marxism carefully, and even taught a course on the subject at Chicago (E. A. Shils, 'The present situation in American sociology', *Pilot Papers*, June 1947, vol. 2, no. 2, p. 14); he also 'attacked the evils of capitalism and monopoly with such vigour that his style sometimes became almost lively'. E. C. Hughes, 'Robert E. Park' in *The Founding Fathers of Social Science*, ed. T. Raison, Penguin Books, Harmondsworth, 1969, p. 170. He and Veblen were contemporaries at Chicago. But this interest was not transmitted to the next generation in the department, and the language of human ecology replaces that of Marxism; the context of research is that of the physical framework of the city, not the relationships of production. The interest in technology remains, 'in human ecology . . . the factor of technology becomes one of major importance', but is considered in terms of 'the means of rapid transport, public utilities, skyscraper

apartment buildings, the new means of communication . . .'. E. W. Burgess, quoted by Odum, H., *op. cit.*, p. 353.

33 G. Simmel (1909), included in R. E. Park and E. W. Burgess, *Introduction to the Science of Society*, University of Chicago Press, 1921, p. 349.

34 Small, *op. cit.*, p. 496.

35 A. Strauss, 'Introduction' to Mead, *op. cit.*, p. xxiv.

36 Quoted by Mills, *Sociology and Pragmatism*, p. 443.

37 For instance, the interpretations of R. A. Nisbet, *The Sociological Tradition* (1966), Heinemann, London, 1967; and H. S. Hughes, *Consciousness and Society*, McGibbon & Kee, London, 1967.

38 James, *op. cit.*; *Essays in Radical Empiricism*, Longmans, London, 1912.

39 James, *Pragmatism*, p. 241.

40 James, *Essays in Radical Empiricism*, p. x.

41 W. James, *The Principles of Psychology*, Holt, New York, 1896, referred to by R. E. Paık, 'New as a form of knowledge' (1940), included in Park, *On Social Control and Collective Behaviour*, pp. 33–52.

42 James, *Essays in Radical Empiricism*, p. 57.

43 *Ibid.*, pp. 65–6.

44 R. E. Park, 'Human nature and collective behaviour' (1927), included in Park, *On Social Control and Collective Behaviour*, p. 190.

45 A. Reiss, 'Introduction' to L. Wirth, *On Cities and Social Life*, University of Chicago Press, 1964, p. xv.

46 Park, 'Human nature and collective behaviour', p. 193.

47 Park, 'The city', p. 5. Over twenty-five years, Park did not waver from this approach, in his methodological papers, his prefatory statements to the books of others, and in his advice to students. In 1921, in his most systematic writings on sociology, he wrote,

> It has been the dream of philosophers that theoretical and abstract science could and perhaps some day would, succeed in putting into formulae and into general terms all that was significant in the concrete facts of life. It has been the tragic mistake of the so-called intellectuals, who have gained their knowledge from textbooks rather than from observation and research to assume that science had already realised its dream. But there is no indication that science has begun to exhaust the sources or significance of concrete experience.

(R. E. Park, and E. W. Burgess, *Introduction to the Science of Society*, University of Chicago Press, 1921, p. 15.)

48 Park, 'The city', p. 7.

49 R. Turner, 'Introduction' to Park, *On Social Control and Collective Behaviour*, p. xv.

50 Mills, *Sociology and Pragmatism*, p. 20.

51 In this respect, note Rex's approach to the city 'as a social organization structured in terms of differential rights in relation to the use of domestic and community buildings', J. Rex, *Race, Colonialism and the City*, Routledge & Kegan Paul, London, 1973, pp. 112–13. Even

the commercial buildings of the city are excluded in this model of the city.

3 A continuing tradition?

1 C. W. Mills, *Sociology and Pragmatism*, Oxford University Press, London, 1966, p. 382.
2 *Ibid.*, p. 380.
3 B. McLoughlin, *Urban and Regional Planning: a Systems Approach*, Faber & Faber, London, 1969, p. 95.
4 M. A. Alihan, *Social Ecology*, Columbia University Press, New York, 1938 (reprinted Cooper Square Publishers, 1964), p. 6.
5 P. Geddes, 'Civics: an applied sociology', *Sociological Papers*, 1904, p. 117.
6 L. Kolakowski, *Positivist Philosophy*, Penguin Books, Harmondsworth, 1972, p. 201.
7 B. Russell, *The History of Western Philosophy*, Allen & Unwin, London, 1946, p 841.
8 *Ibid.*, p. 851.
9 R. J. Chorley and P. Haggett, *Socio-Economic Models in Geography*, Methuen, London, 1967, p. 20.
10 J. Rex and R. Moore, *Race, Community and Conflict*, Oxford University Press, London, 1967.
11 R. Pahl, *Whose City?*, Longman, London, 1970.
12 M. Castells, *La Question urbaine*, Maspero, Paris, 1972.
13 N. Anderson, *The Hobo*, University of Chicago Press, 1923; F. Thrasher, *The Gang*, University of Chicago Press, 1927; W. F. Whyte, *Street Corner Society*, University of Chicago Press, 1943; G. Suttles, *The Social Order of the Slum*, University of Chicago Press, 1969.
14 Redfield responded to Lewis's criticisms of his Rousseauesque presentation of life in a Mexican village with the statement: 'The principal conclusion I can draw from this experience is that we are all better off with two descriptions of Tepoztlan than we would be with only one of them.' R. Redfield, *The Little Community*, University of Chicago Press, 1960, p. 136. Lewis surveyed the same village seventeen years after Redfield, in O. Lewis, *Life in a Mexican Village: Tepoztlan Restudied*, University of Illinois Press, 1951.
15 A. Gouldner, *The Coming Crisis of Western Sociology*, Heinemann, London, 1971, p. 482.
16 The list could include M. Young and P. Willmott, *Family and Kinship in East London* (1957), Penguin Books, Harmondsworth, 1963; B. Jackson and D. Marsden, *Education and the Working Class*, Routledge & Kegan Paul, 1962; Rex and Moore, *op. cit.*; and K. Coates and R. Silburn, *Poverty: the Forgotten Englishmen*, Penguin Books, Harmondsworth, 1969.
17 M. Shaw, 'The coming crisis of radical sociology' in *Ideology and Social Science*, ed. R. Blackburn, Fontana, Collins, London, 1972, p. 34.

18 K. Marx, 'The Paris manuscripts' (1844); quoted by D. McLellan, *Marx Before Marxism*, Penguin Books, Harmondsworth, 1970, p. 260.
19 W. Reich, *What is Class Consciousness?* (1933), Socialist Reproduction Press, London, 1971, p. 23.
20 C. W. Mills, *The Sociological Imagination*, Oxford University Press, London, 1959, p. 5.
21 K. Korsch, *Karl Marx*, Russell & Russell, New York, 1963, p. 79.
22 J. Foster, *Class Consciousness and the Industrial Revolution*, Weidenfeld & Nicolson, London, 1973.
23 R. S. and H. M. Lynd, *Middletown* (1929), Harcourt Brace & World, New York, 1956; and N. Dennis, F. Henriques and C. Slaughter, *Coal is Our Life*, Eyre & Spottiswoode, London, 1956.

Epilogue The city or the town?

1 K. Marx and F. Engels, 'The German Ideology', included in *Pre-capitalist Economic Formations*, ed. E. J. Hobsbawm, Lawrence & Wishart, London, 1964, p. 127.
2 M. Weber, *The Theory of Social and Economic Organization*, Free Press, New York, 1964, p. 113: 'Not every type of contact of human being has a social character. . . . For example, a mere collision of two cyclists may be compared to a natural event. . . .'
3 A. Gouldner, *The Coming Crisis of Western Sociology* (1970), Heinemann, London, 1971, p. 349.
4 Phrases used by the Webbs and G. B. Shaw. See E. J. Hobsbawm, *Labouring Men*, Weidenfeld & Nicolson, London, 1964, chapter 14, 'The Fabians reconsidered', pp. 250–71.
5 Centre of Environmental Studies, *Proceedings of the Conference on Urban Change and Conflict*, London, 1975, CES CP 14, p. 220.
6 K. B. Smellie, *A History of Local Government*, Allen & Unwin, London, 1969, p. 111.
7 Note the discussion by R. Pahl, *Whose City?*, Longmans, London, 1970, pp. 215–25.
8 J. Cornford (ed.), *The Failure of the State*, Croom Helm, London, 1975, p. 9.
9 C. W. Mills, *The Sociological Imagination* (1959), Oxford University Press, London, 1967, p. 136.
10 M. Harloe, 'Inner urban areas—a review article', published as Centre of Environmental Studies, CES WN 356, London, 1973. See also *Proceedings of the Conference on Urban Change and Conflict*, *op. cit.*, p. 7.
11 Gouldner, *op. cit.*, p. 350.
12 M. Shaw, *Marxism and Social Science*, Pluto Press, London, 1975, p. 13.
13 *Ibid.*, p. 14.
14 R. Miliband, *The State in Capitalist Society* (1969), Quartet Books, London, 1973, p. 233.
15 R. Williams, *The Country and the City*, Chatto & Windus, London, 1973, p. 291.

16 R. Pahl, 'Urban processes and social structure', included in *Whose City?*, Penguin Books, Harmondsworth, 1975, p. 234.
17 J. Anderson, 'The political economy of urbanism: a bibliography', Architectural Association, London, unpublished document circulated January 1976, p. 10.
18 Marx and Engels, *op. cit.*, p. 127.
19 B. Russell, quoted by M. Weber, 'The urban place and the non-place urban realm', included in *Explorations in Urban Structure*, ed. M. Webber, University of Pennsylvania Press, Philadelphia, 1964, p. 79.
20 E. Durkheim, *The Rules of Sociological Method*, Free Press, Chicago, 1950, p. 14.

Index

Abercrombie, P., 277n
Abrams, C., 67, 264n
Addams, J., 217, 291, 296n
Alden, P. and Hayward, E., 276n
Alihan, M. A., 211, 212, 292n, 293n, 298n
Allaun, F., 269n
Althusser, L., 256n
Altshuler, A. L., 281n
Amenity, 44, 142, 150; civic, 129; societies, 134, 145, 277n
Amos, F., 280n
Anderson, J., 300n
Anderson, N., 298n
Anti-urbanism, 55, 141, 156, 217–18, 254n
Architects,48,92,95–107,139–40,165
Arnstein, S. R., 206n
Aron, R., 288n
Ash, M., 51, 261n
Ashworth, W., 259n
Ashworth, W., 271n, 278n

Barlow Report, 259n
Barrows, H., 293n
Bauer, C., 62, 264n, 270n
Becker, H., 290n, 291n
Bendix, R., 288n, 289n, 292n
Benevolo, L., 264n
Berger, B., 262n
Bernard, L. and J., 296n
Best, R. H. and Champion, A. G., 260n
Betterment Fund, 279n
Bindoff, S., 257n
Birmingham, 28, 32, 51
Blackman, J., 272n
Blair, T., 267n, 270n, 275n, 281n
Boardman, P., 278n
Booth, C., 28, 68, 72, 170, 220, 226, 228, 249, 251, 263n, 264n
Booth, D., 255n
Bor, W., 279n
Bowley, M., 260n
Bramson, L., 295n
Branson, N. and Heinemann, M., 259
Brett, S., 272n
Briggs, A., 256n, 265n, 291n
British experiment, 43–5, 119, 162, 242
Buchanan, C., 135, 260n, 277n
Buckingham, 131, 276n
Builder's pool, 115
Building societies, 41–2, 58, 110, 114–16, 119, 262n, 273n
Bunge, W., 267n
Bureaucratic guerrilla, 153
Burgess, E. W., 58, 204, 206, 208, 212, 231, 292n
Burney, E., 274n
Burns, 128
Bury, J. P., 268n

Cardoso, F. H., 256n
Carney, J. and Hudson, R., 282n

Castells, M., 2, 7, 64, 126, 202, 233, 253, 254n, 257n, 284n, 285n, 290n, 296n, 298n
Castles, S. and Kosack, G., 275n
Chadwick, G., 280n
Chalklin, C. W., 272n
Chermayeff, S. and Alexander, C., 269n
Chicago School: assumptions, 5, 182; attitudes to urbanization, 217–20, 295n; continuing influence, 167–8, 234–7; community studies, 8, 214, 215, 228, 234–7; environment, 205–6, 276n, 291n; and European sociology, 195–6, 224–6; Golden Age, 195, 199; leadership, 5, 204–5; task for sociology, 66, 216–17, 297n; theoretical innovations, 206–7, 221–30; other references, 23, 45, 194, 241; *see also* Human ecology
Chorley, R. and Haggett, P., 298n
Christaller, W., 12, 256n
Citizenship/citizenry, 1, 121, 132, 138, 190, 193, 197, 248, 288n
City, 248–9; definition of, 30, 172–3, 179, 182, 185, 189–90, 192–3, 194, 201, 288n; in history, 169–70, 171–3, 191–4, 295n; in planning, 65, 137–8, 152; in sociology, 1–2, 7–8, 23, 49, 64–5, 197, 264n; *see also* Regional city, Metropolis, Tönnies, Simmel, Weber, Wirth
Civics, 137–8, 278n
Clark, R. B., 11, 255n
Class/class consciousness, 1, 2, 7, 8, 11–12, 18, 124–6, 172, 254n, 263n
Classical/European sociology, 5–6, 167–8, 169–77, 224–6
Clawson, M., 58, 262n
Cleary, E. J., 260n
Coates, B. and Rawstron, E., 254n
Coates, K. and Silburn, R., 75, 266n, 298n
Cole, G. D. H., 260n
Cole, G. D. H. and Postgate, R., 272n
Columbia University Statement, 264n, 294n
Community, 6, 10, 125, 209, 210–11, 220, 230, 251; community action, 77–8, 124, 154–5, 281; *Community Action*, 273n, 281n; community analysis, 124; community development, 66, 77, 83, 148; community studies, 2, 7, 74–6, 125, 215–16, 228, 234–7, 245
Consensus, in cities, 172, 180–1, 197, 201, 228; in planning process, 150, 151, 156, 160–1, 163
Cornford, J., 299n
'Cost yardsticks', 271n
Cox, A., 270n
Crockford, S. and Fromer, F., 271n
Cullingworth, J. B., 266n
Culture, 10; working-class, 249, suburban, 60
Cybernetics, 150, 231

Darby, H. C., 257n
Darwin, C., 206
Darwinism/social Darwinism, 5, 195, 222–3, 224, 225, 226, 229
Davies, J. G., 135, 266n, 275n, 277n, 281n
Democracy, future of, 190, 192, 197, 214, 217, 220, 221, 229, 241; in urban sociology, 247, 248, 251; *see also* Liberalism
Dennis, N., 14, 73, 78, 157, 256, 265n, 282n
Dennis, N. *et al.*, 237, 251, 256n, 299n
Dependency, 9, 14, 15, 19, 21–2, 256n
Dewey, J., influence on Chicago School, 204, 230; key concepts: action, 235; consensus, 211, 217, 221; inquiry, 165, 229, 232; other references, 222, 231, 236, 290n, 296n

Dickens, C., 49, 69, 89, 249
Dobriner, W. M., 54, 261n, 263n
Donnison, D. V., 269n
Durkheim, É., influence on Chicago School, 201, 209, 215, 217, 219, 225, 226, 293n, 294n, 300n; other references, 5, 117, 171, 181, 196
Dyos, H. J., 57, 262n, 272n

Eden, W. A., 276n
Élites, 151, 159; British, 14, 17–18, 27, 35, 45; provincial, 27–8, 47
Embourgeoisement, 60–1, 263n
Engels, F., expectations of urbanization, 1, 20, 28, 170; and the housing question, 61–2, 86, 92, 123; other references, 30, 49, 69, 253n, 257n, 258n, 269n, 284n
Eversley, D. E. C., 260n, 268n, 282n
Experience, 136–7, 165, 174–5, 203, 227–8, 235–6; urban, 48, 171, 218, 219

Fanon, F., 284n
Fans, R., 296n
Faris, J., 279n
Firey, W., 261n
Foley, D., 134, 143, 275n, 277n
Foster, J., 256n, 299n
France, *bidonvilles*, 86, 107; slum clearance, 267n; urban migration, 258n, 260n; urban sociology, 241
Frank, A. G., 10, 11, 13, 16, 22, 23, 172, 182, 255n, 284n
Frankenberg, R., 263n
Fromm, E., 284n

Galpin, J. C., 293
Gans, H. J., 60, 75, 78, 153, 161, 263, 264n, 281n
Garden cities/movement, 49, 61–2, 64, 129, 130–3, 217, 247
Garden City Association, 127, 130, 131, 132
Gaskell, E., 258
Gauldie, E., 267n, 270n
Gayler, H., 259
Geddes, P., 62, 66, 132, 136–9, 142, 165, 228, 231, 247, 278n, 298n; *see also* Civics, Town planning, Profession
Gemeinschaft, 88, 173, 175, 178–82, 210
Geography, 12, 168, 249; in Chicago, 210, 292n, 293n
George, H., 23, 130, 276n
Gesellschaft, 1, 83, 173, 175, 178–82, 194, 210
Ghetto, 11, 51, 66, 80, 85, 154, 256n, 274
Giddens, A., 286n
Ginsberg, L., 259
Glass, R., 6, 7, 15, 157, 254n, 255n, 279n, 282n
Goffmann, E., 269n
Goffmann, J., 49, 260n
Goist, P., 218, 295n
Goldthorpe, J. and Lockwood, D., 283n
'Golden Circle', 3, 16, 39, 49
Goodman, P., 271n
Goodman, R., 152, 154, 155, 270n
Gouldner, A., 266, 296n, 298n, 299n
Gramsci, A., 172, 283n, 284n
Green ghettoes, 52
Greer, S., 282n
Gurney, S., 277n
Gutkind, E. A., 284n

Haar, C., 279n
Haddon, R., 274n, 275n
Hall, P., 261n
Halsey, A. H., 8, 282n
Harloe, M., 17, 256n, 299n
Harrington, R. L., 262n
Harris, N., 253n, 255n, 258n, 277n, 282n
Harvey, D., 274n
Hauser, P., 290n
Haussmann, G., 31–2
Hebbert, M., 281n
Heberle, R., 286n
Hegemony, 14, 15, 16, 21, 22, 28, 45, 87, 155, 172–3, 180, 246
'Hierarchy', 12, 21, 47

Hill, C., 257n
Hinkle, R. L. and G. J., 294n
Hobsbawm, E. J., 258n, 299n
Hofstadter, R., 294n, 296n
Hoggart, R., 256n
House price spiral, 116–17, 273n
Housing: affluence, 42, 78–9; and class struggle, 121–6, 274–5n; as commodity, 110; dependancy, 91–2, 94–5; design of, 95–107, 164; and enfranchisement, 123; finance, 8, 109–11, 113–14, 119, 272n, 273n; ideology of, 88–90, 119–20, 209; land use planning, effects of, 117–19, 273n; local authority, 104–6, 112–14, 164, 272–3n; in modern economy, 93–4; privacy in, 90–1, 209n; private investment in, 41–2, 111–17, 119; rehabilitation, 45, 77, 79–80, 86, 110, 119–20; as scarce resource, 111–12, 121; self build, 94–5, 107, 119-20; State intervention, 107–9; in urban sociology, 18, 125–6, 249; working class, 95–6, 98, 103–4
Housing and Town Planning Act (1909), 128
Housing classes, 6, 14, 81, 121, 122, 124, 274–5n
Housing Finance Act (1972), 114
Housing question, 86, 88, 120–6; *see also* Engels
Housing Subsistence Act (1956), 105
Howard, E., 28, 36, 49, 62, 128, 129, 130, 131, 136, 142, 217, 247, 249, 259n, 275n, 276n
Hughes, E. C., 291n, 296n
Hughes, H. S., 297n
Hull, 40, 76, 277n, 283n
Human ecology: definition of, 198, 292n; aggregates, 209; community, 210–11; competition, 206; competitive co-operation, 216; derivation of, 221–30; dominance, 120, 212; ecological community, 209; European sociology, 224–6; gains from, 221–2, 229–30, 233; gradient, 292n; influences, 168, 230–4; invasion and succession, 120, 212–14; key concepts, 206–16; moral order, 190, 208, 210–11; natural area, 208–9, 213–16, 220; natural order, 196, 206, 207, 208, 210–11; pragmatism, 226–8, 232; Social Darwinism, 222–4; society, 196, 211; symbiosis, 210; zone, 58, 293n
Hurd, R., 56, 207, 262n, 292n
'Hypothetico deductive method', 279n

Ideology, 59, 120, 131, 134, 136, 143, 149, 151
Images: of cities, 48–50, 64–5, 248; of slums, 68, 75, 79; of suburbs, 54, 59–60, 251; *see also* Myths
Individuality, 156, 177, 186, 187–8, 190, 197, 216, 219–20, 248
Industrial decentralization, 36–8, 39, 43, 62, 63
Industrialization, 46–7, 274
Intellectuals/intelligentsia, 28, 48, 220, 229; intellectual proletariat, 242; weapons of intellect, 246

Jackson, B., 256n
Jackson, B. and Marsden, D., 298n
Jacobs, J., 103, 271n
James, W., influence on Park, 204, 222, 230; key concepts, flux of life, 165; live experience, 227, 234; radical empiricism, 211, 226, 227; theory of knowledge, 226, 227; other references, 221, 224, 236, 249, 255n, 289n, 296n, 297n
Jones, G. S., 12, 29, 254n, 265n
Jencks, C., 270n

Kaye, B., 270n
Keeble, L., 143, 278n

Kidron, M., 272n, 282n
Kiernan, V., 27, 254n
Kolakowski, L., 232, 298n
Kopp, A., 270n
Korsch, K., 257n

'Labour aristocracy', 42, 123
Laclau, E., 256n
Lambert, J., 275n
Land Commission (1967), 118
Land market, 56–7; land reform, 129–30; land values, 208, 211–12
Lawrence, F. W., 265n
Leapfrog, urban game of, 58
Le Corbusier, 99, 100–2, 103–4, 106, 140, 247, 269n, 270n, 271n
Lethaby, W. R., 274n, 280n
Levine, D., 286n, 289n, 295n
Lewis, J. P., 262n
Lewis, O., 265n, 298n
Liberalism, 18, 24, 165–6, 219, 220, 230, 249–50; liberals, 4, 171, 217, 219
Linn, J. W., 291n, 296n
Lloyd George, D., 130
Local authorities, 17, 47, 126, 128, 148, 240, 242–5; local authority housing, 43, 58, 73, 109–10, 112–14; *see also* Local government
Local government, 17, 124, 147–8, 162, 242–5; reorganization, 144, 147–8, 244–6; *see also* Local authorities
Lockwood, D., 263
London, 12, 24–6, 28–9, 30–3, 37, 40, 50, 57, 68, 70–2, 81, 85, 124–5, 128–9, 135, 258n, 268n
Long, N., 264n
Loomis, C. P., 285n
Losch, A., 12, 256n
Lynd, R. and H., 237, 251, 299n

McCrone, D. and Elliott, P., 283n
MacEwan, M., 270n
MacFadyen, D., 276n
McGonaghy, D., 154, 281n
McKenzie, R., 196, 204, 208, 213, 291n, 292n
McKie, R., 266n
Mackinder, 23, 249
McLoughlin, J. B., 279n, 280n, 281n, 298n
McLoughlin, J. B. and Thornley, J., 280n
MacMurray, T., 267n, 277n
Managerialism, 158–9, 161, 243–4
Manchester, 9, 15, 26, 27, 28, 30, 35, 51
Mannheim, K., 277n, 289n
Marcuse, H., 284n
Marr, T., 265n
Martin, L., 100, 270n
Martindale, D., 188, 200, 259n, 286n, 287n
Marx, K., on community, 176; urbanization in capitalism, 1, 20, 170, 176; other references, 24, 176, 177, 181, 225, 235, 253n, 284n, 285n, 290n, 299n
Marxism: discussion of urbanization, 170–1, 172–3; and pragmatism, 235–7; and regional development, 13, 23; and urban sociology, 126, 174, 176–7, 251
Mass society, 10–11, 151, 171–2, 194, 197, 200
Mayhew, H., 29, 68, 71, 251, 264n
Mays, J. B., 265n
Mead, G., 222, 230, 234, 296n
Mearns, A., 68, 71, 266n
Medawar, P. B., 279n
Mellor, R., 266n, 267n
Metropolis: class structure, 12, 29–30, 86–7, 124; community organization, 20, 169, 178–9, 182, 185–7, 198; metropolis/satellite polarization, 10–11, 16, 26, 47, 172; metropolitan dominance, 9, 10, 11, 13, 22, 24, 47, 62, 86–7, 172, 173; metropolitan economy, 14, 30, 84–5, 86; power of, 15, 21, 22, 28, 62, 172–3, 255n
Miliband, R., 299n
Millett, K., 269n

Mills, C. W., 2, 199, 217, 254n, 284n, 290n, 295n, 297n, 298n, 299n
Milner Holland Report, 85, 268n
Modern Movement, 92, 100, 104, 106, 164
Mogey, J., 60, 75, 263n, 269n
Moore, B., 257
Moorhouse, G., 254n, 256
Morris, W., 48–9, 51, 100, 129, 140, 170, 217, 249, 253n, 276n
Morrison, A., 264n
Mumford, L., 49, 52, 103, 131, 137, 140, 171, 189, 247, 271n, 276n, 278n, 283n, 287n
Municipal capitalism, 275n
Muth, R., 267n
Myths, 202, 249, 262n; reformers, 67, 72–3, 79, 140, 251, 267n; of slums, 67, 70, 79; of suburbs, 59, 262n

Neighbourhood planning, 157, 164, 251, 282n
Nevitt, A., 271n, 272n
New towns, 43, 119, 133; *see also* Garden cities
New York, 50, 55, 128
Nisbet, R. A., 284n, 285n, 287n, 289n, 294n, 297n
Norman, P., 283n

Olsen, D. J., 262n
Orleans, P., 295n
Osofsky, G., 213, 261n, 293n
Over-urbanization, 39–40, 45, 69
Owen, R., 28, 49, 131, 133
Oxaal, I., 257n

Pahl, R., 7, 161, 164, 239, 249, 255n, 268n, 282n, 283n, 298n, 299n, 300n
Palmer, J., 83–4, 267n, 277n
Paris, 24, 30, 31–3, 55, 81, 107, 258n, 271n
Park, R.: attitudes to urbanization, 171, 216, 218–20, 295n; career, 199, 204, 291n; city as ecological community, 207–16; debt to Durkheim, 196, 209, 215, 219, 225, 293n, to Dewey, 221, 226; to James, 226–8, to Mead, 220, 296n, to Simmel, 196, 219, 225–6, 293n, 295n, to Small, 207, 223, to Spencer, 224; human ecology, definition, 198; human ecology derivation, 196, 206–10, 221–30; functionalism, use of, 208, 209, 216; social control, emphasis on, 196, 210, 216; statistics, attitude to, 214, 222, 225; theory of community, 219–20, urban methodology, 199, 215–16; other references, 5, 10, 30, 64, 66, 167, 189, 195, 197, 200, 201, 247, 249, 251, 252, 253n, 283n, 289n, 290–7n
Park, R. and Burgess, E., 289n
Parkhill, Sheffield, 101, 271n
Parsons, T., 280n, 285n, 288n
Participation, 78, 79, 144, 145–6
Pawley, M., 270n
Peace and quiet, 53, 89–90
Pennance, F. C. and Gray, H., 262n
Phenomenological empiricism, 228, 249
Pinkney, D., 258n
Planning policies, 43; countryside protection, 44, 52, 107, 115, 118–19, 135; urban dispersal, 43–4, 52, 61, 86–7
Plowden Report, 207n
Political economy, 24, 250–1; of regional development, 23; of urbanization, 45, 250
Pollard, S., 258
Pollins, H. S., 263n
Pragmatism: criticisms of, 230–2; influence on Chicago School, 165, 195, 222, 224–5; and Marxism, 226–8, 230, 232 235–7; and town planning, 164–6, 231–2; and urban sociology, 164–6; other references, 5, 6
Privacy/reserve, 90–1, 102, 185, 209n

Professional middle classes, 4, 28, 71, 89, 141, 216–17, 220, 231, 239, 241; professional proletariat, 242, 248, 280; *see also* Intellectuals, Reformers
Professionalism: definition of, 136–7, 149; in architecture, 98–9, 106; in planning, 134, 152–3, 155; rationality in, 102–3, 141, 149, 159, 165–6, 232
Problematic, 85, 267n
Property speculation, 35, 56, 117–18, 160, 260n, 262n
Provincial community, 14, 46, 47, 63–5, 69, 124, 125
Provincialism, 11, 15, 16, 74

Rainwater, L., 269n
Rasmussen, S. E., 33, 259
Reade, E., 279n, 282n
Redfield, R., 178, 285n, 288n, 296n, 298n
Reform, 70–3, 81–2, 109, 141, 242, 245–6
Reformers, 61, 73, 81–2, 109, 133, 141, 156, 158, 245, 246; *see also* Myths
Regional city/city region, 43, 48, 49, 50, 52, 53, 55, 56, 64–5, 261n
Regional inequalities, *see* Underdevelopment
Rehabilitation, *see* Housing
Reich, C. A., 281n, 283n
Reich, W., 3, 254n, 299n
Reisman, D., 55, 262n
Reissman, L., 141, 278n
Reserve Army, 30, 84
Rex, J., 6–7, 54, 64, 121–2, 157, 239, 267n, 276n, 281n, 282n, 297n, 298n
Rex, J. and Moore, R., 14, 255n
Richardson, H. W., 255n
Richardson, H. W. and Aldcroft D. H., 259n
Rose, E., 274n
Rosebery, A. P. P., 28, 129
Rowntree, 170
Rucker, D., 291n
Ruskin, J., 140, 170, 270n
Russell, B., 232, 283, 298n, 300n

Saville, J., 130, 276n
Schorr, A., 264n
Seabrook, J., 256n
Self, P., 277n, 280n, 283n
Seeley, J., 264, 266n
Seeley *et al.*, 269n
Semmel, B., 276n
Semple, 23
Sennett, R., 218, 264n, 287n, 290n, 295n
'Shanty towns', 86, 107, 272n
Shaw, M., 298n, 299n
Shils, E. M., 290n, 296n
Simmel, G., attitude to urbanization, 64, 169, 171, 173, 198, 247; methodology, 5, 176, 177, 182–5, 188, 192, 194; metropolis, 12, 49, 175, 185–8; urban personality, 185, 188; other references, 3, 167, 189, 191, 195, 196, 197, 201, 202, 204, 211, 216, 219, 225, 226, 232, 254n, 284–7n, 293n, 296n
Skeffington Report, 144, 279n
Slums, 31, 42, 66, 67–72, 73, 85–6, 107; blighted slum, 68, 70; exploitation slum, 68; slum clearance, 69–70, 71–2, 73–4, 76, 111, 113, 128; sociological accounts of, 73–6; urban jungle, 74, 265n; urban village, 74–5, 79, 266n
Small, A., 5, 195–7, 204, 207, 219, 223, 224, 254n, 292n, 296n
Smellie, K., 299n
'Social gatekeepers', 7, 158, 162
'Social polarization', 43, 51, 63
Sociology of constraint, 7–8, 121, 274n
Sorokin, P., 287n
Spencer, H., 131, 183, 222–4, 296n
State, challenge to authority, 16, 21, 22, 47; power of, 24, 158–9, 241; and urban sociology, 241–2, 245–7, 250

State intervention, 17–18; in housing, 96, 108–10, 112, 125; over labour force, 46; over land, 35; on urban problem, 80–1, 86, 123, 128
Status, 11–12, 53, 54, 55, 59, 80, 90, 91, 120
Status decategorization, 261n
Stein, M., 10, 60, 218, 255, 292n, 295n
Stein *et al.*, 255
Strauss, H., 297n
Suburban development, 35, 37–8, 40–3, 45, 128
Suburbanization, 42, 51–2, 53–6
Suburbs: middle-class, 52–5; working-class, 31, 42, 59–61, 263n
Summerson, J., 270n
Sumner, W., 222–3, 224
Suttles, G., 74, 234, 265n, 298n

Tabb, W., 11, 256n
Taylor, W., 260n
Thomas, B., 259n
Thomas, W., 5, 204, 218, 219, 224, 228
Thrasher, F., 298n
Tönnies, I.: attitude to urbanization, 11, 64, 169, 173, 180–2, 188, 200, 220–47; methodology, 174, 176, 177, 178–82; metropolis, 20, 88, 171, 175, 178; urban bourgeoisie, 180, 181–2; other references, 1, 5, 48, 167, 183, 184, 191, 194, 201, 211, 226, 255n, 284–6n
Town and Country Planning Act (1947), 117, 144; (1968), 117, 144; (1971), 145
Town planning methodology: advocacy planning, 153–6; community planning, 148–9, 155–7; corporate planning/management, 147, 148; development plans, 117, 144; environmental planning, 134, 142, 149–50, 156; participation, 78, 79, 144–6, 147, 164, 179n; structure planning, 65, 144, 146–8, 161; systems planning/management, 145, 150–2, 161, 231
Town planning movement: development pressures for, 34, 128–9; ideas, 130–2; political context, 129–33; social reform, 130–3, 161–2 ; *see also* Howard, Garden cities
Town planning profession: founding fathers, 136–43; generation of consensus, 159–61, 163–4; ideological inconsistencies, 133–4, 136, 143; lack of authority, 135, 146, 149–50; new ideologies, *see* Town planning methodology; pragmatic tendencies, 134–5, 150, 163, 165; premises, 141–3, 149–50; recruitment, 35, 136, 277n; reorientation, 134–5, 143–56; role in State, 158–61, 277n
Turner, J., 272n
Turner, R., 297n
Twilight areas, 51, 63, 66–7, 82–5, 121, 122, 124, 159, 267n, 282n; as reception areas, 80–2

Underdevelopment, 19, 22, 36, 40, 45, 47, 69, 84, 172; regional, 36–40
Uneven development, 9, 13, 18, 36, 40, 45, 175
Unwin, R., 35, 100, 128, 132, 136, 139–41, 142, 259n, 278n
Urban bourgeoisie, 1, 17, 35, 47, 172–6, 180–1, 190, 194, 244, 248
Urban community, 1, 6, 7–8, 64, 121–2, 125, 174, 190–3, 196–7
Urban degeneration, 129
Urban jungle, 74, 265n
Urban managers, 7, 67, 79, 120, 162; *see also* Managerialism
Urban problem, 3–4, 20, 65–7, 71, 82–4, 85–7, 159
Urban renewal, 69, 76–7, 81, 101, 111, 113, 159, 164

Urban residuum, 30, 61, 63, 84
Urban social movements, 77, 126, 250
Urban sociology: in America, 5–6, 167–8, 214, 228, 241; in Britain, 4, 6–9, 18, 120–6, 161–2, 238–42; in France, 241, 244–52; in general, 49, 64–5, 156, 165–6, 214, 230, 233–4; redefinition of, 13–16
Urban system, 64, 121, 125, 145, 151–2, 161, 190, 230–1
Urban village: 74–5, 79, 266n
'Urbanism', 4, 5, 15, 16, 173, 175, 177, 194–5, 198–9, 202
Urbanization, 6, 16, 20, 252; in Britain, 9, 17, 24–31, 34–47, 51–2; without industrialization, 25–7, 29, 30–1, 39, 46–7, 259n; sociological account of, *see* Engels, Marx, Park, Simmel, Tönnies, Weber, Wirth
Uthwatt Report, 283n

Wade, R. and Mayer H., 275n, 292n
Ward, S., 279n
Webber, M., 49–50, 150, 260n, 280n
Weber, M., city as community-association, 3, 189–90, 191, 197, 247; critique of urbanization, 170, 171, 173, 189, 192, 193, 220; idea of a city, 190, 193, 248; methodology, 191–2, 193, 195, 225; urbanization in capitalism, 1–2, 174, 191; other references, 5, 7, 9, 64, 122, 158, 167, 175, 184, 201, 202, 210, 217, 226, 232, 233, 253n, 284–99n
Welfare State, 6, 18, 24, 68, 120, 162, 245–8, 277n
Wells, H. G., 253n
White, M. and L., 296n
Whyte, W. F., 265n, 298n
Wilkinson, E., 259n
Williams, R., 202, 248, 254n, 256n, 257n, 290n, 299n
Willmott, P., 58, 262n, 263n
Wirth, L.: attitude to urbanization, 169, 170, 173, 194, 198, 200, 218, 247, to human ecology, 289n, to theory, 198–9; definition of urbanism, 177, 180–2, 193, 194, 196–7; influence of Chicago School, 195, 197, 198; influence of European sociology, 182, 195, 197, 200–2, 286n; other references, 64, 167, 185, 189, 220, 221, 232, 248, 253n, 287–93n
Wiseman, H. V., 254n
Wohl, A. S., 265n
Wolff, K., 286n
Women, 89–90, 91, 102, 104, 294n
Wright, F. C., 99, 269n
Wrigley, E. A., 257n

Young, M. and Willmott, P., 251, 261n, 294n, 298n

'Zone of transition', 83, 112, 120
Zoning, effects of, 117–18, 133
Zorbaugh, H., 206, 264n, 292n

For Product Safety Concerns and Information please contact our EU
representative GPSR@taylorandfrancis.com
Taylor & Francis Verlag GmbH, Kaufingerstraße 24, 80331 München, Germany

www.ingramcontent.com/pod-product-compliance
Ingram Content Group UK Ltd.
Pitfield, Milton Keynes, MK11 3LW, UK
UKHW020153200926
475702UK00016B/2687
9780415851909